Lecture Notes in Mathematics

Edited by A. Dold and B. Eckmann

884

Combinatorial Mathematics VIII

Proceedings of the Eighth Australian Conference on Combinatorial Mathematics Held at Deakin University, Geelong, Australia, August 25 – 29, 1980

Edited by Kevin L. McAvaney

Springer-Verlag
Berlin Heidelberg New York 1981

Editor

Kevin L. McAvaney
Division of Computing and Mathematics, Deakin University
Viktoria 3217, Australia

AMS Subject Classifications (1980): 05-06, 05 A 10, 05 A 15, 05 B xx, 05 C xx, 06 D 05, 08 B 99, 10 A 25, 51 E 15, 68 D 35, 68 E 05, 90 B 10

ISBN 3-540-10883-1 Springer-Verlag Berlin Heidelberg New York
ISBN 0-387-10883-1 Springer-Verlag New York Heidelberg Berlin

Library of Congress Cataloging in Publication Data
Australian Conference on Combinatorial Mathematics (8th: 1980: Deakin University) Combinatorial mathematics VIII. (Lecture notes in mathematics; 884) Bibliography: p. Includes index. 1. Combinatorial analysis--Congresses. I. McAvaney, Kevin L. (Kevin Lawrence), 1946- II. Title. III. Series: Lecture notes in mathematics (Springer-Verlag); 884. QA3.L28 vol. 884 [QA164] 510s [511'.6] 81-18255 AACR2
ISBN 0-387-10883-1 (U.S.)

Printed in Germany

Printing and binding: Beltz Offsetdruck, Hemsbach/Bergstr.
2141/3140-543210

PREFACE

The Eighth Australian Conference on Combinatorial Mathematics was conducted by the Combinatorial Mathematics Society of Australasia and held at Deakin University, Geelong, from 25 to 29 August 1980. It heard 37 speakers deliver 42 addresses: 3 expository, 9 invited, and 30 contributed. This refereed volume contains 31 of these papers; the remainder are listed by title.

Ten countries were represented at the conference by the 47 participants. There were 9 invitees:

Dr. Brian R. Alspach (Simon Fraser University)
Dr. Chuan-Chong Chen (National University of Singapore)
Dr. Ronald L. Graham (Bell Laboratories)
Dr. Jun-Shung Hwang (Academia Sinica)
Prof. Peter J. Lorimer (University of Auckland)
Prof. Ronald C. Read (University of Waterloo)
Prof. Johan J. Seidel (University of Technology, Eindhoven)
Dr. John Sheehan (University of Aberdeen)
Prof. Ralph G. Stanton (University of Manitoba)

Grateful acknowledgement is due to the following for their generous financial support of the conference:

Division of Computing and Mathematics, Deakin University
Department of Mathematics, Royal Melbourne Institute of Technology
Australia and New Zealand Banking Group Limited
Blue Circle Southern Cement Limited
Australian Mathematical Society
National Mutual Life Association of Australasia Limited
Ian Potter Foundation
Trans-Australia Airlines

Indeed I thank all those who helped with the conference organisation and the publication of this volume, in particular: divisional chairman Alan Parish and secretaries Jenny Sayers and Betty Worland for their co-operation; Deakin University for the use of its facilities; Deakin University Union for housing and feeding the participants; those who chaired sessions; the referees; the mini-bus drivers; D.A. Book Depot P/L., Harcourt Brace Jovanovich Group (Aust.) P/L., and Gordon Breach Science Publishers Ltd. for lending display books and journals; the participants for coming; the contributors for making it worthwhile; Springer-Verlag for disseminating the proceedings to the rest of the world; and not least Joy who remains my wife in spite of it all.

- K.L. McAvaney

TABLE OF CONTENTS

* denotes speaker

EXPOSITORY PAPERS

INVITED PAPERS

CONTRIBUTED PAPERS

REMAINING ADDRESSES

32
19
20
9
22
34
11
42
26
47
1
37
38
21
39
44
18
46
4
5
41
12
43
24
13
10
29
40
36
27
45
2
15
33
14
25
28
3

PARTICIPANTS

Brian ALSPACH	1	Department of Mathematics, Simon Fraser University, Burnaby, British Columbia V5A 1S6, Canada.
Rosemary BAILEY	2	Faculty of Mathematics, The Open University, Walton Hall, Milton Keynes MK7 6AA, England.
Uday BARUA	3	Department of Mathematics, University of Melbourne, Parkville, Victoria 3052.
David BILLINGTON	4	Department of Mathematics, University of Queensland, St. Lucia, Queensland 4067.
Elizabeth BILLINGTON	5	Department of Mathematics, University of Queensland, St. Lucia, Queensland 4067.
Stephen BOURN	6	Department of Pure Mathematics, University of Adelaide, G.P.O. Box 498, Adelaide, South Australia 5001.
Alan BRACE	7	School of Information Sciences, Canberra College of Advanced Education, P.O. Box 381, Canberra City, Australian Capital Territory 2601.
Rey CASSE	8	Department of Pure Mathematics, University of Adelaide, G.P.O. Box 498, Adelaide, South Australia 5001.
Chuan-Chong CHEN	9	Department of Mathematics, National University of Singapore, Bukit Timah Road, Singapore 1025.
Jeremy DAWSON	10	Division of Mathematics and Statistics, C.S.I.R.O., P.O. Box 218, Lindfield, New South Wales 2070.
Erich DURNBERGER	11	Department of Mathematics, Simon Fraser University, Burnaby, British Columbia V5A 1S6, Canada.
Peter EADES	12	Department of Computer Science, University of Queensland, St. Lucia, Queensland 4067.
Roger EGGLETON	13	Department of Mathematics, University of Newcastle, New South Wales 2308.

Les FOULDS	14	Department of Economics, University of Canterbury, Christchurch, New Zealand.
Ron GRAHAM	15	Bell Laboratories, 600 Mountain Avenue, Murray Hill, New Jersey 07974, U.S.A.
Bill HAEBICH	16	Operations Research Department, National Mutual Life Association, 447 Collins Street, Melbourne, Victoria 3000.
Bob HALE	17	Division of Computing and Mathematics, Deakin University, Victoria 3217.
Alan HARTMAN	18	Department of Combinatorics and Optimization, University of Waterloo, Waterloo, Ontario N2L 3G1, Canada.
Irith HARTMAN	19	Department of Combinatorics and Optimization, University of Waterloo, Waterloo, Ontario N2L 3G1, Canada.
Katherine HEINRICH	20	Department of Mathematics, Simon Fraser University, Burnaby, British Columbia V5A 1S6, Canada.
Derek HOLTON	21	Department of Mathematics, University of Melbourne, Parkville, Victoria 3052.
Jun Shung HWANG	22	Institute of Mathematics, Academia Sinica, Nankang, Taipei, Taiwan, Republic of China.
Tony KLEMM	23	Division of Computing and Mathematics, Deakin University, Victoria 3217.
Mordechai LEWIN	24	Department of Mathematics, Israel Institute of Technology, Technion City 3200, Haifa, Israel.
Charles LITTLE	25	Department of Mathematics, Royal Melbourne Institute of Technology Limited, G.P.O. Box 2476V, Melbourne, Victoria 3001.
Peter LORIMER	26	Department of Mathematics, University of Auckland, Private Bag, Auckland, New Zealand.
Brian MARTIN	27	Department of Computer Science, University of Manitoba, Winnipeg, Manitoba R3T 2N2, Canada.

Kevin McAVANEY	28	Division of Computing and Mathematics, Deakin University, Victoria 3217.
Bernhard NEUMANN	29	Department of Mathematics, Institute of Advanced Studies, Australian National University, P.O. Box 4, Canberra, Australian Capital Territory 2600.
Ian PARBERRY	30	Department of Computer Science, University of Queensland, St. Lucia, Queensland 4067.
Alan RAHILLY	31	School of Applied Science, Gippsland Institute of Advanced Education, Switchback Road, Churchill, Victoria 3842.
Ron READ	32	Department of Combinatorics and Optimization, University of Waterloo, Waterloo, Ontario N2L 3G1, Canada.
Bob ROBINSON	33	Department of Mathematics, University of Newcastle, New South Wales 2308.
Douglas ROGERS	34	68 Liverpool Road, Watford, Hertfordshire WD1 8DN, England.
Robert ROSSENBERG	35	431 Cardigan Street, Carlton, Victoria 3054.
Chris ROWLEY	36	The Open University, London Region, Parsifal College, 527 Finchley Road, London NW3 7BE, England.
Johan SEIDEL	37	Department of Mathematics, University of Technology, P.O. Box 513, 5600 MB Eindhoven, The Netherlands.
John SHEEHAN	38	Department of Mathematics, University of Aberdeen, Dunbar Street, Aberdeen AB9 2TY, Scotland.
Ralph STANTON	39	Department of Computer Science, University of Manitoba, Winnipeg, Manitoba R3T 2N2, Canada.
Deborah STREET	40	Department of Mathematics, University of Queensland, St. Lucia, Queensland 4067.
Don TAYLOR	41	Department of Pure Mathematics, University of Sydney, New South Wales 2006.
Richard TAYLOR	42	Department of Mathematics, University of Melbourne, Parkville, Victoria 3052.

Helge TVERBERG	43	Department of Mathematics, University of Bergen, 5014 Bergen, Norway.
John VAN REES	44	Department of Computer Science, University of Manitoba, Winnipeg, Manitoba R3T 2N2, Canada.
Wal WALLIS	45	Department of Mathematics, University of Newcastle, New South Wales 2308.
Peter WILD	46	Department of Pure Mathematics, University of Adelaide, G.P.O. Box 498, Adelaide, South Australia 5001.
Sheila WILLIAMS	47	Department of Mathematics, University of Queensland, St. Lucia, Queensland 4067.

SOME PROPERTIES OF H-DESIGNS

R.G. STANTON AND R.C. MULLIN

1. *Introduction.* In a very innovative paper [3] dealing with minimal coverings of triples by quadruples, W. H. Mills introduced two types of design, known as G-systems and H-systems. These designs are the topic of this survey, and we begin by recapitulating the definitions given by Mills.

Let m and r be positive integers. Let T be a collection $\{T_1, T_2, \ldots, T_m\}$ of disjoint r-element sets whose union is S; S is thus a set with mr elements. By a *transverse* of T, we mean a subset of S that meets each T_i in at most one point.

Definition 1. An (m,r,k,t) group-divisible system or $G(m,r,k,t)$ system on T is a collection $K_1, K_2, \ldots, K_u$, of k-element subsets (blocks) of S such that each t-element subset of S is contained in exactly one of the $m+u$ subsets $T_1, \ldots, T_m, K_1, \ldots, K_u$.

Definition 2. An $H(m,r,k,t)$ system on T is a collection $K_1, K_2, \ldots, K_u$, of k-element transverses of T such that each t-element transverse is contained in exactly one of the u subsets K_i.

It is important to note the fact that the $G(m,r,k,2)$ and the $H(m,r,k,2)$ systems are identical.

In order to clarify these important concepts, we now introduce two examples.

Example 1. $G(2,6,4,3)$.

Take two sets 1 2 3 4 5 6 and 1'2'3'4'5'6'.

Write one-factors

F_1: 12	34	56	F_1'
F_2: 13	25	46	F_2'
F_3: 14	26	35	F_3'
F_4: 15	24	36	F_4'
F_5: 16	23	45	F_5'

The Cartesian products $F_i \times F_i'$ generate $5(3 \times 3) = 45$ quadruples, and every triple occurs in one of the original sextuples or in one of these 45 quadruples (this particular G-system was used in [5]).

Example 2. H(4,2,4,3).

Take the sets 12 34 56 78. The number of elements in the required H-system must be $\binom{4}{3} 2^3/4 = 8$. The construction is easy, namely,

1 3 5 7	1 3 6 8
1 4 5 8	1 4 6 7
2 3 5 8	2 3 6 7
2 4 5 7	2 4 6 8

In the paper cited, Mills determines N(3,4,v), the minimum number of quadruples from a v-element set S required in order to ensure that each triple from S occurs in at least one of the quadruples, for all values of v except for $v \equiv 7$ modulo 12. In particular, some of the systems G(m,6,4,3), H(m,6,4,3), and H(m,6,3,2) are used to establish the value of the covering number N(3,4,v) for $v \equiv 1$ mod 12. Mills also proved the important result that a G(m,6,4,3) exists for all $m \geq 1$. Brouwer [1] later used this result to obtain the packing number D(3,4,v) for $v \equiv 0$ mod 6. (D(3,4,v) is the maximum number of quadruples in a v-element set with the property that no repeated triple appears.)

In [4] Stanton and Mullin showed that a generalized type of G system could be used in certain instances to obtain good estimates for $N(3,4,v)$ in the case $v \equiv 7 \bmod 12$. Their construction relied heavily on the existence of systems $H(m,6,4,3)$ in the case when m is an odd integer; it is these systems that we shall survey here. The Mills paper constructed a system $H(5,6,4,3)$, and it was very useful in [4]. Mills also proved the following powerful lemma.

LEMMA 1. *If an* $H(m,r,k,k-1)$ system exists, then an $H(m,rw,k,k-1)$ *system exists for every positive integer* w.

Proof. We are given m sets of r elements; basically what we do is replace each element by w elements. It will suffice to illustrate the procedure by taking the system $H(4,2,4,3)$ given in Example 2 and expanding it to give a system $H(4,4,4,3)$.

The 4 sets of elements given now become 4 sets of elements $1_1\ 1_2\ 2_1\ 2_2$, $3_1\ 3_2\ 4_1\ 4_2$, $5_1\ 5_2\ 6_1\ 6_2$, $7_1\ 7_2\ 8_1\ 8_2$. Each of the original sets K_i is replaced by 8 new sets. For example, the set 1 3 5 7 is replaced by the 8 sets $1_1\ 3_1\ 5_1\ 7_1$, $1_2\ 3_2\ 5_2\ 7_2$, and $1_a\ 2_b\ 3_c\ 4_d$ (six sets obtained by taking two subscripts as 1 and two subscripts as 2). Clearly all 3-element transverses occur in this system.

If we are looking for systems $H(2s+1,6,4,3)$, then this suggests the device of looking for an $H(2s+1,t,4,3)$, where $t|6$, and enlarging the system by the appropriate factor.

To help decide when this might be a useful approach, we introduce the idea of a *derived system*.

Let H be an $H(m,r,k,t)$ system, and let x be a point of the system which occurs in the set T_i of T. Consider those blocks of H which contain x.

These contain no other member of T_i. If we take those sets of T other than T_i as a new system T', and take as blocks of a new system H' those blocks of H which contain x and then remove x, it is clear that the resulting system is an $H(m-1,r,k-1,t-1)$ system on T'. This system is called the derived system of H (with respect to x).

THEOREM 1. *Let* $H(m,t,4,3)$ *be an* H *system in which* $t|6$ *and* m *is odd. If* $m \equiv 0 \bmod 3$, *then* $t = 6$; *otherwise* $t = 2$ *or* $t = 6$.

Proof. Let us first show that t can not be one. If $t = 1$, then each triple of points occurs in precisely one block, and the blocks of H form a Steiner quadruple system on the points of H; this is only possible for $m \equiv 2$ or $4 \bmod 6$.

Now let us assume that $t = 3$. In this case, any derived system H' is an $H(m-1,3,3,2)$ system, or equivalently a $G(m-1,3,3,2)$ system. Clearly the sets T' and the blocks of H must form a Steiner Triple System on the points of H'. However, since the total number of points in H' is $3(m-1)$ (an even number), this is impossible, since Steiner triple systems only exist on v points for $v \equiv 1$ or $3 \bmod 6$. Hence $t = 2$ or $t = 6$. If $t = 2$, consider any derived system $H(m-1,2,3,2)$. This is a $G(m-1,2,3,2)$; thus, if we add a new point, say ∞, to each of the pairs of T', we obtain a collection of triples on $2(m-1)+1$ points. These triples cover all pairs of the points, that is, we again have a Steiner Triple System. Hence $2(m-1)+1 = 2m-1 \equiv 1 \bmod 6$ or $2m-1 \equiv 3 \bmod 6$, that is $m \equiv 1$ or $m \equiv 2 \bmod 3$.

2. *Small H-Systems.*

In this section, we consider H(m,6,4,3) systems for odd values of $m \leq 15$.

For $m = 1$, whether such a system exists is a matter of individual preference (cf. the well-known paper, "Is the Null Graph a Pointless Concept?").

For $m = 3$, there is a universal agreement that no such system exists.

For $m = 5$, as noted, Mills had displayed such a system. One notes that such a system might possibly come from an H(5,2,4,3) system. This is not possible, as the following lemma shows.

LEMMA 2. *There is no* H(5,2,4,3).

Proof. If such a system exists, then any derived system must be a Steiner Triple System of order 9, with one deleted point (the affine plane of order 3, less a point).

Represent the affine plane by the usual schema

5 7 3

8 6 4

1 2 9

The lines not containing 9 are 357, 468, 158, 267, 471,238, 245, 136. Hence the H(5,2,4,3) contains 8 blocks T357, T468, T158, T267, T471, T238, T245, T136; 8 blocks of the form 9---; 4 blocks on 1,2,...,8. These last 4 blocks contain 1 element from each of 12, 34, 56, 78; it is easy to try 13--, 14--, 23--, 24--, and see that none of them completes. Hence, we do not have an H(5,2,4,3).

For $m = 7$, one can obtain a system by modifying a construction for quadruple systems used by various authors (see, for example, Hartman [2]). Indeed the blocks

$0_0\ 1_0\ 2_0\ 4_0$ $\qquad$ $0_1\ 3_1\ 5_1\ 6_1$

$0_1\ 3_0\ 5_0\ 6_0$ $\qquad$ $0_0\ 1_1\ 2_1\ 4_1$

$0_0\ 1_0\ 2_1\ 6_1$ $\qquad$ $0_1\ 1_0\ 3_1\ 6_1$

$0_0\ 2_0\ 4_1\ 5_1$ $\qquad$ $0_0\ 2_0\ 5_1\ 6_1$

$0_0\ 4_0\ 1_1\ 3_1$ $\qquad$ $0_0\ 4_0\ 3_1\ 5_1$

developed modulo 7 produce such a design with 70 blocks.

The cases $m = 9$ and $m = 15$ can not be built up by inflating smaller systems as noted in Theorem 1. These seem to be difficult to construct.

The case $m = 11$ could conceivably be obtained from an $H(11,2,4,3)$, but at the present it appears that neither this nor the larger system $H(11,6,4,3)$ are known.

The author has recently heard that Mills has constructed an $H(13,2,4,3)$ with computer assistance. Also, Alan Hartman has constructed $H(v,2,4,3)$ for $v = 19,31,43$.

3. *Conclusion.*

It is apparent that H-systems are important in covering problems and in other problems involving triples. However, because G-systems and H-systems agree for $t = 2$, but are separate entities for $t \geq 3$, we gain insight into the problem of extending the powerful theory of R. M. Wilson [6] for pairwise balanced designs to t-wise balanced designs for $t \geq 3$. Some of the properties of pairwise balanced designs used by Wilson are those basic to H-systems, while others are basic to G-systems. For example, the device of "breaking up blocks" is a property of G-systems, not of H-systems. For example, if one has a G-system $G(m,rn,k,t)$ on a set $T = \{T_1,\ldots,T_m\}$ and a G-system $G(m,r,k,t)$, then one can form a G-system $G(mn,r,k,t)$ by the simple device of replacing each of the sets T_i by a copy of a $G(n,r,k,t)$ system on that same set of points. For $t > 2$, however, this device does not work for H-systems. Thus any extension of Wilson's theory will rely on the interplay of G and H systems, with possibly other structure required as well.

At this stage, we should point out the need for more study of systems $G_\lambda(m,r,k,t)$ and $H_\lambda(m,r,k,t)$. These are generalizations of G and H systems where the phrase "exactly one of the subsets" is replaced by "exactly λ of the subsets".

REFERENCES

[1] A. Brouwer, *On the packing of quadruples without common triples*, Ars Combinatoria *5* (1978), 3-6.

[2] A. Hartman, *Kirkman's trombone player problem*, Ars Combinatoria (to appear).

[3] W. H. Mills, *On the covering of triples by quadruples*, Congressus Numerantium *10* (1974), 563-581.

[4] R. G. Stanton and R. C. Mullin, *Some new results on the covering numbers* N(t,k,v), Combinatorial Mathematics VII (Springer-Verlag), to appear.

[5] R. G. Stanton and J. G. Kalbfleisch, *The λ-μ problem: $\lambda = 1$ and $\mu = 3$*, Proc. 2nd Chapel Hill Conference on Combinatorial Mathematics, Univ. North Carolina (1970), 451-462.

[6] R. M. Wilson, *Construction and uses of pairwise balanced designs*, Proc. NATO Advanced Study Inst. on Combinatorics, Nijenrode Castle, Bruekelen, The Netherlands (1974), 19-42.

Dept. of Computer Science, University of Manitoba, Winnipeg, Canada, R3T 2N2.

Dept of Combinatorics and Optimization, University of Waterloo, Waterloo, Ontario, Canada, N2L3G1.

COMPUTATION OF SOME NUMBER-THEORETIC COVERINGS

R. G. STANTON AND H. C. WILLIAMS

In this expository lecture, we give a survey of the Polignac problem concerning the primality of $k-2^n$ *and the Sierpinski problem concerning the primality of* $1+k.2^n$. *Various numerical results are given related to the problem of determining the smallest* k *for which* $1+k.2^n$ *is always composite.*

1. INTRODUCTION

We shall first give an historical survey of a problem that does not appear to have any pronounced combinatorial aspects; subsequently, we shall show that it can be made to undergo a metamorphosis into a combinatorial problem. This problem was first formulated by A. Polignac in 1849; Polignac made the conjecture that, for any odd integer $k > 1$, k can be expressed in the form

$$k = 2^n + p,$$

where p is a prime (p=1 was, of course, permissible in 1849).

2. THE ERDÖS DISPROOF OF THE POLIGNAC CONJECTURE

We sketch the disproof of the Polignac conjecture given by Erdös [3]. Let n be any natural number; then, by considering congruence classes, mod 24, we note that any n falls into at least one of six classes. Then certain exponential congruences follow, as tabulated.

I	$n \equiv 0\ (2)$	$2^2 \equiv 1\ (3)$	$2^n \equiv 1\ (3)$
II	$n \equiv 0\ (3)$	$2^3 \equiv 1\ (7)$	$2^n \equiv 1\ (7)$
III	$n \equiv 1\ (4)$	$2^4 \equiv 1\ (5)$	$2^n \equiv 2\ (5)$
IV	$n \equiv 3\ (8)$	$2^8 \equiv 1(17)$	$2^n \equiv 2^3(17)$
V	$n \equiv 7(12)$	$2^{12} \equiv 1(13)$	$2^n \equiv 2^7(13)$
VI	$n \equiv 23(24)$	$2^{24} \equiv 1(241)$	$2^n \equiv 2^{23}(241)$

Now use the Chinese Remainder Theorem to determine a number a satisfying the following congruences.

$a \equiv 1(2)$, $a \equiv 1(3)$, $a \equiv 1(7)$, $a \equiv 2(5)$, $a \equiv 2^3(17)$, $a \equiv 2^7(13)$, $a \equiv 2^{23}(241)$, $a \equiv 3(31 = 2^5-1)$.

Then $a - 2^n \equiv 0$ for one of the primes 3,7,5,17,13,241, and so is divisible by one of these primes.

But modulo 31, $2^n \equiv 1$ or 2 or 4 or 8 or 16. Thus $a - 2^n$, modulo 31, $\equiv$ 2 or 1 or -1 or -5 or -13. But this shows that $a - 2^n$ is not a member of {3,7,5,17,13,241}; so $a - 2^n$ is a composite number for all n.

So this particular a (and hence an infinite AP of a's) violates the Polignac conjecture.

3. THE SIERPINSKI PROBLEM

The numbers $1 + k2^n$ have always been of considerable interest; for instance, they include the Fermat numbers. R. M. Robinson gave an extensive table of primes of this form [5], and Sierpinski showed that there existed values of k for which $1 + k2^n$ was never prime for any n [6].

Sierpinski's method was to use the Fermat primes 3,5,17,257,65537, and the two factors of F_5, namely, 641 and 6700417. He then used the Chinese Remainder Theorem to find a k such that $1 + k2^n$ is always composite (his k was rather large!).

Actually, the Sierpinski problem and the Polignac-Erdős problem are equivalent. According to Sierpinski, this fact was pointed out by Schinzel. Suppose that one has determined a particular value a such that $a - 2^n$ is composite for all n, and divisible by at least one of 3,5,7,13,17,241. Then there is an infinite AP of solutions; in particular, there is a negative solution $a = -k$; hence $-k-2^n$, and thus $k + 2^n$, is divisible by one of 3,5,7,13,17,241.

Let $P = (3)(5)(7)(13)(17)(241)$; then

$$k + 2^{n(\phi(P)-1)}$$

is divisible by one of the prime factors of P, say p.

But

$$k + 2^{n(\phi(P)-1)} \equiv k + 2^{n\phi(P)}2^{-n} \equiv 0 \bmod p,$$

and $\quad 2^{\phi(P)} \equiv 1 \bmod P$ (hence, $2^{\phi(P)} \equiv 1 \bmod p$).

Thus $\quad k + 2^{n(\phi(P)-1)} \equiv k + 2^{-n} \equiv 0 \bmod p;$

multiply by 2^n, and we have

$$1 + k2^n \equiv 0 \bmod p.$$

Thus, a solution for the Erdős problem gives a solution for the Sierpinski problem, and vice versa.

4. THE SMALLEST VALUE OF k.

In 1963, Oystein Ore [4] posed an interesting problem in the American Mathematical Monthly. He pointed out that $\phi(x) = 14$ has no solution, and asked for the smallest odd k_α such that the equation

$$\phi(x) = 2^\alpha k_\alpha$$

has no solution.

One of the solutions given was by John Selfridge. He used the fact that

$$1 + k2^n$$

can, for suitable k, be always composite. Take k as a prime (the values of k form an AP, and so this is possible by Dirichlet's Theorem). Then

$$\phi(x) = 2^\alpha k$$

has no solution.

Selfridge pointed out that the Robinson table gave $k_0 = 3$, $k_1 = 7$, $k_2 = 17$, $k_3 = k_4 = k_5 = 19$, $k_6 = k_7 = 31$, $k_\alpha = 47$ for $8 \le \alpha < 583$, $k_\alpha = 383$ for $583 \le \alpha < 2313$.

Clearly, the k_α are bounded above by the smallest prime k such that $1 + k2^n$ is always composite. The first reference we have found to an explicit formulation of the question "What is the smallest number k for which all numbers $1 + k2^n$ ($n = 0, 1, 2, \ldots$) are composite?" is in Sierpinski [7].

5. SOME VERY LARGE PRIMES.

Selfridge had verified that

$$1 + k2^n$$

always contained at least one prime value for $k < 383$. For $k = 383$, he verified that $1 + k2^n$ was composite for $n < 2313$; later, in 1976, N. S. Mendelsohn and B. Wolk pushed this value up to $n \le 4017$. Finally, using the prime-testing methods described in [2], Baillie, Cormack, and Williams [1] established that

$$1 + 383\,(2^n)$$

first assumes a prime value for $n = 6393$. In the same paper, they give those values of $n \ge 3000$ for $k < 10000$ as follows.

k	n	k	n
383	6393	7957	5064
2897	9715	8543	5793
6313	4606	9323	3013
7493	5249		

Selfridge (unpublished) had shown in 1962 that $k = 78557$ produces a set of numbers that are always composite and, indeed, divisible by one of 3, 5, 7, 13, 19, 37, 73. Baillie, Cormack, and Williams tested all k up to 78557, and were able to find a prime in $\{1 + k2^n\}$ for all but 118 values of k. The smallest such k now is 3061, and no prime occurs in $1 + k2^n$ for $n \leq 16000$. It is highly likely that $k = 78557$ is the least value of k for which all values $1 + k2^n$ are composite.

In the succeeding lecture, we look at this problem from an alternative point of view which shows that the problem is basically more combinatorial than number-theoretic.

REFERENCES

[1] R. Baillie, G. V. Cormack, H. C. Williams, *Some Results Concerning a Problem of Sierpinski*, submitted, Math. Comp.

[2] G. V. Cormack and H. C. Williams, *Some Very Large Primes of the Form* $k.2^n+1$, Math. Comp. 35 (1980), 1419-1421.

[3] P. Erdös, *On Integers of the Form* $2^n + p$ *and Some Related Problems*, Summa Brasiliense Mathematicae II - 8(1950), p.119.

[4] O. Ore, cf. *Solution to Problem 4995*, Amer. Math. Monthly 70 (1963), p. 101.

[5] R. M. Robinson, *A Report on Primes and on Factors of Fermat Numbers*, Proc. Amer. Math. Soc. 9 (1958), pp. 673-681.

[6] W. Sierpinski, *250 Problems in Elementary Number Theory,* Elsevier, New York, (1970), p. 10 and p. 64.

[7] W. Sierpinski, *Sur un problème concernant les nombres* $k.2^n+1$, Elemente der Mathematik 15 (1960), pp. 73-74 (cf. also p. 85).

Department of Computer Science, The University of Manitoba, Winnipeg, Canada, R3T 2N2.

THE SEARCH FOR LONG PATHS AND CYCLES IN VERTEX-TRANSITIVE GRAPHS AND DIGRAPHS

BRIAN ALSPACH*

§1. Introduction

We shall assume the reader is familiar with basic graph theory terminology and refer to [6] as a reference. If X is a graph, we use V(X) and E(X) to denote the vertex-set and edge-set of X, respectively. We use Aut(X) to denote the group of automorphisms of X. We call a graph X <u>vertex-transitive</u> if Aut(X) acts transitively on V(X).

In 1969, Lovász [14] posed the question of whether or not every connected vertex-transitive graph has a hamiltonian path. In the intervening time not a great deal of progress has been made towards the resolution of this question. We shall discuss the progress as well as some related questions and the work done on them.

One striking feature of the observations on Lovász's fundamental question is that we realize in general it is asking 'more' of a graph to possess a hamiltonian cycle as compared to a hamiltonian path. Yet there are only four known connected vertex-transitive graphs that do not have hamiltonian cycles. All four have hamiltonian paths. These four graphs are the Petersen graph [6, p.236] , the Coxeter graph [6, p.241] and the graph obtained from each of the preceding two by replacing each vertex with a K_3 and joining corresponding vertices as indicated in Figure 1.

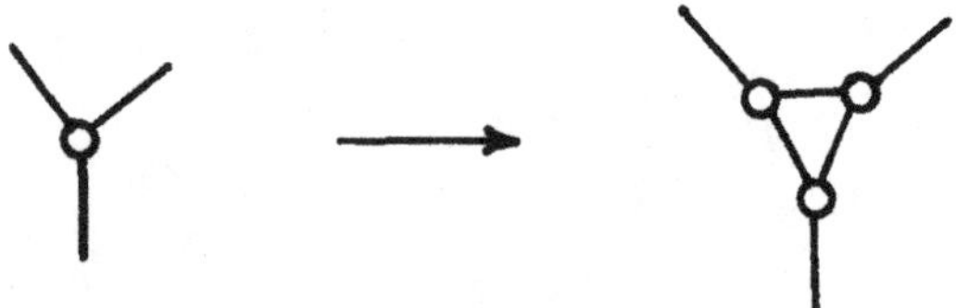

figure 1

The Petersen graph and the Coxeter graph both have cycles of length n-1 where n is the number of vertices. The two graphs derived from them have longest cycles of length n-3.

The above considerations lead to the following two questions.

*This research was supported by the Natural Sciences and Engineering Research Council of Canada under Grant A-4792.

Question 1. Are there an infinite number of connected vertex-transitive graphs that do not have a hamiltonian cycle?

Question 2. If d is a positive integer, does there exist a connected vertex-transitive graph with n vertices and longest cycle of length r such that $n-r > d$?

When one is trying to prove that every object in a class has a certain property and is unable to do so, then typically two approaches are taken. One approach is to restrict the class of objects and prove the property holds over the restricted class while the other approach is to restrict the property and prove that every object in the class satisfies the restricted property. We now consider these two approaches in the next three sections.

§2. Universal bound

We are interested in finding a lower bound on path or cycle lengths in connected vertex-transitive graphs with n vertices. In [3] Babai has proved the following result.

THEOREM 1. <u>Every connected vertex-transitive graph with</u> $n \geq 4$ <u>vertices has a cycle of length greater than</u> $(3n)^{\frac{1}{2}}$.

We now outline a proof of Theorem 1. A result proved independently by Mader [15] and Watkins [25] states that if the degree of regularity of the connected vertex-transitive graph X is 3 or more, then X is 3-connected. It is easy to show that any two longest cycles in a 3-connected graph have at least three vertices in common. If the length of a longest cycle in X is r, one defines an r-uniform hypergraph H with vertex-set V(X) whose edges are the various r-subsets of V(X) that are the vertices of r-cycles in X. We see that H is regular because X is vertex-transitive. Any two edges of H have at least three elements in common so that by Lemma 3.4 of [4] we have $r^2 > 3n$. □

The simplicity of the proof of the preceding result might lead one to think that better lower bounds are known. However, presently this is not the case and, more than that, Babai himself in [3] points out that the best lower bound he could get for paths is the bound implicit in Theorem 1.

Using Theorem 1 in the case of cubic graphs and observing that when n is at least 6 any two longest cycles must have at least four vertices in common, the same proof as used for Theorem 1 yields the following result.

COROLLARY 1.1. <u>Every connected vertex-transitive graph regular of degree</u> 3 <u>with</u> $n \geq 6$ <u>vertices has a cycle of length greater than</u> $2n^{\frac{1}{2}}$.

Question 3. Does there exist a constant c such that every connected vertex-transitive graph with n vertices has a cycle of length at least cn?

Question 4. The same as Question 3 with the word 'path' replacing the word 'cycle'.

§3. Vertex-transitive graphs of special orders

An approach that looks for classes of graphs that contain hamiltonian paths or cycles must be based on sufficiently understanding the structure of the graphs in the classes. For certain orders of vertex-transitive graphs this has been done. First we give a few necessary definitions.

Let $S \subseteq \{1,2,\ldots,n\}$ and satisfy $i \in S$ if and only if $n - i \in S$. The _circulant_ graph $X(n,S)$ has vertex-set $\{u_1,u_2,\ldots,u_n\}$ and edge-set E with $u_iu_j \in E$ if and only if $j - i \in S$ where all arithmetic is done modulo n using the appropriate residues.

If X is a graph that is regular of degree d, then a _hamiltonian partition_ of X is a partition of E(X) into $\frac{d}{2}$ hamiltonian cycles when d is even or $\frac{d-1}{2}$ hamiltonian cycles and a 1-factor when d is odd.

A graph X is said to be _hamiltonian connected_ if for every pair of vertices u and v of V(X) there is a hamiltonian path whose terminal vertices are u and v.

There are two results that have proved to be useful for the classes of graphs under discussion in this section. The first is a complete characterization of vertex-transitive graphs with a prime number of vertices given by Turner [24].

THEOREM 2. _A graph X with a prime number p of vertices is vertex-transitive if and only if it is a circulant_ $X(p,S)$.

The second result deals with strong hamiltonicity of circulant graphs with a prime number of vertices. It was first given in [1].

LEMMA 3. _Let_ $p \geq 5$ _be a prime. The circulant graph_ $X(p,S)$ _is hamiltonian connected if and only if_ $|S| \geq 4$.

In the following theorem, we summarize the results about hamiltonian properties of vertex-transitive graphs that have been derived from our knowledge about vertex-transitive graphs with a prime number of vertices. We follow the statement of the theorem with a brief discussion about the general approach to proving the various parts of the theorem.

THEOREM 4. *Let* p *be any prime. If* X *is a connected vertex-transitive graph with* n *vertices, then*

(i) X *has a hamiltonian path if* n *is of the form* p, p^2, p^3, 2p, 3p, 4p *or* 5p;

(ii) X *has a hamiltonian cycle if* n *is of the form* p, p^2, p^3, 2p *or* 3p *with the sole exception of the Petersen graph; and*

(iii) X *has a hamiltonian partition if* n *is of the form* p *or* 2p *where* $p \equiv 3 \pmod 4$ *in the latter case*.

From Theorem 2 we see that a vertex-transitive graph with p vertices either has no edges (the graph being denoted $\overline{K}_p$) or is connected whenever $S \neq \emptyset$. In the latter case, each i in S gives rise to the hamiltonian cycle $u_1 u_{i+1} u_{2i+1} \cdots u_{-i+1} u_1$ so that each of the statements in Theorem 4 is true when n = p. These results are implicitly contained in Turner's paper [24].

When n is 2p, 3p, 4p or 5p, the various results are proved in a similar fashion with varying details required to handle special cases. For example, let us consider the case when n = 2p. The case that p = 2 is easily disposed of separately. So we assume p is at least 3 and consider elements of Aut(X) of order p. If such an element has only one orbit of cardinality p, then it is easy to find a hamiltonian cycle in X. Thus we may assume an element of order p has two orbits of cardinality p with X_1 and X_2 denoting the two subgraphs induced by the orbits. If X_1 and X_2 are regular of degree 4 or more, then by Lemma 3 it is easy to find a hamiltonian cycle in X. Meanwhile, if each vertex of X_1 is adjacent to at least two vertices of X_2, it is again easy to find a hamiltonian cycle in X. In other words, unless the subgraphs X_1 and X_2 have small degree of regularity and the number of edges joining vertices of X_1 with vertices of X_2 is also small, it is easy to find a hamiltonian cycle. In the case when n = 2p which we are considering, we are left with the case that X is a generalized Petersen graph in addition to being vertex-transitive. The main result of [5] completes the proof. The case when n = 2p was first proved in [1].

The cases of n equal to 3p, 4p or 5p are done in a similar fashion but the details for exceptional cases become more complicated. Indeed, at this time the 4p and 5p cases only claim the existence of a hamiltonian path. The case of n = 3p was done by Marušič in [18] while the n = 4p and n = 5p cases are done by Marušič and Parsons [19, 20].

The result that X has a hamiltonian cycle if it is vertex-transitive and has either p^2 or p^3 vertices and is connected was an oral communication from Marušič. We shall say a few words about the hamiltonian partition result for vertex-transitive graphs with 2p vertices when

$p \equiv 3 \pmod 4$. This result depends heavily on a construction for vertex-transitive graphs with 2p vertices. This construction is given in [2] and [16].

It is not known whether or not the construction produces all vertex-transitive graphs with 2p vertices. The problem is that there may exist vertex-transitive graphs with 2p vertices that are neither complete nor have an imprimitive automorphism group. The Petersen graph and its complement are the only known such graphs and if another exists it would have at least 626 vertices. Since there are no primitive permutation groups of degree 2p that are not doubly transitive when $p \equiv 3 \pmod 4$ by Theorem 31.2 of [21], the above mentioned construction produces all vertex-transitive graphs with 2p vertices in this case. In particular, as pointed out in [2], all such graphs may be constructed by taking two identical copies of a circulant graph X(p,S) with p vertices and cyclically joining vertices in the two copies; that is, vertex u_i in one copy is adjacent to vertex v_j in the other copy if and only if u_{i+k} is adjacent to v_{j+k} for all $k = 1, 2, \ldots, p-1$ with subscripts reduced modulo p. This makes the proof of the second part of (iii) in Theorem 4 straight forward.

§4. Cayley graphs and digraphs

Another class of vertex-transitive graphs is the Cayley graphs which are defined in the following manner. Let G be a finite group and $H \subseteq G$ satisfy $1 \notin H$ and $y \in H$ if and only if $y^{-1} \in H$. The Cayley graph X(G,H) has the elements of G for its vertices and an edge joining x and y if and only if $x^{-1}y \in H$. Notice that circulant graphs are Cayley graphs on the cyclic group. A Cayley digraph X(G,H) is defined in a similar fashion except that $y \in H$ need not imply that $y^{-1} \in H$ and there is an arc from x to y if and only if $x^{-1}y \in H$. We emphasize that the following discussion pertains only to finite groups.

Each of the four graphs mentioned in Section 1 that do not have a hamiltonian cycle is also not a Cayley graph. This naturally suggests the following question.

Question 5. Does every connected Cayley graph have a hamiltonian cycle?

The class of groups with the simplest structure is the abelian groups. It is not surprising that the question has been completely settled for this class. The following result has been independently proved by Chen and Quimpo [7], Lee [13], Marušič [17] and probably others.

THEOREM 5. *Every connected Cayley graph of an abelian group of order at least 3 has a hamiltonian cycle.*

Chen and Quimpo [8] have in fact proved a much stronger result than Theorem 5. This result also includes Lemma 3 as a special case.

THEOREM 6. *A connected Cayley graph* X(G,H) *of an abelian group G of order at least 3 is hamiltonian connected if and only if it is not bipartite and it is regular of degree at least 3. In the case the Cayley graph is connected, bipartite and regular of degree at least 3, there is a hamiltonian path joining any two vertices in different bipartition sets.*

COROLLARY 6.1. *Every edge of a connected Cayley graph of an abelian group of order at least 3 is contained in a hamiltonian cycle.*

Corollary 6.1 follows immediately from Theorem 6 when the graph is regular of degree at least 3 because of the graph being hamiltonian connected if it is not bipartite; while if it is bipartite, every edge has its endvertices in different bipartition sets. When it is regular of degree 2, the graph itself is a hamiltonian cycle.

The proofs of Theorems 5 and 6 are most easily done using induction on $|H|$. The details will be omitted here but an outline of the essential ideas follows. In Theorem 5 the initial condition for the induction occurs when $|H| = 2$. It is easy to see that X(G,H) is a hamiltonian cycle no matter whether H contains two elements of order 2 or an element of order 3 or more and its inverse. In the induction step when we choose $H' = H - \{g,g^{-1}\}$ for some g in H, we must consider whether or not H' also generates all of G. When $G = \langle H' \rangle$, we have a hamiltonian cycle by the induction hypothesis so that the interesting case is when the group G' generated by H' is a proper subgroup of G. We know the Cayley graph X(G',H') contains a hamiltonian cycle by the induction hypothesis and that the Cayley graph X(G,H) is made up of vertex-disjoint isomorphic copies of X(G',H') with edges defined by g and g^{-1} between some of the copies. It is then possible to join together the hamiltonian cycles in each of the isomorphic copies using the edges defined by g to obtain a hamiltonian cycle in all of X(G,H). For Theorem 6 the initial condition occurs with $|H| = 3$ and in the induction step more care must be exercised over how edges joining disjoint isomorphic copies behave. The details are more intricate but not overly difficult.

If K and L are groups, then a *semidirect product* of K by L is a group G such that K is a normal subgroup of G, L is a subgroup of G, $K \cap L$ is the identity element of G and $K \cup L$ generates G. Marušič has proved the following result [17].

THEOREM 7. *Every connected Cayley graph of a semidirect product of a cyclic group of prime order by an abelian group of odd order has a hamiltonian cycle.*

The next result has been proved by Quimpo [22].

THEOREM 8. *Every edge of the connected Cayley graph* X(G,H) *is contained in a hamiltonian cycle in each of the following cases.*

(1) G *has order* pq *where* p *and* q *are primes.*

(2) $H = \{x,x^{-1},y,y^{-1}\}$ *where* $\langle x\rangle$ *is a normal subgroup of* G, $\frac{G}{\langle x\rangle}$ *is cyclic and generated by* $y\langle x\rangle$.

(3) G *is metacyclic and* $H = \{x,x^{-1},y,y^{-1}\}$.

(4) G *is a hamiltonian group.*

(5) *The elements of* H *can be arranged in a sequence* $x_1,x_2,\ldots,x_h$ *so that* $x_i^{-1}H_{i-1}x_i$ *is a subgroup of* H_{i-1} *for* $i = 2,\ldots,h$ *where* H_i *is the subgroup generated by* $x_1,\ldots,x_i$ *and* $h = |H|$.

Part (4) of Theorem 8 can be strengthened to X(G,H) is hamiltonian connected when H contains an element of odd order. Also, notice that there is some overlap between Theorems 7 and 8.

We say that a digraph is *weakly connected* if the graph we obtain upon making u and v adjacent if and only if there is an arc in the digraph with u and v as terminal vertices is itself connected. A digraph is *strongly connected* if for every pair of vertices u and v there is a directed path from u to v.

The overall problem of finding directed hamiltonian paths and cycles in Cayley digraphs is more complicated than the corresponding problem for Cayley graphs. This can be seen by comparing the following results with their counterparts for Cayley graphs. For example, the following theorem of Holsztyński Nathanson [9, Theorem 3.1] should be compared with Theorem 5 and Theorem 8(4).

THEOREM 9. *Every strongly connected Cayley digraph of a hamiltonian group or an abelian group has a directed hamiltonian path.*

The preceding result is best possible in that the Cayley digraph X(G,H) with $G = C_2 \times C_6$ and $H = \{(1,x), (y,x^2)\}$, where y generates C_2 and x generates C_6, has no directed hamiltonian cycle. This is the digraph of Figure 2 in [9]. On the other hand, Klerlein [10] has shown that every finite abelian group G has a minimal generating set H such that X(G,H) has a directed hamiltonian cycle.

In [9], Holsztyński and Strube proved that every strongly connected circulant digraph on n vertices has a directed hamiltonian cycle if and only if $n \geq 3$ is a prime power.

Semidirect products of two cyclic groups have been investigated in several papers. Trotter and Erdös [23] considered the special case of

$G = C_m \times C_n$ and $H = \{(1,0), (0,1)\}$ and proved that $X(G,H)$ has a directed hamiltonian cycle if and only if $\gcd(m,n) = d \geqq 2$ and there exist positive integers d_1, d_2 such that $d_1 + d_2 = d$ and $\gcd(m,d_1) = \gcd(n,d_2) = 1$. Klerlein and Starling [11, 12] considered arbitrary semidirect products of two cyclic groups and found sufficient conditions for such Cayley digraphs to have a directed hamiltonian cycle.

Nijenhuis and Wilf [21] showed that if G is the symmetric group on 5 letters and H consists of a 5-cycle and a transposition, then X(G,H) does not even contain a directed hamiltonian path. Thus, the situation for digraphs is evidently more complicated. It is natural to rephrase Question 2 for digraphs.

Question 6. If d is a positive integer, does there exist a vertex-transitive digraph with n vertices and longest directed cycle of length r such that $n - r > d$?

Question 7. The same question as number 6 except that the word 'path' is replaced by the word 'cycle' is the last question of this paper.

REFERENCES

1. Brian Alspach, Hamiltonian cycles in vertex-transitive graphs of order 2p, Proc. Tenth Southeastern Conf. Combinatorics, Graph Theory and Computing, Congress. Num. XXIII, Utilitas Math., Winnipeg, 1979, 131-139.

2. Brian Alspach and Richard J. Sutcliffe, Vertex-transitive graphs of order 2p, Annals N. Y. Acad. Sci., 319(1979), 19-27.

3. Lászlό Babai, Long cycles in vertex-transitive graphs, J. Graph Theory, 3(1979), 301-304.

4. Lászlό Babai, On the complexity of canonical labeling of strongly regular graphs, SIAM J. Comput., 9(1980), 212-216.

5. Kozo Bannai, Hamiltonian cycles in generalized Petersen graphs, J. Combinatorial Theory Ser. B, 24(1978), 181-188.

6. J. A. Bondy and U. S. R. Murty, Graph theory with applications, American Elsevier, New York, 1976.

7. C. C. Chen and N. F. Quimpo, On some classes of hamiltonian graphs, Southeast Asian Bull. Math., Special issue(1979), 252-258.

8. C. C. Chen and N. F. Quimpo, On strongly hamiltonian abelian group graphs, Lecture Notes in Mathematics, this volume, 1981, Springer-Verlag, Berlin.

9. W. Holsztyński and R. F. E. Strube, Paths and circuits in finite groups, Discrete Math., 22(1978), 263-272.

10. Joseph B. Klerlein, Hamiltonian cycles in Cayley color graphs, J. Graph Theory, 2(1978), 65-68.

11. Joseph B. Klerlein and A. Gregory Starling, Hamiltonian cycles in Cayley color graphs of semi-direct products, Proc. Ninth Southeastern Conf. Combinatorics, Graph Theory and Computing, Congress. Num. XXI, Utilitas Math., Winnipeg, 1978, 411-435.

12. Joseph B. Klerlein and A. Gregory Starling, Hamiltonian cycles in Cayley color graphs of some special groups, preprint.

13. Lawrence Lee, On the hamiltonian connectedness of graphs, M. Sc. Thesis, Simon Fraser University, 1980.

14. László Lovász, Problem 11, Combinatorial structures and their applications, Gordon and Breach, New York, 1970.

15. W. Mader, Ein Eigenschaft der Atome endlicher Graphen, Arch. Math. (Basel), 22(1971), 333-336.

16. Dragan Marušič, On vertex symmetric digraphs, Discrete Math., to appear.

17. Dragan Marušič, Hamiltonian circuits in Cayley graphs, preprint.

18. Dragan Marušič, Hamiltonian circuits in vertex-symmetric graphs of order 3p, preprint.

19. Dragan Marušič and T. D. Parsons, Hamiltonian paths in vertex-symmetric graphs of order 4p, preprint.

20. Dragan Marušič and T. D. Parsons, Hamiltonian paths in vertex-symmetric graphs of order 5p, preprint.

21. A. Nijenhuis and H. Wilf, Combinatorial algorithms, second edition, NEXPER Chap., Academic Press, New York, 1978.

22. Norman F. Quimpo, Hamiltonian properties of group graphs, Ph. D. Thesis, Ateneo de Manila Univ., Philippines, 1980.

23. William T. Trotter, Jr. and Paul Erdös, When the cartesian product of directed cycles is hamiltonian, J. Graph Theory, 2(1978), 137-142.

24. James M. Turner, On point symmetric graphs with a prime number of points, J. Combinatorial Theory, 3(1967), 136-145.

25. M. E. Watkins, Connectivity of transitive graphs, J. Combinatorial Theory, 8(1970), 23-29.

26. Helmut Wielandt, Finite permutation groups, Academic Press, New York, 1964.

Department of Mathematics
Simon Fraser University
Burnaby, British Columbia, V5A 1S6, Canada

ON STRONGLY HAMILTONIAN ABELIAN GROUP GRAPHS

C.C. CHEN AND N.F. QUIMPO

ABSTRACT.

Let G be a graph containing a spanning subgraph H isomorphic to $C_m \times L_n$ where C_m denotes the cycle with m vertices ($m \geq 3$) and L_n the path with n vertices ($n \geq 2$). Then H, in turn, contains a spanning subgraph H' isomorphic to $L_m \times L_n$. Vertices in H' can thus be coloured by two colours, say blue and red, so that no two adjacent vertices in H' are of the same colour. Then any two vertices in G are connected by a hamiltonian path if and only if G contains an edge joining two blue vertices and an edge joining two red vertices. This result enables us to characterize abelian group graphs G in which any two vertices are connected by a hamiltonian path.

1. Introduction

In this paper, we consider only finite simple undirected graphs without loops. For any two vertices a,b in a graph G, we shall write $a \sim b(G)$, or simply $a \sim b$, if a and b are adjacent in G. A *path* p from a to b in G is a sequence $a_0 a_1 \ldots a_n$ of distinct vertices in G such that $a_0=a$, $a_n=b$ and $a_i \sim a_{i+1}$ for $i=0,1,\ldots,n-1$. If the path p is such that $a_0 \sim a_n$, then we obtain a *cycle* in G denoted by $a_0 a_1 \ldots a_n a_0$. The path p is called a *hamiltonian path* if $G = \{a_0, a_1, \ldots, a_n\}$. The graph G is said to be *strongly hamiltonian* if, for any two vertices a,b in G, there exists a hamiltonian path from a to b.

Now, let G be a graph containing a spanning subgraph H isomorphic to $C_m \times L_n$ ($n \geq 3$, $n \geq 2$) where C_m denotes the cycle with m vertices and L_n the path with n vertices. We shall represent vertices in H by ordered pairs (i,j) with $0 \leq i \leq m-1$ and $0 \leq j \leq n-1$ in such a way that two ordered pairs (i,j), (h,k) are adjacent if, either i=h and $|j-k|=1$, or j=k and $i-h \equiv \pm 1 \pmod m$. The subgraph of H obtained from H by deleting the edges joining (0,j) to (m-1,j), ($j=0,1,\ldots,n-1$), will be denoted by H'. Evidently, H' is isomorphic to $L_m \times L_n$, and hence H' is a bipartite graph. We shall colour vertices in G by two colours, say blue and red, in such a way that no two adjacent vertices in H' are of the same colour; a vertex x in G is coloured blue if the distance from x to the vertex (0,0) in H' is even; whereas x is coloured red if this distance is odd. An edge joining two vertices a,b in G is called a *red-edge* if a,b are both coloured red; a *blue-edge* if both a,b are coloured blue; and a *purple edge* if a,b are differently coloured. Note that all edges in H' are purple; whereas, for even m, all edges in H are also purple.

In this paper, it will be proved that the graph G is strongly hamiltonian if and only if G has a blue edge and a red edge. This result enables us to characterize strongly hamiltonian abelian group graphs as follows.

Let G be a group and A a set of generators of G not containing the identity e. We shall denote by G(A) the graphs whose vertices are elements of G and for any a,b in G, $a \sim b(G(A))$ if and only if $a^{-1}b \in A$. We call G(A) a *group graph*. If G is an abelian group, then G(A) is called an *abelian group graph*. We shall prove that an abelian group graph G(A) is strongly hamiltonian if and only if it is neither a cycle nor a bipartite graph. Equivalently, G(A) is strongly hamiltonian if and only if A has two elements a,b with $a \neq b^{-1}$ and there exists a sequence $(a_1, a_2, \ldots, a_n)$ of elements of A such that n is odd and $a_1 a_2 \ldots a_n = e$.

2. Some basic lemmas

Let K be isomorphic to $L_m \times L_n$ $(m, n \geq 2)$. We may represent vertices in K by ordered pairs (i,j); $0 \leq i < m-1$, $0 \leq j < n-1$ and colour them blue or red as described in the **previous** section. We shall call a vertex x in K a *corner-vertex* if it is of degree 2; a *side-vertex* if it is of degree 3; and an *interior-vertex* if it is of degree 4. We have the following results which are useful for subsequent discussions.

Lemma 1. If mn is even, then there exists a hamiltonian path from any corner-vertex x to any vertex in K whose colour is different from that of x.

Proof. By symmetry, we may take x = (0,0) which is coloured blue and assume that m is even. First, let us consider the case when m = 2. We shall prove by induction on n that there exists a hamiltonian path from x to any red vertex y in K. If n = 2, then y = (1,0) or (0,1). We may then choose the path p to be (0,0)(0,1)(1,1)(1,0) or (0,0)(1,0)(1,1)(0,1) respectively. Hence Lemma 1 holds. Assume that the result holds for n < k (k > 2) and consider the case when n = k. If y = (0,1), then we have the following hamiltonian path from x to y :

$$p = (0,0)(1,0)(1,1)\ldots(1,k-1)(0,k-1)(0,k-2)\ldots(0,1).$$

If y = (1,0), then the required path is :

$$p = (0,0)(0,1)(0,2)\ldots(0,k-1)(1,k-1)(1,k-2)\ldots(1,0).$$

Next, if y = (0,j) where j = 3,5,..., let $K_1 = \{(i,h) \mid i=0,1 ; 0 \leq h < j\}$ and $K_2 = K - K_1$. Applying induction hypothesis to the section graphs K_1 and K_2, there exist a hamiltonian path p_1 from (0,0) to (1,j-1) in K_1 ; and a hamiltonian path p_2 from (1,j) to (0,j) in K_2. (See Fig.1). Then $p = p_1 p_2$ will be a hamiltonian path from (0,0) to (0,j) in K.

Finally, if y = (1,j), j = 2,4,6,..., we let $K_1 = \{(i,h) \mid i=0,1; 0 \leq h < j\}$ and $K_2 = K - K_1$. As before, we may apply the induction hypothesis to the section graphs K_1 and K_2 to get two hamiltonian paths p_1 from (0,0) to (0, j-1) and p_2 from (0,j) to y of K_1 and K_2 respectively such that $p_1 p_2$ will be a hamiltonian path of K from (0,0) to (1,j). This shows that Lemma 1 holds for m = 2.

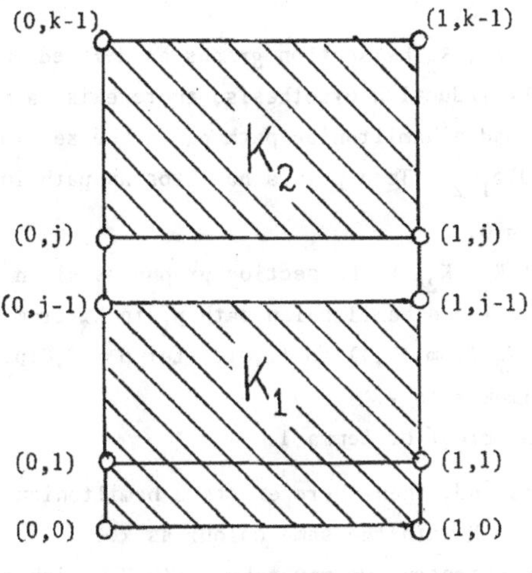

FIG. 1

We shall now prove that our result holds by induction on m. We have just dealt with the case $m = 2$. Assume that Lemma 1 is valid for all even $m < k$ where k is an even integer greater than 2 and consider the case when $m = k$. Let $y = (i,j)$ be any red vertex in K. We have the following three cases :

Case 1. $i \geq 2$.

In this case, let $K_1 = \{(s,t) \in K | s < 2\}$ and $K_2 = K - K_1$. By induction hypothesis, there exist a hamiltonian path p_1 in the section graph K_1 from (0,0) to (1,0); and a hamiltonian path p_2 in the section graph K_2 from (2,0) to y. Then $p = p_1p_2$ will be a hamiltonian path in K from x to y. (See Fig.2)

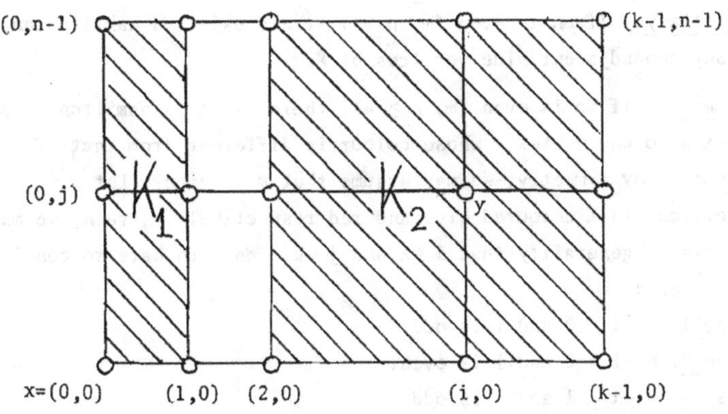

FIG. 2

Case 2. $i < 2$ and $j > 0$.

In this case, let K_1, K_2 be section graphs as defined in Case 1 and let $K_3 = K_1 - \{(0,0),(1,0)\}$. By induction hypothesis, there exist a hamiltonian path p_1 in K_2 from (2,0) to (2,1) and a hamiltonian path p_2 in the section graph K_3 from (1,1) to y. Let $p = (0,0)(1,0)p_1p_2$. Then p is a hamiltonian path in K from x to y.

Case 3. $i = 1$, $j = 0$.

In this case, let K_1, K_2, K_3 be section graphs as given in Case 2. By induction hypothesis, there exist a hamiltonian path p_1 in K_3 from (0,1) to (1,1) and a hamiltonian path p_2 in K_2 from (2,1) to (2,0). Let $p = (0,0)p_1p_2(1,0)$. Then p is a hamiltonian path in K from x to y.

This completes the proof of Lemma 1.

Lemma 2. If mn is odd, then there exists a hamiltonian path from any corner-vertex x to any vertex in K with the same colour as x.

Proof. Again, by symmetry, we may take $x = (0,0)$ which is coloured blue. Let $y = (i,j)$ be any blue vertex in K. We have the following two cases :

Case 1. $i > 0$.

In this case, let $K_1 = \{(0,0),(0,1),\ldots,(0,n-1)\}$ and $K_2 = K - K_1$. By Lemma 1, there exists a hamiltonian path p_1 in K_2 from (1,n-1) to y. Let $p = (0,0)(0,1)\ldots(0,n-1)p_1$. Then p will be a hamiltonian path in K from x to y.

Case 2. $i = 0$.

In this case, let $K_1 = (0,0),(1,0),\ldots,(m-1,0)$ and $K_2 = K - K_1$. By Lemma 1, there exists a hamiltonian path p_1 in K_2 from (m-1,1) to y. Let $p = (0,0)(1,0)\ldots(m-1,0)p_1$. Then p is a hamiltonian path in K from x to y.

The proof of Lemma 2 is complete.

We have the following immediate consequence of Lemmas 1 and 2.

Corollary. If mn is even ($m, n \geq 4$), there exists a hamiltonian path in K connecting any two adjacent side-vertices of K.

Lemma 3. If mn is even ($m, n \geq 4$), there exists a hamiltonian path in K from any vertex x to any vertex y whose colour is different from that of x.

Proof. By symmetry, we may assume that m is even. Let $x = (i,j)$, $y = (h,k)$ be two vertices in K coloured blue and red respectively. Again, we may assume without loss of generality that $i \leq h$ and $j \leq k$. We then have to consider the following four cases :

Case 1. $h - i = 0$ and i is odd.

Case 2. $h - i = 0$ and i is even.

Case 3. $h - i \geq 1$ and i is odd.

Case 4. $h - i \geq 1$ and i is even.

In case 1, let $K_1 = \{(s,t) \in K \mid t \leq j\}$ and $K_2 = K - K_1$. Then by Lemma 1, there exist a hamiltonian path p_1 in K_1 from x to (0,j) and a hamiltonian path p_2 in K_2

from (0, j+1) to y. Thus p_1p_2 will be a hamiltonian path in K from x to y.

In case 2, if $j > 0$, then we may define K_1, K_2 as above. Again, by Lemma 1, there exist a hamiltonian path p_1 in K_1 from x to (m-1, j) and a hamiltonian path p_2 in K_2 from (m-1, j+1) to y. Then p_1p_2 will be a hamiltonian path in K from x to y. On the other hand, if $j = 0$, then we let $K_3 = \{(s,t) \in K \mid i \leq s, 1 \leq t\}$, and $K_4 = \{(s,t) \in K \mid s < i\}$. (Note that K_4 may be empty). By Lemma 1, there exists a hamiltonian path p_1 in K_3 from (m-1, 1) to y. The path p_1 must contain an edge (i,d)(i, d+1) for some $d = 1,2,\ldots, n-2$. By the corollary to Lemmas 1 and 2, there exists a hamiltonian path p_2 in K_4 from (i-1, d) to (i-1, d+1). Let p_3 be the path obtained from p_1 by replacing the edge (i,d)(i, d+1) by p_2. Then (i,0)(i+1,0)... (m-1, 0)p_3 will be a hamiltonian path in K from x to y.

In case 3, let $K_5 = \{(s,t) \in K \mid s \leq i\}$ and $K_6 = K - K_5$. Then by Lemma 1, there exist a hamiltonian path p_1 in K_5 from x to (i,0) and a hamiltonian path in K_6 from (i+1, 0) to y. Then p_1p_2 will be a hamiltonian path in K from x to y.

In case 4, let $K_7 = \{(s,t) \in K \mid s \geq i, t \geq j\}$. By Lemma 1, there exists a hamiltonian path p in K_7 from x to y. The path p must contain an edge (d,j)(d+1, j) for some $d = i, i+1,\ldots, m-1$ or an edge (i,e)(i, e+1) for some $e = 0, 1,\ldots, n-1$ (say the former). Then by Lemma 1 again, there exists a hamiltonian path p_1 in K_8 from (d,j-1) to (d+1, j-1) where $K_8 = \{(s,t) \in K \mid s \geq i, t < j\}$. Let p_2 be the path obtained from p by replacing the edge (d,j)(d+1,j) by p_1. Then p_2 is a hamiltonian path in $K_7 \cup K_8$ from x to y. We can then extend p_2 to a hamiltonian path in K from x to y. (Note that if $K_8 = \emptyset$, then we can extend p immediately to a hamiltonian path in K from x to y.) This complete the proof.

Lemma 4. If mn is odd ($m, n \geq 3$), then there exists a hamiltonian path in K from any blue vertex x to any other blue vertex different from x.

Proof. Let x = (i,j), y = (h,k) be two different blue vertices of K. Without loss of generality, we may assume that $i \leq h$ and $j \leq k$. We then have the following three cases :

Case 1. i is even and $i < m-1$.

Case 2. i is even and $i = m-1$.

Case 3. i is odd.

In case 1, let $K_1 = \{(s,t) \in K \mid s \geq i, t \geq j\}$. Then by Lemma 2, there exists a hamiltonian path p in K_1 from x to y. This path can then be extended to a path in K from x to y (as in the proof of Lemma 3).

In case 2, if $j = 0$, then we just apply Lemma 2 to get a hamiltonian path from x to y. If $j > 0$, we let $K_2 = \{(s,t) \in K \mid t \geq j\}$. By Lemma 2, there exists a hamiltonian path p in K_2 from x to y. We may then extend p to get a hamiltonian path in K from x to y.

In case 3, if $i = m-2$ and $j = n-2$, then $y = (m-1, n-1)$, and so we can apply Lemma 2 to get a hamiltonian path in K from x to y. Hence we may assume either

$i < m-2$ or $j < n-2$, say the former. Now, if $i < h$, then we let $K_3 = \{(s,t) \in K | s \leq i\}$ and $K_4 = K - K_3$. By Lemma 1, there exists a hamiltonian path p_1 in K_3 from x to (i,0) and by Lemma 2, there is a hamiltonian path in K_4 from (i+1,0) to y. Then p_1p_2 will be a required hamiltonian path from x to y in K. On the other hand, if $j < k$, we let $K_5 = \{(s,t) \in K | t \leq j\}$ and $K_6 = K - K_5$. Then by Lemma 1 there exists a hamiltonian path p_3 in K_5 from x to (0,j) and by Lemma 2, there is a hamiltonian path p_4 in K_6 from (0,j+1) to y. Hence p_3p_4 will be a hamiltonian path in K from x to y, completing the proof.

<u>Lemma 5</u>. If $m \geq 3$ is odd and $n \geq 2$, then $C_m \times L_n$ is strongly hamiltonian.

<u>Proof</u>. Let $K = C_m \times L_n$. We first represent vertices of K by ordered pairs (s,t), $s = 0,1,\ldots,m-1$; $t = 0,1,\ldots,n-1$, and colour (s,t) blue or red according as i+j is even or odd. Let $x = (i,j)$, $y = (h,k)$ be two distinct vertices in K. Without loss of generality, we may assume that $i \leq 1$, $i \leq h$ and $j \leq k$ and that x is coloured blue. We have the following four cases :

<u>Case 1</u>. n is even and $i < h$.

<u>Case 2</u>. n is even and $i = h$.

<u>Case 3</u>. n is odd and $i < h$.

<u>Case 4</u>. n is odd and $i = h$.

In Case 1, we may assume that y is coloured blue. Let $K_1 = \{(s,t) \in K | s \leq i, t \geq j\}$, $K_2 = \{(s,t) \in K | s > i, t \geq j\}$. Then there exist a hamiltonian path p_1 in K_1 from x to the red vertex (0,n-1) and a hamiltonian path p_2 in K_2 from (m-1, n-1) to y. Then p_1p_2 will be a hamiltonian path in $K_1 \cup K_2$ from x to y, which can be extended to a hamiltonian path in K from x to y.

In Case 2, again we may assume that y is coloured blue. Let $K_3 = \{(s,t) \in K | t \leq j\}$ and $K_4 = \{(s,t) \in K | s < m-1, t > j\}$. Then there is a hamiltonian path p_1 in K_3 from x to (m-1,j) and a hamiltonian path in K_4 from (0, n - 1) to y. Then p_1(m-1, j+1)(m-1, j+2)...(m-1, n-1)p_2 will be a hamiltonian path in K from x to y.

In Case 3, we may assume that y is coloured red. Let $K_5 = \{(s,t) \in K | s \leq i, t \geq j\}$, $K_6 = \{(s,t) \in K | s > i, t \geq j\}$. If i is even, then there exist a hamiltonian path p_1 in K_5 from x to (0, n-1) and a hamiltonian path p_2 in K_6 from (m-1,n-1) to y. Then p_1p_2 will be a hamiltonian path in $K_5 \cup K_6$ from x to y which can be extended to a hamiltonian path in K from x to y. On the other hand, if i is odd, there exist a hamiltonian path p_3 in K_5 from x to (i, n-1) and a hamiltonian path p_4 in K_6 from (i+1, n-1) to y. Then p_3p_4 will be a hamiltonian path in $K_5 \cup K_6$ from x to y which can be extended to a hamiltonian path in K from x to y.

In Case 4, again we assume that y is coloured red. Let $K_7 = \{(s,t) \in K | t \leq j\}$ and $K_8 = \{(s,t) \in K | s < m-1, t > j\}$. Then there exist a hamiltonian path p_1 in K_7 from x to (m-1,j) and a hamiltonian path p_2 in K_8 from (0,n-1) to y. Then p_1(m-1,j+1)(m-1,j+2)...(m-1,n-1)p_2 will be a hamiltonian path in K from x to y.

The proof is complete.

Note that in the above proof, some small cases for m and n must be checked separately since there are restrictions on m and n in earlier lemmas.

Corollary. $C_m \times L_n$ is hamiltonian for any $m \geq 3$, $n \geq 1$.

3. Strongly hamiltonian graphs spanned by $C_m \times L_n$.

Let G be a graph containing a spanning subgraph H isomorphic to $C_m \times L_n$ ($m \geq 3$, $n \geq 2$). We shall represent vertices in H by ordered pairs (i,j) with $0 \leq i < m$, $0 \leq j < n$ and colour vertices in G by blue and red as described in section 1. We have :

Theorem 1. G is strongly hamiltonian if and only if G contains a blue-edge and a red-edge.

Proof. If m is odd, then by lemma 5, G is strongly hamiltonian. G contains also a blue edge (0,0)(m-1,0) and a red edge (0,1)(m-1,1). Hence we need only to consider the case when m is even.

Assume first that there exist a blue-edge b_1b_2 and a red-edge r_1r_2 in G. To prove that G is strongly hamiltonian, let x = (i,j), y=(h,k) be any two vertices in G. If x,y are differently coloured, then by Lemmas 1 and 3, x,y are connected by a hamiltonian path in H. Thus we may assume that x,y are of the same colour. Again, without loss of generality, we may assume that i = 0, and $j \leq k$. We assume also that x,y are both blue. The case that x,y are both coloured red can be settled in a similar manner. We can then prove by induction on k-j that there exists a hamiltonian path in H from x to y. First, consider the case k-j = 0. In this case, let $H_1 = \{(s,t) \in H \mid s < h \text{ and } t \geq j\}$, $H_2 = \{(s,t) \in H \mid s \geq h \text{ and } t \geq j\}$, $H_3 = \{(s,t) \in H \mid s < h \text{ and } t < j\}$ and $H_4 = \{(s,t) \in H \mid s \geq h \text{ and } t < j\}$ (See Fig. 3).

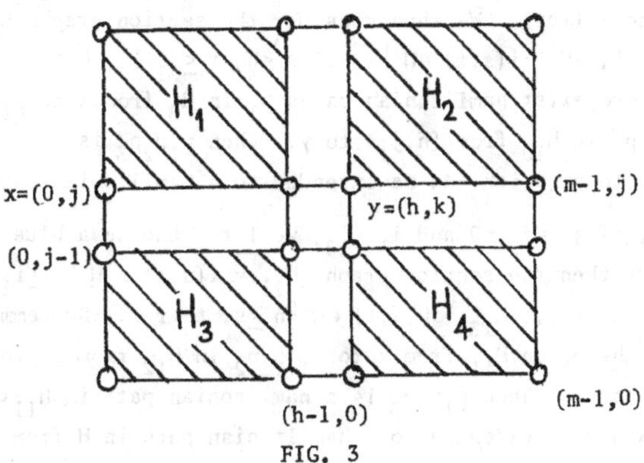

FIG. 3

Consider then the red-edge r_1r_2, with $r_1 = (i_1,j_1)$ and $r_2 = (i_2,j_2)$. We need to consider the following four cases :

Case 1. $r_1 \in H_1$ and $r_2 \in H_2$,

Case 2. $r_1 \in H_1$ and $r_2 \in H_4$,

Case 3. $r_1, r_2 \in H_1$; and

Case 4. $r_1 \in H_1$ and $r_2 \in H_3$.

In Case 1, we have by Lemma 1, a hamiltonian path p_1 in H_1 from x to r_1 and a hamiltonian path p_2 in H_2 from r_2 to y. Then p_1p_2 will be a hamiltonian path in $H_1 \cup H_2$ from x to y. If this path contains an edge $(r,j)(r+1,j)$ for some $0 \leq r < m$, by the Corollary to Lemmas 1 and 2, there exists a path q in $H_3 \cup H_4$ from $(r,j-1)$ to $(r+1, j-1)$. Replacing the edge $(r,j)(r+1,j)$ in p_1p_2 by the path $(r,j)q(r+1,j)$, we would obtain a hamiltonian path in H from x to y. If p_1p_2 contains no such edge $(r,j)(r+1,j)$, then we must have $m=4$, $r_1=(1,j)$ and $r_2=(3,j)$. Then we choose hamiltonian paths p_1 in $H_3 \cup H_4$ from $(0,j-1)$ to $(1,j-1)$, p_2 in $(H_1 \cup H_2) - \{(i,j) \mid i=0,1,2,3\}$ from $(3,j+1)$ to $(2,j+1)$. We thus obtain a hamiltonian path $xp_1r_1r_2p_2y$ in H from x to y.

In case 2, we have by Lemma 1 and its corollary, a hamiltonian path p_1 in H_1 from x to r_1 ; a hamiltonian path p_2 in $H_3 \cup H_4$ from r_2 to $(h+1,k-1)$; and a hamiltonian path p_3 in H_2 from $(h+1,k)$ to y. Then $p_1p_2p_3$ will be a hamiltonian path in H from x to y.

Let us now consider Case 3. Since any hamiltonian path in $H_1 \cup H_2$ from x to y can be extended to a hamiltonian path in H from x to y (see Case 1), we may assume that $H_3 \cup H_4 = \emptyset$ (i.e. $j=0$).

If $j_1 \neq j_2$ (say $j_1 < j_2$), then we consider the section graphs $H_5 = \{(s,t) \in H \mid s < h, t \leq j_1\}$, $H_6 = \{(s,t) \in H \mid t > j_1\}$ and $H_7 = H - (H_5 \cup H_6)$. Let z be any blue vertex represented by (r, j_1+1) for some $r \geq h$ and let $w = (r,j_1) \in H_7$. By Lemma 1 and its corollary, there exist hamiltonian paths p_1 in H_5 from x to r_1; p_2 in H_6 from r_2 to z ; and p_3 in H_7 from w to y. Then $p_1p_2p_3$ will be a hamiltonian path in H from x to y.

On the other hand, if $j_1 = j_2 \neq 0$, then $i_1 \neq i_2$ (say $i_1 < i_2$). Either i_1 or i_1+1 is odd (say the latter). We then consider the section graphs $H_8 = \{(s,t) \in H \mid s \leq i_1+1, t \leq j_1\}$, $H_9 = \{(s,t) \in H \mid i_1+1 < s < h, t \leq j_1\}$, $H_{10} = \{(s,t) \in H \mid s \geq h, t \leq j_1\}$. By Lemma 1, there exist hamiltonian paths p_1 in H_8 from x to r_1; p_2 in H_9 from r_2 to $(h-1,j_1)$; and p_3 in H_{10} from (h,j_1) to y. Then $p_1p_2p_3$ is a hamiltonian path in $H_8 \cup H_9 \cup H_{10}$ from x to y, which can be extended to a hamiltonian path in H from x to y.

Finally, if $j_1 = j_2 = 0$ and $i_1 < i_2$, we first choose a blue vertex $(h-1,r)$, with $r > 0$. Consider then the section graphs $H_{11} = \{(s,t) \in H \mid s \leq i_1, t \leq r\}$, $H_{12} = \{(s,t) \in H \mid i_1 < s < h, t \leq r\}$, $H_{13} = \{(s,t) \in H \mid h \leq s, t \leq r\}$. By Lemma 1, there exist hamiltonian paths p_1 in H_{11} from x to r_1 ; p_2 in H_{12} from r_2 to $(h-1,r)$ and p_3 in H_{13} from (h,r) to y. Then $p_1p_2p_3$ is a hamiltonian path in $H_{11} \cup H_{12} \cup H_{13}$ from x to y, which can again be extended to a hamiltonian path in H from x to y.

This settles Case 3.

Consider now Case 4. By Lemma 1 and its corollary, there exist hamiltonian paths p_1 in H_1 from x to r_1 ; p_2 in H_3 from r_2 to $(h-1,j-1)$ and p_3 in $H_2 \cup H_4$ from $(h,j-1)$ to y. Then $p_1p_2p_3$ will be a hamiltonian path in H from x to y.

Now assume that there exists a hamiltonian path from x to y in H whenever $k-j = v$ and consider the case when $k-j = v+1$. Let $H_{14} = \{(s,t) \in H \mid t \leq j\}$, $H_{15} =$

$\{(s,t) \in H \mid j < t < k\}$ and $H_{16} = \{(s,t) \in H \mid k \le t\}$. If $r_1, r_2 \in H_{14} \cup H_{15}$, by induction hypothesis there exists a hamiltonian path p_1 in $H_{14} \cup H_{15}$ from x to any blue vertex $z = (s,t)$ with $t = k-1$. By the corollary to Lemmas 1 and 2, there exists a hamiltonian path p_2 in H_{16} from (s,k) to y. Then p_1p_2 will be a hamiltonian path p in H from x to y. Similarly, if $r_1, r_2 \in H_{15} \cup H_{16}$, we would also obtain such a hamiltonian path. It remains to consider the case when $r_1 \in H_{14}$ and $r_2 \in H_{16}$. We then apply Lemma 1 to get a hamiltonian path p_1 in H_{14} from x to r_1. As in Case 1, we can extend p_1 to a hamiltonian path p_2 in $H_{14} \cup H_{15}$ from x to r_1. Again, there exists a hamiltonian path p_3 in H_{16} from r_2 to y. Then p_2p_3 will be a hamiltonian path in H from x to y. This proves the sufficiency of Theorem 1.

To establish the necessity of the theorem, assume that G is strongly hamiltonian. Since m is even, H contains only purple-edges. Moreover, exactly half of its vertices are blue and half are red. Let $a_1a_2 \ldots a_{mn}$ be any hamiltonian path in G from a blue vertex a_1 to another blue vertex a_{mn}. Then in the path $p = a_2a_3 \ldots a_{mn-1}$ with mn-2 vertices, there are $\frac{mn}{2}$ vertices which are red. Hence p must contain a red-edge a_ia_{i+1} ($i = 2, \ldots, mn-2$). This shows that G contains a red-edge. Similarly, it also contains a blue-edge. This completes the proof of Theorem 1.

4. Strongly hamiltonian abelian group graphs

Let G be any finite abelian group of order at least 3 and A a set of generators of G not containing the identity e. We first establish the following :

Lemma 6. G(A) contains a spanning subgraph isomorphic to $C_m \times L_n$ for some $m \ge 3$, $n \ge 1$.

Proof. We first choose a minimal subset A' of A which generates G and list out elements of A' as : $A' = \{a_1, a_2, \ldots, a_k\}$. Let $A_i = \{a_1, \ldots, a_i\}$ and G_i be the subgroup of G generated by A_i ($i = 1,2,\ldots,k$). If $G_1 = G$, then G(A) contains a spanning subgraph $C_m \times L_n$ where $m = O(G)$, $n = 1$. On the other hand, if $G_1 \ne G$, then $a_2 \notin G_1$. Let n_2 be the smallest positive integer such that $a_2^{n_2} \in G_1$. It is easy to check that $G_1 \cup a_2G_1 \cup \ldots \cup a_2^{n_2-1} G_1$ is a subgroup of G and so $G_2 = G_1 \cup a_2G_1 \cup \ldots \cup a_2^{n_2-1} G_1$. Moreover, each $a_2^j G_1$ is isomorphic to the graph $G_1(A_1)$ under the isomorphism $f : G_1 \to a_2^j G_1$ defined by $f(x) = a_2^j x$ and also each $a_2^j a_1^i$ is adjacent to $a_2^{j+1} a_1^i$. Therefore $G_2(A_2)$ contains a spanning subgraph H isomorphic to $C_{n_1} \times L_{n_2}$ where $n_1 = O(a_1)$.

Now, if $G_2 = G$, then we have nothing to prove; whereas, if $G_2 \ne G$, then G_2 is, by the corollary to Lemma 5, hamiltonian. Hence G_2 contains a spanning cycle C_{m_2} with $m_2 = n_1n_2$ vertices. By the same argument as before, $G_3(A)$ contains a spanning subgraph $C_{m_2} \times L_{n_3}$ where n_3 is the smallest positive integer such that $a_3^{n_3} \in G_2$. The validity of Lemma 6 thus follows by induction.

Corollary. Every abelian group graph of order ≥ 3 is hamiltonian.

Theorem 2. Let G be a finite abelian group and A a set of generators of G such that the subgroup $[a]$ generated by a is not G for some $a \in A$. Then G(A) is strongly hamiltonian if and only if G(A) is not bipartite.

Proof. It is easy to see that bipartite graphs are never strongly hamiltonian. Hence the necessity is obvious. To prove the sufficiency, assume that G(A) is not bipartite. Since $[a] \neq G$, by Lemma 6, G(A) contains a spanning subgraph $C_m \times L_n$, $m \geq 3$, $n \geq 2$. [Note that if G(A) has only four elements, then G(A) is a complete graph and hence strongly hamiltonian]. If m is odd, then G(A) is strongly hamiltonian by Lemma 5. We may thus assume that m is even. As indicated in Section 1, vertices in G(A) may be coloured blue or red. Let B, R be the sets of all blue vertices and red vertices respectively. Then $G(A) = B \cup R$ and $|B| = |R|$. Since G(A) is not bipartite, it must contain a blue or a red edge (say a blue edge b_1b_2 with $b_2 = b_1a_1$ for some a_1 in A). Define a mapping $f : G \to G$ by putting $f(x) = xa_1$ for all x in G. Then f is one : one and onto. Thus $f(R) \neq B$ since $b_2 = f(b_1) \notin f(R)$. Therefore $R \cap f(R) \neq \emptyset$. Hence $f(r_1) = r_2$ for some r_1, r_2 in R, and so r_1r_2 is a red edge in G(A). This shows that G(A) always contains a blue edge and a red edge. By Theorem 1, G(A) is thus strongly hamiltonian.

Theorem 3. Let G be a finite abelian group of order ≥ 3 and A a set of generators of G such that $[a] = G$ for all a in G and $a \neq b^{-1}$ for some distinct a,b in A. Then G(A) is strongly hamiltonian if and only if $|G|$ is odd.

Proof. Let $G = Z_n$ where Z_n denotes the addition cyclic group $\{0,1,\ldots,n-1\}$ of order n. Assume that $1 \in A$. If n is even, we may colour vertices x in G blue or red according as x is even or red. Then G is bipartite and so not strongly hamiltonian.

Conversely, assume that n is odd. Since $|A| > 1$, $A \cup A^{-1}$ contains an integer $k = 0,1,\ldots,[\frac{n}{2}]$ different from 1 and relatively prime with n. Then $n = km + r$ for some positive integer m,r with $0 < r < k$. It is easy to see that G(A) contains the graph in Figure 4 as a spanning subgraph.

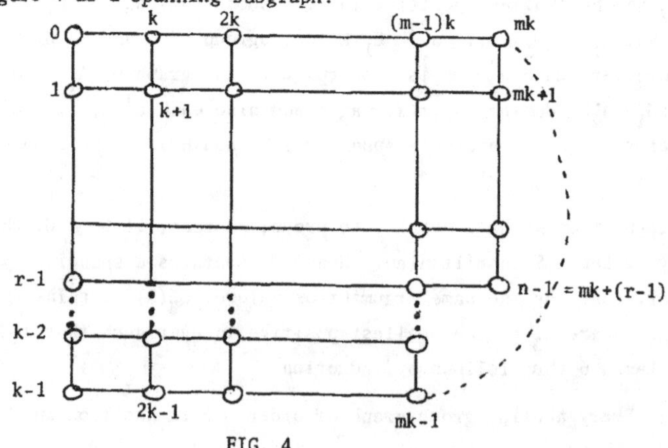

FIG. 4

Claim. There exists a hamiltonian path in G(A) from $n-1$ to any other vertex.

Indeed, we first let $H=\{0,1,\ldots,mk-1\}$, $K=\{ik+j \mid i=0,1,\ldots m; j=0,1,\ldots,r-2\}$, $I=G(A)-K$, and p be the path $(n-1)(n-2)\ldots(mk)$. We have to consider the following three cases :

Case 1. k is even.

Case 2. k is odd and m is odd.

Case 3. k is odd and m is even.

In Case 1, let x be any vertex in H. Then the colour of x must be different from $(m-1)k$ or $mk-1$ (say the former). By Lemma 1, there exists a hamiltonian path p_1 in H from $(m-1)k$ to x. Then pp_1 will be a hamiltonian path in G(A) from n-1 to x. This proves that there exists a hamiltonian path in G(A) from n-1 to any vertex $x=0,1,\ldots,[\frac{n}{2}]$. For $y=[\frac{n}{2}]+1,\ldots,n-2$, we let $x=n-2-y$. Then $x \leq [\frac{n}{2}]$ and so there is a hamiltonian path p_2 in G(A) from n-1 to x. If we replace each vertex z in p_2 by $(n-2)-z$, then we would obtain a hamiltonian path p_3 [called the *dual path* of p_2] in G(A) from n-1 to y. This settles Case 1.

In Case 2, for any blue vertex x in H, there is a hamiltonian path p_1 in H from either $(m-1)k$ or $mk-1$ to x. Then pp_1 will be a hamiltonian path in G(A) from n-1 to x. Further, if x is a blue vertex in G(A) -H, then let p_2 be a hamiltonian path in K from mk to x, and p_3 be a hamiltonian path in I from r to mk-1. Then $(n-1)(n-1-k)(n-1-2k)\ldots(r-1)p_3p_2$ will be a hamiltonian path in G(A) from n-1 to x. This proves that for any blue vertex x in G(A), there is a hamiltonian path from n-1 to x. By taking dual paths as in Case 1, every red vertex is also connected to n-1 by a hamiltonian path. This settles Case 2.

In Case 3, for any blue vertex x in H, there exists a hamiltonian path p_1 in H from $(m-1)k$ to x. Then pp_1 will be a hamiltonian path in G(A) from n-1 to x. Further, if x is a blue vertex in G(A) -H, then there exists a hamiltonian path in $K \cup \{r-1, r-1+k, r-1+2k, \ldots, n-1\}$ from n-1 to x, which can then be extended to a hamiltonian path in G(A) from n-1 to x. This shows that all blue vertices are connected to n-1 by hamiltonian paths. Again by taking dual paths as in Case 1, all red vertices are also connected to n-1 by hamiltonian paths. The proof is complete, since G(A) is vertex-transitive.

Theorem 4. A finite abelian group graph G(A) is strongly hamiltonian if and only if it is neither a cyclic nor a bipartite graph.

Proof. The necessity is clear. To prove the sufficiency, assume that G(A) is neither a cycle nor bipartite. Then there exist distinct $a,b \in A$ with $a \neq b^{-1}$. If $[x]=G$ for all $x \in A$, then $|G|$ is odd and so by Theorem 3, G(A) is strongly hamiltonian. On the other hand, if $[x] \neq G$ for some $x \in G$, then by Theorem 2, G(A) is strongly hamiltonian.

Note that a graph is not bipartite if and only if it contains an odd cycle (or equivalently, an odd closed walk). Note also that if $a_1a_2\ldots a_t = e$ for some $a_1, a_2, \ldots, a_t$ in A, then $ex_1x_2\ldots x_t$ is a closed walk in G(A), where $x_i = a_1a_2\ldots a_i$ $(i = 1, 2, \ldots, t)$. With these, we may rephrase Theorem 4 as follows :

Theorem 5. A finite abelian group graph G(A) is strongly hamiltonian if and only if it is not a cycle and $a_1^{t_1} a_2^{t_2} \ldots a_n^{t_n} = e$ for some integers $t_1, \ldots, t_n$ with an odd sum $t_1 + \ldots + t_n$ and some elements $a_1, \ldots, a_n$ in A.

5. Conclusion.

To end this paper, we would like to point out that the technique employed in characterizing strongly hamiltonian abelian group graphs here can also be used to characterize some other classes of strongly hamiltonian group graphs. (See [2] ,[3]). Also, we find that the study of hamiltonian properties of group graphs is important, because the construction of graphs from groups provides us with a big class of vertex-transitive graphs; and, since 1968, it was conjectured by L. Lovász that every vertex-transitive connected graph has a hamiltonian path. (See [1] , Problem 20). The validity of this conjecture is still far from being determined. It is hoped that group-theoretical approach may provide some partial (if not the whole) solution to the problem.

References

[1] J. A. Bondy and U. S. R. Murty, Graph Theory with Applications, (1976), The MacMillan Press Ltd., London.

[2] C. C. Chen and N. F. Quimpo, On Some Classes of Hamiltonian Graphs, SEA Bull. Math. Special Issue (1979), 252-258.

[3] C. C. Chen and N. F. Quimpo, On a Class of Strongly Hamiltonian Group Graphs, to appear.

Department of Mathematics,
National University of Singapore,
Bukit Timah Road,
Singapore 1025.

Ateneo de Manila University
P.O.B. 154,
Manila,
Republic of the Phillippines

MONOCHROMATIC LINES IN PARTITIONS OF Z^N

R.L. GRAHAM, WEN-CHING WINNIE LI,
AND
J.L. PAUL

INTRODUCTION

It is well known that perfect play in the familiar game of Tic-Tac-Toe* always results in a draw. It is less well known (but equally true) that no draw is possible for the 3-dimensional analogue of Tic-Tac-Toe. More precisely, if the set of integer points $I_3^3 = \{(x_1,x_2,x_3):x_i \in \{0,1,2\}, 1 \le i \le 3\}$ is arbitrarily partitioned into two classes, say $I_3^3 = C_1 \cup C_2$, then at least one of the classes must contain three collinear points. However, it is easy to construct partitions of $I_4^3 = \{(x_1,x_2,x_3):x_i \in \{0,1,2,3\}$ into two classes, neither of which contains four collinear points.**

In this paper we would like to describe recent work directed towards sharpening the known bounds for the n-dimensional generalizations of this problem. In particular, we show that in n dimensions, there exists a board of size $O(n)$ for which no draw is possible.

Because of space requirements, and keeping more in line with the talk on which this paper is based, we will furnish detailed proofs for very few of the assertions made. Rather we will indicate the underlying ideas needed and the general techniques used. Full details for these assertions (as well as their generalizations to the rather more difficult case of partitions into q classes for an arbitrary prime power q) can be found in [11].

* Also known as Noughts and Crosses in some parts of the world.

** The 4 × 4 × 4 analogue of Tic-Tac-Toe, marketed under the name of Qubic, has recently been shown by O. Patashnik to be a win for the first player. For an interesting account of this difficult computation, the reader should consult [16].

PRELIMINARIES

We first formulate the problem we study more precisely. For an arbitrary (fixed) integer $n \geq 2$, let $\mathbb{Z}^n$ denote the set of integer points in Euclidean n-space. A geometric line L of length ℓ in $\mathbb{Z}^n$ is defined to be a set of points described by

$$L = \{(x_1,\ldots,x_n): x_i = c_i + d_i u, u = 1,2,\ldots,\ell\}$$

where

(1) $$\text{g.c.d.}\{d_1, d_2, \ldots, d_n\} = 1.$$

The special lines in Tic-Tac-Toe have all $d_i = 0$ or ± 1. Condition (1) just guarantees that any lattice point in the convex hull of L is also in L.

By a 2-coloring χ of $\mathbb{Z}^n$, we just mean a map $\chi: \mathbb{Z}^n \to \{0,1\}$. A subset $X \subseteq \mathbb{Z}^n$ is said to monochromatic* under χ if for some $i \in \{0,1\}$,

$$X \subseteq \chi^{-1}\{i\}.$$

For a 2-coloring χ of $\mathbb{Z}^n$, let $\ell(\chi)$ denote the length of the longest monochromatic line in $\mathbb{Z}^n$. Finally, define

(2) $$\rho(n) = \inf_{\chi} \ell(\chi)$$

where χ ranges over all 2-colorings of $\mathbb{Z}^n$.

It follows from the fundamental result of Hales and Jewett [12], [10] that

(3) $$\rho(n) \to \infty \text{ as } n \to \infty.$$

Essentially, this theorem asserts the following: For any integers r and t, there is an integer $N = N(r,t)$ so that in any r-coloring of $I_t^N = \{(x_1,\ldots,x_N): x_i = 0,1,\ldots,t-1, 1 \leq i \leq N\}$ there is always a monochromatic line of length t with all $d_i = 0$ or 1.

The best bounds currently known for $N(r,t)$, as well as related corollaries such as van der Waerden's theorem for arithmetic progressions, are extremely weak. We will discuss these more fully at the end of the paper.

* Also often called homogeneous.

Our goal will be to bound $\rho(n)$ from above.

THE LINEAR UPPER BOUND

It turns out the basic functions we will use in our proofs depend on very old and fundamental quantities in combinatorics, namely, the binomial coefficients. However, we will derive several (what we believe to be) new results concerning them which are of interest* in themselves.

Let $\mathbb{Z}_2 = \{0,1\}$ denote the field of two elements.

Definition: For $a \geq 0$, define g_a: $\mathbb{Z} \to \mathbb{Z}_2$ by

$$g_a(x) \equiv \binom{x}{a} \equiv \frac{x(x-1)\ldots(x-a+1)}{a!} \pmod 2 \tag{4}$$

In Table 1 we list some of the initial values of the g_a.

a \ x	0	1	2	3	4	5	6	7
0	1	1	1	1	1	1	1	1
1	0	1	0	1	0	1	0	1
2	0	0	1	1	0	0	1	1
3	0	0	0	1	0	0	0	1
4	0	0	0	0	1	1	1	1
5	0	0	0	0	0	1	0	1
6	0	0	0	0	0	0	1	1
7	0	0	0	0	0	0	0	1

$g_a(x)$

Table 1

We next list various facts concerning the g_a. Let us write

$$a = \sum_{i \geq 0} a_i 2^i, \quad x = \sum_{i \geq 0} x_i 2^i, \text{ etc.,}$$

in their binary expansions.

Fact 1. $g_a(x) = 1$ if and only if $x_i \geq a_i$ for all i.

* In fact, perhaps of more interest than the main results of the paper.

Proof: Since the exact power of 2 which divides m! is $\sum_{k\geq 1}\left[\frac{m}{2^k}\right]$ then $\binom{a+b}{b}$ is odd if and only if

$$(5) \qquad \sum_{k\geq 1}\left[\frac{a+b}{2^k}\right] = \sum_{k\geq 1}\left[\frac{a}{2^k}\right] + \sum_{k\geq 1}\left[\frac{b}{2^k}\right].$$

But

$$[\alpha+\beta] \geq [\alpha] + [\beta]$$

implies that (5) holds if and only if

$$\left[\frac{a+b}{2^k}\right] = \left[\frac{a}{2^k}\right] + \left[\frac{b}{2^k}\right] \text{ for all } k.$$

Thus, (5) holds iff there is <u>no carrying</u> when adding a and b written base 2. Therefore, $g_a(x) = 1$ iff $x_i \geq a_i$ for all i. ∎

From Fact 1, a number of very useful results follow.

<u>Fact 2</u>. If $2^t \leq a < 2^{t+1}$ then g_a has period 2^{t+1}, i.e., $g(x+2^{t+1}) = g(x)$ for all $x \in \mathbf{Z}$.

<u>Fact 3</u>.

$$g_a(x) = \begin{cases} 0 \text{ for } x = 0,1,\ldots,a-1, \\ 1 \text{ for } x = a. \end{cases}$$

It follows from Facts 2 and 3 that the g_a, $0 \leq a < 2^{t+1}$, are <u>independent</u> over $\mathbf{Z}_2$ and, in fact, form a <u>basis</u> for functions $f\colon \mathbf{Z} \to \mathbf{Z}_2$ which have period 2^{t+1}.

It is clear that $g_a(x+1)$ has the same period as $g_a(x)$. More precise information is given in the following.

<u>Fact 4</u>.

$$g_a(x+1) \equiv g_a(x) + \sum_{i<a} \varepsilon_i g_i(x) \pmod 2$$

for a suitable choice of $\varepsilon_i = \varepsilon_i(a) \in \mathbf{Z}_2$.

Similarly, the period of $g_a(dx)$ divides the period of $g_a(x)$. It is not too difficult to prove the following crucial result.

Fact 5.

$$g_a(dx) \equiv \begin{cases} g_a(x) + \sum_{i<a} \varepsilon_i(d)g_i(x), & d \text{ odd} \\ \sum_{i\leq a/2} \varepsilon_i(d)g_i(x), & d \text{ even} \end{cases}$$

for suitable $\varepsilon_i(d) \in \mathbb{Z}_2$.

Finally, we have the very useful product formula.

Fact 6 (orthogonality). If $a + b < 2^{t+1}$ then

$$\sum_{x=0}^{2^{t+1}-1} g_a(x)g_b(x) \equiv \begin{cases} 1 \text{ if } a + b = 2^{t+1} - 1, \\ 0 \text{ otherwise.} \end{cases}$$

The preceding facts can now be used to prove the following result.

Lemma. If $\varepsilon_a \not\equiv 0 \pmod 2$ then

$$f(x) \equiv \sum_{i\leq a} \varepsilon_i g_i(x) \pmod 2$$

can have at most a consecutive equal values.

Proof: Assume the contrary and suppose without loss of generality (by Fact 4) that

$$(6) \qquad f(2^{t+1}-a-1) = \ldots = f(2^{t+1}-2) = f(2^{t+1}-1) = 0.$$

By Fact 3,

$$g_b(0) = g_b(1) = \ldots = g_b(b-1) = 0.$$

Thus, for $2^{t+1} - a - 1 = b$,

$$0 = \sum_{x=0}^{2^{t+1}-1} f(x)g_b(x) = \sum_{x=0}^{2^{t+1}-1} \sum_{i=0}^{a} \varepsilon_i g_i(x)g_b(x)$$

$$= \sum_{i=0}^{a} \varepsilon_i \sum_{x=0}^{2^{t+1}-1} g_i(x) g_b(x)$$

$$= \varepsilon_a$$

by Fact 6, which contradicts the initial hypothesis on ε_a. ∎

We are now in a position to prove a linear upper bound on $\rho(n)$. This result first appeared in [18]. Define a 2-coloring χ of the points $\bar{x} = (x_1,\ldots,x_n) \in \mathbf{Z}^n$ by

$$(6) \qquad \chi(\bar{x}) = \sum_{k=1}^{n} g_{n-1+k}(x_k) \pmod 2$$

Then

$$(7) \qquad \ell(\chi) \leq 2n-1.$$

To see this, consider a line

$$L: x_i(u) = c_i + d_i u, \quad u = 0,1,2,\ldots$$

where $\text{g.c.d.}\{d_1,d_2,\ldots,d_n\} = 1$. Thus, some d_i is odd. Let k be the largest index with d_k odd. Then

$$\chi(\bar{x}(u)) = \sum_{i=1}^{n} g_{n-1+i}(x_i(u)) = \sum_{i=1}^{n} g_{n-1+i}(c_i+d_i u)$$

$$= g_{n-1+k} + \sum_{i<n-1-k} \varepsilon_i g_i(u)$$

for suitable $\varepsilon_i \in \mathbf{Z}_2$ since by Fact 5, the $g_{n-1+i}(c_i+d_i u)$ with d_i even collapse into $\mathbb{Z}_2$-linear combinations of $g_j(u)$'s with $j \leq \frac{1}{2}(n-1+i) < n \leq n-1+k$. Thus, by the Lemma, $\chi(\bar{x}(u))$ has at most

$n - 1 + k \leq 2n - 1$ consecutive equal values. This proves (7). It follows from (7) that

(8) $$\rho(n) \leq 2n - 1$$

which is the upper bound promised for this section. Note that (7) actually holds when g.c.d.$\{d_1,\ldots,d_n\}$ is odd.

A SUBLINEAR UPPER BOUND

The basic idea we will use in reducing the bound in (8) is to replace the terms $g_j(x)$ in the definition of χ by functions $h_j(x,y,\ldots,z)$ of many variables which behave in certain ways like $g_j(x)$. The key result for such a substitution is the following extension of Fact 6.

Fact 7 (generalized orthogonality). If $0 \leq a,b,\ldots,c < 2^{t+1}$ then

$$\sum_{x=0}^{2^{t+1}-1} g_a(x)g_b(x)\ldots g_c(x) \equiv \begin{cases} 0 \text{ if } a_i+b_i+\ldots+c_i = 0 \text{ for some } i, \\ 1 \text{ if } a_i+b_i+\ldots+c_i = 1 \text{ for all } i. \end{cases}$$

As an example of the type of substitution we have in mind, consider the function h_5 defined by

$$h_5(x_1,x_2) \equiv g_5(x_1) + g_5(x_2) + g_1(x_1)g_4(x_2) \pmod 2.$$

Suppose $\bar{x}(u) = (x_1(u),\ldots,x_N(u))$ with

$$x_i(u) = c_i + d_iu, \quad \text{g.c.d.}\{d_1,\ldots,d_N\} = 1$$

and

$$\chi(\bar{x}) = \ldots + h_5(x_1,x_2) + \ldots .$$

Then

$$\chi(\bar{x}(u)) = \ldots + g_5(c_1+d_1u) + g_5(c_2+d_2u)$$
$$+ g_1(c_1+d_1u)g_4(c_2+d_2u) + \ldots .$$

Therefore,

$$\sum_{u=0}^{2^{t+1}-1} \chi(\bar{x}(u))g_2(u)$$

$$= \ldots + \sum_{u=0}^{2^{t+1}-1} h_5(c_1+d_1u, c_2+d_2u)g_2(u) + \ldots$$

$$= \ldots + \sum_{u=0}^{2^{t+1}-1} g_5(c_1+d_1u)g_2(u) + \sum_{u=0}^{2^{t+1}-1} g_5(c_2+d_2u)g_2(u)$$

$$+ \sum_{u=0}^{2^{t+1}-1} g_1(c_1+d_1u)g_4(c_2+d_2u)g_2(u) + \ldots .$$

By Fact 7, the sum of the three displayed terms is 1 modulo 2 <u>unless</u> $d_1 \equiv d_2 \equiv 0 \pmod 2$. Thus, if the line $\bar{x}(u)$ moves non-trivially in either the x_1 or the x_2 direction then $g_5(u)$ occurs in the expansion of $\chi(\bar{x}(u))$ with a coefficient of 1. It is in this sense that the use of h_5 in the definition of the coloring χ is equivalent to the use of g_5. However, h_5 "uses up" <u>two</u> coordinates where as g_5 only uses up one. For $k = \sum_{i\geq 0} k_i 2^i$, let $w(k)$ (called the <u>weight</u> of k) be defined by $w(k) = \sum_{i\geq 0} k_i$. Define $h_k : \mathbf{Z}^{w(k)} \to \mathbb{Z}_2$ by

$$h_k(z_1,\ldots,z_{w(k)}) = \sum_I G_I$$

where I ranges over all nonempty subsets of $\{1,\ldots,w(k)\}$ and $G_I = \prod_{i\in I} g_{a_i}(z_i)$ with $\sum_{i\in I} a_i = k$ having no carries when performed to the base 2. It is not hard to show in this case (using most of the preceding Facts) that if

$$z_i(u) = c_i + d_i u \text{ and } 2^t \leq k < 2^{t+1}$$

then

$$\sum_{u=0}^{2^{t+1}-1} h_k(z_1(u),\ldots,z_{w(k)}(u))g_{2^{t+1}-1-k}(u) \equiv 0 \pmod 2)$$

if and only if all the d_i are <u>even</u>.

The coloring χ^* we use for the sublinear bound can now be described. Let $W(n)$ denote $\sum_{j=1}^{n} w(j)$ and set $N(n) = W(2n-1) - W(n-1)$. Define

$$\chi^*: \mathbb{Z}^{N(n)} \to Z_2 \text{ by}$$

$$\chi^*(z_1,\ldots,z_{N(n)}) = \sum_{k=n}^{2n-1} h_k(z_{W(k-1)-W(n-1)+1},\ldots,z_{W(k)-W(n-1)})$$

It follows from the preceding Facts and the definition of h_k (in much the same way as in the proof of (7)) that if g.c.d.$\{d_1,\ldots,d_{N(n)}\} = 1$ then in the expansion of $\chi^*(\bar{z}(u))$ as a $\mathbb{Z}_2$-linear combination of g_k's, some g_i with $i \le 2n - 1$ occurs non-trivially. The Lemma then implies that $\chi^*(\bar{z}(u))$ has at most $2n - 1$ equal values. Thus,

$$\ell(\chi^*) \le 2n - 1$$

and consequently,

$$(9) \qquad \rho(N(n)) \le 2n - 1.$$

Finally, since it can be shown that (cf. [3])

$$W(m) = (1+o(1)) \frac{m \log m}{2 \log 2}$$

then straightforward computation using (9) implies the main result of the paper:

<u>Theorem</u>.

$$(10) \qquad \rho(n) \le (1+o(1)) \frac{2n \log 2}{\log n}$$

CONCLUDING REMARKS

As we remarked earlier, a similar but more complicated analysis can be carried out (see [11]) with an arbitrary prime power q replacing 2. These results imply the following upper bound on $\rho_r(n)$, the quantity which corresponds to $\rho(n)$ when r colors are used in coloring $\mathbb{Z}^n$.

$$\rho_r(n) \leq (1+o(1)) \frac{2 \log r}{r-1} \cdot \frac{n}{\log n}.$$

It is interesting to note that the properties of the g_a expressed in Facts 2, 3 and 4 actually characterize $\binom{x}{a}$ (mod 2). In fact, for an arbitrary prime p, suppose $f_a : \mathbb{Z} \to \mathbb{Z}_p$, $a = 0,1,2,\ldots$, is a sequence of functions satisfying:

(i) $f_a(x)$ has period p^{t+1} where $p^t \leq a < p^{t+1}$; $f_0(x)$ has period 1;

(ii) $f_a(x) = \begin{cases} 0 \text{ for } x = 0,1,\ldots,a-1, \\ 1 \text{ for } x = a, \end{cases}$

(iii) $f_a(x+1)$ is a $\mathbb{Z}_p$-linear combination of $f_i(x)$, $0 \leq i \leq a$.

Then (see [11])

$$f_a(x) \equiv \binom{x}{a} \pmod{p}.$$

This is not the case when p is composite.

A fundamental question in this subject which at present remains completely unanswered is whether or not the <u>density</u> version of the Hales-Jewett theorem holds. To explain what we mean by this, consider the well known theorem of van der Waerden on arithmetic progressions (see [22], [9], [10]):

> For all integers k and r, there is a number W(k,r) so that in any r-coloring of $\{1,2,\ldots,W(k,r)\}$ there is always a monochromatic arithmetic progression of k terms.

Observe that this result follows from the Hales-Jewett theorem - simply associate the integers in $[0,k^n-1]$ with $\mathbb{Z}_k^n$ by $x = \sum_{i=1}^{n} x_i k^{i-1} \leftrightarrow (x_1,x_2,\ldots,x_n)$. Nearly 50 years ago, Erdös and Turán [4] raised the question of determining <u>which</u> color class contains the long arithmetic progressions. In particular, they conjectured that the "most frequently occurring" color should have this property. This was shown to be the case for 3-term progressions by Roth [19] and finally, in 1974, Szemerédi managed to prove the general result.

<u>Theorem</u>* (Szemerédi [21]). For all k and $\varepsilon > 0$ there is a number

* Recently, Furstenburg and others (see [5], [6]) have succeeded in proving Szemerédi's theorem using newly developed results from ergodic theory.

$S(k,\varepsilon)$ such that if $R \subseteq \{1,2,\ldots,S(k,\varepsilon)\}$ and $|R| \geq \varepsilon\{S(k,\varepsilon)\}$ then R must contain a k-term arithmetic progression.

Van der Waerden's theorem is an example of a Ramsey (or partition) theorem. Szemerédi's theorem (which clearly implies van der Waerden's result) is the stronger density version of it.

It is very tempting to believe that the corresponding density version of the Hales-Jewett theorem should hold.

Conjecture*. For all t and $\varepsilon > 0$, there exists a number $C(t,\varepsilon)$ so that if $n \geq C(t,\varepsilon)$ and $R \subseteq \{(x_1,\ldots,x_n):x\varepsilon\{0,1,\ldots,t-1\}\}$ has $|R| \geq \varepsilon t^n$ then R contains a line of length t of the form $x_i = c_i + d_i u$ with all $d_i = 0$ or 1.

Even the case $t = 2$, the only non-trivial case for which the conjecture is known to be true, requires an argument. In this case, we are required to find a "line" which consists of two points of the form:

$$\begin{cases} x = (\ldots,a,\ldots,0,\ldots,b,\ldots,0,\ldots,c,\ldots), \\ \\ y = (\ldots,a,\ldots,1,\ldots,b,\ldots,1,\ldots,c,\ldots). \end{cases}$$

However we can associate to each point $z = (z_1,z_2,\ldots,z_n)$, $z_i = 0$ or 1, a subset $Z = \{1,2,\ldots,n\}$ in the usual way; namely, $i \varepsilon Z$ iff $z_i = 1$. Under this association, our "line" is just a pair of subsets X, Y with X a proper subset of Y. But a theorem of Sperner [20] shows that the largest family of subsets of $\{1,2,\ldots,n\}$ having no member properly contained in another has at most $\binom{n}{\left[\frac{n}{2}\right]} \sim \frac{2^n}{\sqrt{\pi n}} = o(2^n)$ members. Thus, if $|R| > \varepsilon 2^n$ then it must contain a line of the desired form.

Very recently, progress has been made by T. C. Brown and J. P. Buhler (see [2]) for a weakened version of the case $t = 3$.

Finally, we make a few remarks concerning what we believe to be the "truth" concerning the actual values of $\rho(n)$. From above it seems likely that $\rho(n) = o(\log n)$ or perhaps even $\rho(n) = o(\log\log n)$ should be true. However, the lower bound for $\rho(n)$, which depends on the known lower bounds for the Hales-Jewett theorem, is embarrassingly weak (it is not even primitive recursive). As an example of a measure of our ignorance in this area, consider the following related Ramsey-type theorem.

* One of the authors is currently offering US \$1000 for a resolution of this conjecture. It is actually a strengthened version of an earlier conjecture of Moser [15].

Theorem (see [8]). Let $Q_n = \{(x_1,\ldots,x_n): x_i = 0 \text{ or } 1\}$. Then there is a number N_0 so that for any 2-coloring of the line segments joining pairs of points in Q_n, there always exist four coplanar points of Q_n spanning 6 line segments all having the same color.

The best estimate from above currently available for N_0 can be described as follows (also see [1], [7], [14]).

Following Knuth [13], define

$$a\uparrow n = a^n,$$

$$a\uparrow\uparrow n = \underbrace{a\uparrow(a\uparrow(\ldots(a\uparrow a))\ldots)}_{n\ a\text{'s}},$$

$$= a^{a^{\cdot^{\cdot^{\cdot^{a}}}}}\Big\}\,n$$

and, in general,

$$a\underbrace{\uparrow\uparrow\ldots\uparrow}_{t+1} n = a\underbrace{\uparrow\ldots\uparrow}_{t}(a\underbrace{\uparrow\ldots\uparrow}_{t}(\ldots(a\underbrace{\uparrow\ldots\uparrow}_{t} a)\ldots)).$$

where n a's occur on the right-hand side. (The reader is invited to something as simple as $3\uparrow\uparrow\uparrow\uparrow 3$ into normal notation). Then it has been shown that

$$N_0 \le \left.\begin{matrix} 3\uparrow\uparrow\uparrow\uparrow 3 \\ 3\uparrow\uparrow\ldots\uparrow 3 \\ 3\uparrow\uparrow\uparrow \quad \uparrow\uparrow 3 \\ \cdots\cdots\cdots \\ 3\uparrow\uparrow\uparrow\uparrow\ldots\ldots\ldots\uparrow\uparrow\uparrow\uparrow 3 \end{matrix}\right\}\ 64 \text{ layers}$$

where each number represents the number of arrows in the expression below it.

The best lower bound known for N_0 is:

$$N_0 \ge 6.$$

Probably, $N_0 = 6$.

REFERENCES

[1] G. R. Blakely and I. Borosh, Knuth's iterated powers, Advances in Math. 34 (1979), 109-136.

[2] T. C. Brown and J. P. Buhler, A density version of a geometric Ramsey theorem (to appear).

[3] L. E. Bush, An asymptotic formula for the average sum of digits of integers, Amer. Math. Monthly, 47 (1940), 154-156.

[4] P. Erdös and P. Turán, On some sequences of integers, J. London Math. Soc., 11 (1936), 261-264.

[5] H. Furstenberg, Recurrence in ergodic theory and combinatorial number theory, Princeton Univ. Pr. (to appear).

[6] H. Furstenberg and B. Weiss, Topological dynamics and combinatorial number theory, J. Anal. Math., 34 (1978), 61-85.

[7] M. Gardner, Mathematical games in which joining sets of points by lines leads into diverse (and diverting) paths, Sci. American 237 (1977), 18-28.

[8] R. L. Graham and B. L. Rothschild, Ramsey's theorem for n-parameter sets, Trans. Amer. Math. Soc. 159 (1971), 257-292.

[9] R. L. Graham and B. L. Rothschild, A short proof of van der Waerden's theorem on arithmetic progressions, Proc. Amer. Math. Soc. 42 (1974), 385-386.

[10] R. L. Graham, B. L. Rothschild, J. H. Spencer, Ramsey Theory, John Wiley and Sons, New York, 1980.

[11] R. L. Graham, Wen-Ching Winnie Li and J. L. Paul, Homogeneous collinear sets in partitions of $\mathbb{Z}^n$ (to appear).

[12] A. W. Hales and R. I. Jewett, Regularity and positional games, Trans. Amer. Math. Soc., 106 (1963), 222-229.

[13] D. E. Knuth, Mathematics and computer science: Coping with finiteness, Science 194 (1976), 1235-1242.

[14] N. McWhirter, Guiness book of world records, Bantam, New York, 1980, p. 193.

[15] L. Moser, Problem 170, Canad. Math. Bull. 13 (1970, p. 268.

[16] O. Patashnik, Computer-aided problem solving and 4 × 4 × 4 Tic-Tac-Toe (to appear in Math. Mag.).

[17] J. L. Paul, On the partitioning of sets of lattice point paths, Proc. 5th British Comb. Conf. 1975, Congressus Numerantum XV, Utilitas Mathematica Pub. Inc., Winnipeg, 1976, pp. 497-502.

[18] J. L. Paul, Partitioning the lattice points in $\mathbb{R}^n$, J. Comb. Th. (A), 26 (1979), 238-248.

[19] K. F. Roth, On certain sets of integers, J. London Math. Soc., 28 (1953), 104-109.

[20] E. Sperner, Ein Satz über Untermenge einer endlichen Menge, Math. Z., 27 (1928), 544-548.

[21] E. Szemerédi, On sets of integers containing no k elements in arithmetic progression, Acta Arith., 27 (1975), 199-245.

[22] B. L. van der Waerden, Beweis einer Baudetschen Vermutung, Nieuw Arch. Wisk., 15 (1927), 212-216.

Bell Laboratories
600 Mountain Ave
Murray Hill
New Jersey 07974
U.S.A.

Department of Mathematics
Pennsylvania State University
University Park
Pennsylvania 16802
U.S.A.

Department of Mathematics
University of Cincinnati
Cincinnati
Ohio 45221
U.S.A.

COMPLETE STABLE MARRIAGES AND SYSTEMS OF I-M PREFERENCES

J. S. HWANG

Recently, we have introduced the notion of stable permutations in Latin squares. In this paper, we introduce the systems of I-M preferences in the marriage theory and we prove that in such a system, the study of stable marriages in two matrices is equivalent to the study of stable permutations in one matrix.

1. INTRODUCTION

The notion of stable marriages has been introduced by D. Gale and L. S. Shapley [2]. Let M be the set of n men $A_1, A_2, \ldots, A_n$ and W be the set of n women $a_1, a_2, \ldots, a_n$. A marriage m is a bijection from M onto W, for instance, if each man A_i gets married to each woman a_i, than we have a marriage

$$(1) \qquad m = \begin{pmatrix} A_1 A_2 \ldots A_n \\ a_1 a_2 \ldots a_n \end{pmatrix} = ((a_1, a_2 \ldots a_n)),$$

where the order of A_i is fixed, for each $i = 1, 2, \ldots n$.

A system of preference of order n

$$(2) \qquad P(n) = \{A_i : (a_{ij}) ; a_i : (A_{ij})\}, \quad i, j = 1, 2, \ldots, n,$$

is complete if and only if each man A_i has an order of preference a_{ij}, $j = 1, 2, \ldots, n$, for all women, and each women a_i also has an order of preference A_{ij}, $j = 1, 2, \ldots, n$, for all men, see D. E. Knuth [5].

For simplicity, we shall use the following notation

$$a_1 A a_2 \Leftrightarrow A \text{ prefers } a_1 \text{ to } a_2,$$

$$A_1 a A_2 \Leftrightarrow a \text{ prefers } A_1 \text{ to } A_2.$$

Let m be a marriage defined by (1), we say that m is unstable if and only if there exist an unstable pair (A_i, a_j) such that

(3) $$a_j A_i a_i \quad \text{and} \quad A_i a_j A_j .$$

We call a marriage m stable if and only if m is not unstable. The meaning of unstability (3) is consistent with our common life, namely, there is a pair (A_i, a_j), of a man and a woman, such that they are not married but each loves the other more than their spouse.

Having defined the notion of complete systems of preferences, we shall now define the systems of ideal maximum (I-M) preferences. By the very meaning of completeness, we can see each man A_i has the order of preference on each woman a_j which we denote by $A_i(a_j)$, and each woman a_j also has the order of preference on each man A_i which we denote by $a_j(A_i)$. A preference $P(n)$ of order n is called an I-M preference if it satisfies

(4) $$A_i(a_j) + a_j(A_i) = n + 1, \quad \text{for all } i, j = 1, 2, \ldots, n.$$

With the above notations, (3) can also be represented by

(3)* $$A_i(a_j) < A_i(a_i) \text{ and } a_j(a_i) < a_j(A_j), \text{ for some } i \text{ and } j.$$

The starting point for us to study the systems of I-M preferences essentially comes from the following Knuth system [5, Problem 5].

Example 1. *Let* P(4) *be a system of preference defined by*

#:	1	2	3	4		#:	1	2	3	4
A:	a	b	c	d		a:	D	C	B	A
B:	b	a	d	c		b:	C	D	A	B
C:	c	d	a	b		c:	B	A	D	C
D:	d	c	b	a	,	d:	A	B	C	D

Then clearly, we have $A(a) + a(A) = 1 + 4$, $A(b) + b(A) = 2 + 3$, ..., $D(a) + a(D) = 4 + 1$. *This shows that* P(4) *is an* I-M *preference.*

We now explain the choice of the term I-M preference by two reasonings. First, it is computable. The number of all stable marriages can be computed for each I-M preference by connecting stable marriages and stable permutations. Second, it is constructable. For each number n of the form $n = 2^p$, there can be constructed an I-M preference for which the number of all stable marriages is reasonably large and by which we can construct an ideal maximum number of stable marriages for each positive integer, where the "ideal maximum" means close to the maximum, but itself might not be the maximum.

It is known that the theory of stable marriages studies the behaviour between two matrices. In general, this is much more difficult than to study one matrix. The importance of the system of I-M preferences is to reduce the marriage theory to the study of one matrix in terms of stable permutations (see Theorem 8).

2. SYSTEMS OF I-M PREFERENCES

The notion of I-M preferences has already been defined in (4). From this property, instead of (2), we shall represent a preference of order n by

$$(5) \qquad P(n) = \left\{ \begin{matrix} \# : & j & \# : & j \\ A_i : & (a_{ij}), & a_i : & (A_{ij}) \end{matrix} \right\}, \quad i,j = 1, 2, \ldots, n.$$

For simplicity, $P(n)$ will be called a Latin square preference if both (a_{ij}) and (A_{ij}) are Latin squares, i.e. no row or column contains the same element more than once, refer to the book of J. Dénes and A. D. Keedwell [1]. For convenience, we shall call the above two matrices pair matrices. The following theorem describes the relation between the I-M preferences and the Latin squares.

<u>Theorem 1</u>. *If a preference* $P(n)$ *is an* I-M *preference then both pair matrices are Latin squares and one of them uniquely determines the other.*

<u>Proof</u>. Suppose on the contrary that one of them, say, (a_{ij}) is not a Latin square. Then there is some $i_1 \neq i_2$ and some j such that

$$a_{i_1 j} = a_{i_2 j} = a_{j*}, \; 1 \leq i_1 \neq i_2, j, j^* \leq n,$$

which gives

$$A_{i_1}(a_{j*}) = A_{i_2}(a_{j*}) = j .$$

Owing to (4), this implies that

$$a_{j*}(A_{i_1}) = a_{j*}(A_{i_2}) = n+1-j ,$$

which is absurd due to $i_1 \neq i_2$.

To prove that one of them, again (a_{ij}), determines uniquely the other, we may start from the first column of (a_{ij}). By virtue of (4) again, this uniquely determines the last column of (A_{ij}) . Similarly, each j-th column of (a_{ij}) uniquely determines each $(n+1-j)$-th column of (A_{ij}). This completes the proof.

The significance of the above theorem is that it shows us that for I-M preferences, the study of stable marriages between two matrices can be reduced to study what we call stable permutations in one matrix [3]. This will be explained more clearly in Section 6. We also notice that from Theorem 1 together with a well-known result [1, p.146], we can see the total number of I-M preferences of order n is at least $n!(n-1)!\ \ldots\ 2!1!$. Also notice that a preference $P(n)$ whose pair matrices are both Latin squares, is not necessarily an I-M preference. One such example follows.

<u>Example 2.</u> *Let* $P(3)$ *be a preference defined by*

#:	1	2	3		#:	1	2	3
A:	a	b	c		a:	C	B	A
B:	b	c	a		b:	A	C	B
C:	c	a	b	,	c:	B	A	C .

Since $A(c) + c(A) = 3+2 > 4$, $P(3)$ *is not an* I-M *preference. In this case, there are only two stable marriages* ((abc)) *and* ((bca)). *Certainly, one matrix can not determine the other.*

3. FUNDAMENTAL MARRIAGES

In this section, we shall prove a necessary and sufficient condition for a preference $P(n)$ to be an I-M preference. For this, we let $P(n)$ be a Latin square preference and let it be written as (5). Then both of the pair matrices of (5) are Latin squares. From this Latinness, we can define the j-th fundamental marriage of $P(n)$ by the following marriage

$$(6) \qquad F_j = \begin{pmatrix} A_1 & A_2 & \ldots & A_n \\ a_{1j} & a_{2j} & \ldots & a_{nj} \end{pmatrix}, \quad j = 1,2,\ \ldots,\ n.$$

We call such marriages fundamental as they correspond to those fundamental permutations defined in [3], which are stable in one matrix. With the help of this definition, we shall prove the following equivalent relation between the I-M preferences and the stability of fundamental marriages.

<u>Theorem 2.</u> *Let* $P(n)$ *be a Latin square preference and* F_j *be*

the j-th *fundamental marriage, then* P(n) *is an* I-M *preference if and only if all* F_j *are stable,* $j = 1,2, \ldots, n$.

Proof. Let P(n) be an I-M preference. We shall prove the stability of all fundamental marriages. We first observe from the definitions (5) and (6) that for each j ,

(7) $$A_i(a_{ij}) = j, \quad i = 1,2, \ldots, n .$$

It follows from (4) and (7) that for each j , we have

(8) $$a_{ij}(A_i) = n+1-j, \quad i = 1,2, \ldots, n .$$

Now suppose that F_{j_1} is unstable for some j_1, $1 \le j_1 \le n$. Then by (3), there is an unstable pair (A_p, a_{qj_1}) such that

(9) $$a_{qj_1} A_p a_{pj_1} \quad \text{and} \quad A_p a_{qj_1} A_q .$$

Let $A_p(a_{qj_1}) = j_0$, then by choosing $i = p$ and $j = j_1$ in (7) and the first part of (9), we must have

(10) $$j_0 < j_1 \quad \text{and} \quad a_{qj_1}(A_p) = n+1-j_0 .$$

By choosing $i = q$ and $j = j_1$ from (8), we obtain

(11) $$a_{qj_1}(A_q) = n+1-j_1 .$$

Owing to (11) together with the second part of (9) and (10), we must have

$$n+1-j_0 = a_{qj_1}(A_p) < a_{qj_1}(A_q) = n+1-j_1 ,$$

which contradicts the first part of (10). We thus establish the stability of all F_j, $j = 1,2, \ldots, n$.

Conversely, if all F_j are stable, we want to prove that P(n) must be an I-M preference. We start from the last one

$$F_n = \begin{pmatrix} A_1 & \ldots A_p & \ldots A_q & \ldots A_n \\ a_{1n} & \ldots a_{pn} & \ldots a_{qn} & \ldots a_{nn} \end{pmatrix} .$$

Since $A_i(a_{in}) = n$, by virtue of (4), it is sufficient to show that $a_{in}(A_i) = 1$, for all $i = 1,2, \ldots, n$. Suppose on the contrary that

(12) $$a_{qn}(A_q) > 1 = a_{qn}(A_p) , \quad \text{for some } p \text{ and } q .$$

By considering the preference of A_p, we find that

(13) $$A_p(a_{qn}) < n = A_p(a_{pn}) .$$

It follows from (13) and (12) that

$$A_p(a_{qn}) < A_p(a_{pn}) \quad \text{and} \quad a_{qn}(A_p) < a_{qn}(A_q) ,$$

which contradicts the stability of F_n due to (3)*.

For the general case, we shall verify by induction. Assume that for all $j > k$, we have

(14) $$a_{ij}(A_i) = n+1-j, \; i = 1,2, \ldots, n .$$

We shall prove that (14) holds too for $j = k$. Again, suppose on the contrary that for some p and q, $p \neq q$,

(15) $$a_{qk}(A_q) > n+1-k = a_{qk}(A_p) .$$

Since $p \neq q$, there are only two cases to be considered, namely, either $A_p(a_{qk}) > k$ or $< k$. In the first case, we can represent $a_{qk} = a_{pr}$ for some $r \geq k+1$, and therefore by (14) we obtain

$$a_{qk}(A_p) = a_{pr}(A_p) = n+1-A_p(a_{pr}) = n+1-A_p(a_{qk}) < n+1-k ,$$

which contradicts (15). For the second case, we have from (7)

(16) $$A_p(a_{qk}) < k = A_p(a_{pk}) .$$

It follows from (15) and (16) that

$$A_p(a_{qk}) < A_p(a_{pk}) \quad \text{and} \quad a_{qk}(A_p) < a_{qk}(A_q) ,$$

which again contradicts the stability of F_k due to (3)*. This shows that (14) holds for $j = k$ and the proof is complete.

In the second example, P(3) is not an I-M preference and therefore by using the above theorem we can see some fundamental marriage must be unstable, in fact, F_3 = ((cab)) is the only one unstable.

Notice that it has been shown by Gale and Shapley [2, Theorem 1] that there always exists a stable marriage in any system of preference and in general, the number of stable marriages can not be expected to be greater than one, see [2, Example 2]. However, for I-M preference, we do have more than one and in fact from the above Theorem 2 we obtain immediately the following

Corollary 1. If P(n) is a system of I-M preference, then the number of stable marriages is at least n .

4. OPTIMALITY OF I-M PREFERENCES.

For each marriage m defined by (1), there can be associated with it a sequence of n numbers defined by

(17) $D(m) = (d_1, d_2, \ldots, d_n)$, where $d_j = A_j(a_j) + a_j(A_j)$, $j = 1,2,\ldots,n$.

It is clear that P(n) is an I-M preference if and only if for each marriage m , we always have

(18) $D(m) = (n+1, n+1, \ldots, n+1)$, or $d_j = n+1$, $j = 1,2,\ldots, n$.

With respect to (17) and (18), we shall prove the following optimal property of I-M preferences.

Theorem 3. *Let* P(n) *be a Latin square preference and let* m *be a marriage for which the numbers* d_j *defined by (17) satisfy*

(19) $$d_j \geq n + 2, \; j = 1,2, \ldots, n.$$

Then m *is unstable.*

Proof. We first let

(20) $1 < b = \min A_j(a_j), \;\; 1 \leq j \leq n,$ and $A_k(a_k) = b$ for some k.

We consider the first matrix (a_{ij}) of $P(n)$. Owing to the Latin property, we find that this minimum element a_k has to occur in each column. We denote by A_{n_j} the man whose j-th preference is this woman a_k, $j = 1,2, \ldots, b-1$. Remember the man A_{n_j} gets married with the woman a_{n_j} due to the notation (1). By virtue of (20), we find that

(21) $$A_{n_j}(a_k) < A_{n_j}(a_{n_j}), \; j = 1,2, \ldots, b-1.$$

If we can show that there exists a man A_{n_ℓ} such that

(22) $$a_k(A_{n_\ell}) < a_k(A_k), \;\; \text{where} \;\; 1 \leq \ell \leq b-1 \; ,$$

then the unstability of m will follow from (21) and (22).

Owing to (17), (19), and (20), we obtain

$$a_k(A_k) \geq n+2-A_k(a_k) = n+2-b \; .$$

This yields

(23) $$n - a_k(A_k) \leq b-2 \; .$$

Suppose now that there doesn't exist a man A_{n_ℓ} satisfying (22), then we have

$$a_k(A_{n_j}) \geq a_k(A_k), \quad j = 1,2, \ldots, b-1.$$

By considering the row $A_{k1}, A_{k2}, \ldots, A_{kn}$ in the preference of the woman a_k, we can see that there are $n-a_k(A_k)$ men after the man A_k in this preference of a_k. It follows that

$$n - a_k(A_k) \geq b-1 .$$

This contradicts (23) and shows the unstability of m.

As a corollary, we shall explain a geometric meaning of the above theorem for some particular case. To do this, we consider the pair matrices (a_{ij}) and (A_{ij}) as two lattice planes. We divide them into left and right open half-planes which we denote by $L(a)$, $R(a)$; $L(A)$, $R(A)$ respectively. With these notations, Theorem 3 yields immediately the following

Corollary 2. *Let* $P(n)$ *be a Latin square preference associated with* $L(a)$, $R(a)$; $L(A)$, $R(A)$. *Let* m *be a marriage defined by (1). If all women* $a_j \in R(a)$ *and all men* $A_j \in R(A)$, *then* m *is unstable.*

We notice that the situation of Corollary 2 can never occur in an I-M preference because of (18). We also notice that Theorem 3 is best possible in the sense that condition (18) can not be replaced by some d_{j_0} instead of all d_j. This will be seen from the following

Example 3. *Let* $P(4)$ *be defined by*

#:	1	2	3	4		#:	1	2	3	4
A:	a	b	d	c		a:	A	B	C	D
B:	b	c	a	d		b:	D	C	B	A
C:	c	d	b	a		c:	B	A	D	C
D:	d	a	c	b	,	d:	C	D	A	B .

Then the marriage $m = \begin{pmatrix} A & B & C & D \\ a & b & d & c \end{pmatrix}$ is stable, however, $D(c) + c(D) = 3 + 3 = 6$.

5. STABLE PERMUTATIONS

The notion of stable permutations in Latin squares has recently been introduced in [3, 4]. The main purpose in this section is to prove that in a system of I-M preference the study of stable marriages in two matrices is equivalent to the study of stable permutations in one matrix. As described in Corollary 2, what we need is to consider the pair matrices as two lattice planes. Since we deal only with I-M preferences, it is sufficient to regard one of them, say, (a_{ij}) to be a lattice plane. In this case, each woman $\{a_{ij}\}$ is considered to be a point set on the plane (a_{ij}). We fix the order of men, then a marriage can be written as in (1), i.e. $m = ((a_1 a_2 \ldots a_n))$. We consider m as a polygon on the plane by joining all a_j together with this order

(24) $$m = \overline{a_1 a_2 \ldots a_n}, \quad \text{where the cardinality } |m| = n.$$

This polygon m divides the plane into two open parts, left and right which we denote by $L(m)$ and $R(m)$ respectively. We notice that it is sufficient to consider the open parts due to the fact that the pair matrices of an I-M preference are Latin squares, see Theorem 1.

Similarly, for each point $\{a_i\} \in m$, the vertical line passing through $\{a_i\}$ divides the whole plane into the left and the right open half-plane which we denote by $L(a_i)$ and $R(a_i)$ respectively. With the help of those geometric notions, we are now able to state and prove the following graph of unstable marriages for I-M preferences.

Note that for ease of printing, in the sequel, the graph of polygons will be sketched on the right handside of the elements.

Theorem 4. *Let* $P(n)$ *be a system of* I-M *preference and* m *be a marriage denoted by*

(25) $$m = \begin{pmatrix} A_1 \ldots A_p \ldots A_q \ldots A_n \\ a_1 \ldots a_p \ldots a_q \ldots a_n \end{pmatrix}.$$

Then m *is unstable if and only if it satisfies*

(26) $$\{a_p\} \in R(a_p) \cap L(m) \quad \text{for some } p,\ 1 \le p \le n.$$

Equivalently, m *is stable if and only if it satisfies*

(27) $\{a_i\} \cap R(a_i) \cap L(m) = \phi$, for all $i = 1,2, \ldots, n$.

Proof. Suppose that (26) holds for some p, $1 \le p \le n$. Sinee the preferences are not changed, we may, without loss of generality, assume that the polygon m starts from the left monotonically decreasing to the right in the plane (a_{ij}). We notice that the same point set $\{a_p\}$ on both sides of (26) does not occur in the same position, in fact $\{a_p\}$ occurs in each column of (a_{ij}). The one on the right hand side of (26) lies on m but the left one lies inside the region $R(a_p) \cap L(m)$. To avoid the confusion, let us explain by the following graph.

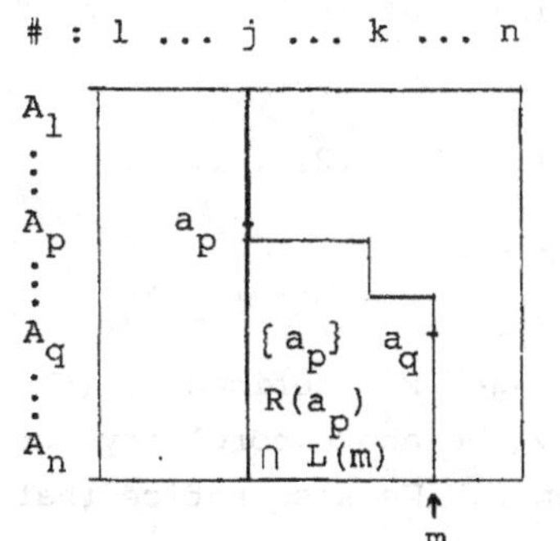

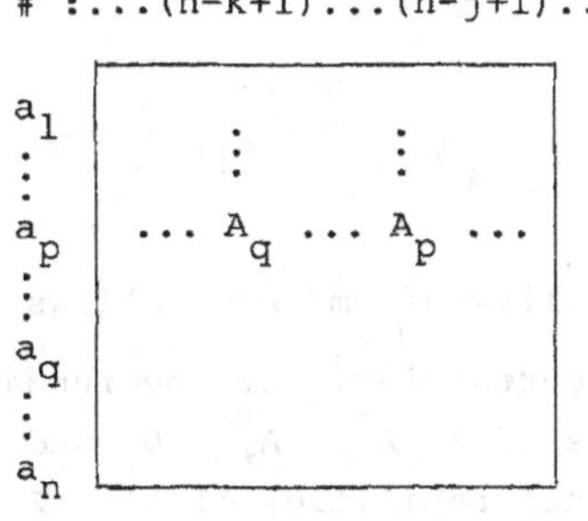

Let the right $\{a_p\} = a_{pj}$ and the left $\{a_p\} = a_{qk}$, then by (25) and (26), we can see that

(28) $$A_p(a_p) = j < k = A_q(a_p) < A_q(a_q) .$$

It follows from (4) that

(29) $$a_p(A_q) = n-k+1 < n-j+1 = a_p(A_p) .$$

The unstability of m is a consequence of $(3)^*$, (28), and (29).

Conversely, if m is unstable, then there is an unstable pair (A_q, a_p) which satisfies (29) together with the last part of (28). Again, by (4), we find that the first part of (28) holds which yields (26).

Finally, it is obvious that (27) is equivalent to (26). This completes the proof.

We shall now prove the following useful corollary which is the same as [3, Theorem 2] for stable permutations.

Corollary 3. *Under the hypothesis of Theorem 4, if in addition, we have*

$$(30) \qquad A^* - A_* \le 1 \;, \text{ where } A^* = \max_{1\le i\le n} A_i(a_i) \,,\; A_* = \min_{1\le i\le n} A_i(a_i) \,,$$

then m *is stable.*

Proof. According to our definition of the open sets of $R(a_i)$ and $L(m)$, we can see that condition (30) yields immediately

$$R(a_i) \cap L(m) = \phi \;, \text{ for all } i = 1,2, \ldots, n.$$

The stability of m now follows from (27).

In particular, for any fundamental marriage F_j defined by (6), we always have $A^* - A_* = 0$ and therefore by the above corollary, we obtain the stabilities of all F_j in Theorem 2. We also notice that the condition of I-M preferences in the above theorem is necessary, by the following

Example 4. *Let* $P(3)$ *be a Latin square preference defined by*

```
#: 1 2 3       #: 1 2 3
A: a|b c       a: A B C
B: c (a) b|    b: B C A
C: b c|a   ,   c: C A B .
```

Then the marriage $m = ((abc))$ *is stable, however, we have*

$$\{a\} \in R(a) \cap L(m).$$

Clearly, $P(3)$ *is not an* I-M *preference. We notice that Example 2 is not an* I-M *preference either, but the behaviours of fundamental marriages in these two examples are different. There is only the first one stable in Example 4.*

Theorem 4 is the most important theorem we have found due to the fact that it allows us to define what we call stable permutations in a single matrix in terms of (27). Starting from this definition, we have systematically developed the whole theory in Latin squares and the computation of such stable permutations, see [3, 4]. Certainly, our original definition of (27) is called the left stable which, for Latin squares, can be unified with another kind of right stable [3, Theorem 1]. The notion of right stable will be discussed in the next section.

To finish this section, we shall point out a relation between stable marriages, stable permutations, and transversals. It was shown in [4, Theorem 8] that all transversals are unstable permutations in one matrix. However, as a marriage, there are some transversals which are stable marriages. For instance, the marriage $m = ((abc))$ in Example 4 denotes a transversal in the Latin square (a_{ij}), which is stable. This phenomenon occurs because $P(3)$ is not an I-M preference.

6. INVERSE PREFERENCES

Let $\vec{P}(n)$ be a preference defined by (5). We call $\overleftarrow{P}(n)$ the inverse preference of $\vec{P}(n)$ if the orders of preferences of men and women in $\overleftarrow{P}(n)$ are inverse to those in $\vec{P}(n)$. This can be represented by

$$(31) \qquad \overleftarrow{P}(n) = \left\{ \begin{matrix} j & : \# & j & : \# \\ (a_{ij}) & : A_i, & (A_{ij}) & : a_i \end{matrix} \right\}$$

where

$$\vec{P}(n) = \left\{ \begin{matrix} \# : & j & \# : & j \\ A_i : & (a_{ij}), & a_i : & (A_{ij}) \end{matrix} \right\}, \qquad i,j, = 1,2, \ldots, n.$$

In order to have the same type of representation, we write the inverse preference by

$$(32) \qquad \overleftarrow{P}(n) = \left\{ \begin{matrix} \# & : & j & \# & : & j \\ A_i^* & : & (a_{ij}^*), & a_i^* & : & (A_{ij}^*) \end{matrix} \right\}$$

where $a_{ij}^* = a_{i(n+1-j)}$ and $A_{ij}^* = A_{i(n+1-j)}$. Clearly, we have

(33) $A_i^*(a_j^*) = n + 1 - A_i(a_j)$ and $a_j^*(A_i^*) = n + 1 - a_j(A_i)$.

We notice that the definition (31) yields immediately the following recursive formula

$$\left\{\overleftarrow{P}(n)\right\}^2 = \overleftarrow{\overleftarrow{P}}(n) = \vec{P}(n), \quad \left\{\overleftarrow{P}(n)\right\}^3 = \overleftarrow{P}(n), \ \ldots$$

From this, we can say that $\vec{P}(n)$ and $\overleftarrow{P}(n)$ are each other's inverse. With those definitions, we shall prove the following invariance of I-M preferences between an inverse pair.

Theorem 5. *Let* $\vec{P}(n)$ *and* $\overleftarrow{P}(n)$ *be an inverse pair, then* $\vec{P}(n)$ *is an* I-M *preference if and only if* $\overleftarrow{P}(n)$ *is also an* I-M *preference.*

Proof. As represented by (32), we have the two equalities of (33). By adding these two equalities, we obtain

$$A_i^*(a_j^*) + a_j^*(A_i^*) = 2(n+1) - \left(A_i(a_j) + a_j(A_i)\right) .$$

This yields the invariance of I-M preference between an inverse pair due to (4).

By applying Theorem 5 together with the argument in Theorem 4, we obtain the following graph of unstable marriages for the inverse preference.

Theorem 6. *Let* $\vec{P}(n)$ *be a system of* I-M *preference,* $\overleftarrow{P}(n)$ *the inverse preference, and* m *a marriage. Then* m *is unstable in* $\overleftarrow{P}(n)$ *if and only if it satisfies*

(34) $\{a_p\} \in L(a_p) \cap R(m)$ for some $p, 1 \leq p \leq n$.

Equivalently, m *is stable in* $\overleftarrow{P}(n)$ *if and only if it satisfies*

(35) $\{a_i\} \cap L(a_i) \cap R(m) = \phi$, for all $i = 1,2, \ldots, n$.

It is now easy to prove the following invariance of unstable marriages between an inverse pair of I-M preferences.

Theorem 7. *Let* $\vec{P}(n)$ *and* $\overleftarrow{P}(n)$ *be an inverse pair of* I-M *preferences, then a marriage* m *is unstable in* $\vec{P}(n)$ *if and only if* m *is also unstable in* P(n) .

Proof. Let m be a marriage unstable in $\vec{P}(n)$, then by Theorem 4, we find that (26) holds for some woman a_p. Owing to [3, Theorem 1], we can see that (26) in turn implies (34) for some woman $a_p^* \neq a_p$. It follows from Theorem 6 that m is also unstable in $\overleftarrow{P}(n)$ and vice versa.

For an I-M preference, the above theorem allows us to call a marriage unstable if it satisfies either (26) or (34). This is the same as the left or the right unstable permutation in one matrix. From this remark, we finally obtain the following important result.

Theorem 8. *For any* I-M *preference, the notion of stable marriages can be unified with that of stable permutations.*

We notice that the condition of I-M preference in the above Theorems 7 and 8 is necessary, for instance, in Example 4, we can see that the marriage m is stable, but the identity permutation is unstable. Meanwhile, m is stable in $\vec{P}(3)$, but unstable in $\overleftarrow{P}(3)$.

References

[1] J. Dénes and A.D. Keedwell, Latin Squares and Their Applications. Academic Press. New York and London 1974.

[2] D. Gale and L.S. Shapley, College admissions and the stability of marriages. Amer. Math. Monthly 69(1962), 9-15.

[3] J.S. Hwang, Stable permutations in Latin squares. Soochow J. Math. 4(1978), 63-72.

[4] ———————, On the invariance of stable permutations in Latin rectangles (to appear in ARS Combinatoria).

[5] D.E. Knuth, Mariages stables et leurs relations avec d'autres problèmes combinatoires. Les Presses de l'Université de Montréal 1976.

Institute of Mathematics
Academia Sinica
Taipei, Taiwan.

THE CONSTRUCTION OF FINITE PROJECTIVE PLANES

PETER LORIMER

I propose to discuss this topic in a general way, keeping to those aspects which are easily presented in a talk.

I acknowledge the generous financial support of the Australia and New Zealand Banking Group Limited.

Three axioms serve to define projective planes: a projective plane is a set of points and lines which satisfy

Axiom 1. *Each pair of points is joined by one line.*

Axiom 2. *Each pair of lines meets in one point.*

Axiom 3. *There is a configuration of four points, no three of which are collinear.*

The first two are the effective axioms, the third serving to exclude some "degenerate" configurations.

In a projective plane there is a 1-1 correspondence between the points on any line and the lines through any point. If this common number is $n+1$, n being finite, then n is called the *order* of the plane. In a plane of order n there are n^2+n+1 points and n^2+n+1 lines. Thus a projective plane of order n is a configuration of n^2+n+1 points and n^2+n+1 lines, there being $n+1$ points on every line and $n+1$ lines through every point.

The smallest plane is the Fano plane which has order 2: it is a configuration of 7 points and 7 lines with 3 points on each line and 3 lines through each point. Figure 1 is a drawing of this plane, the circle being one of the lines: as a matter of fact, this configuration is not realisable in the Euclidean plane with each line being represented by a Euclidean line.

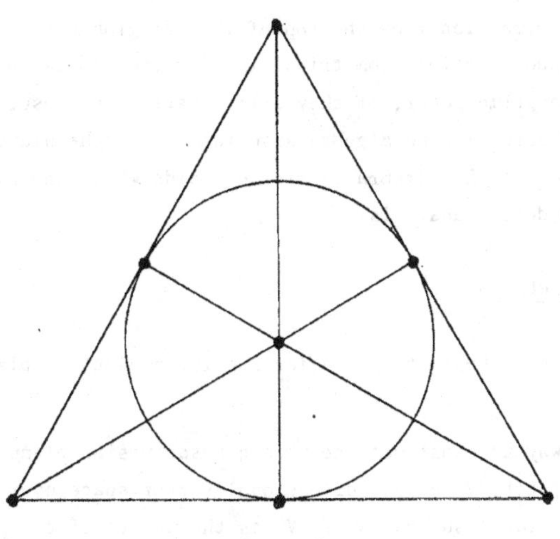

FIGURE 1

Finite projective planes fall into four main types

Desarguesian Planes

|

Translation Planes

|

Planes of prime power order which are not translation planes.

|

Planes of order not a prime power.

The Desarguesian planes are completely known, there being one of each prime power order. They are included among the translation planes which necessarily have prime power order, but cannot have prime order unless they are Desarguesian. Infinite classes of planes which are not translation planes are known: every one has prime power but not prime order. No plane of order not a prime power is known.

There is a progression from the top of this diagram where the Desarguesian planes can be described as being geometrical in character to the bottom where the planes of order not a prime power, if they exist, have to be described as combinatorial. This is reflected in the algebra associated with the planes: the algebra of Desarguesian planes is the algebra of finite fields while the algebra of the planes at the bottom defies analysis.

1. DESARGUESIAN PLANES

The best place to begin to get a feeling for projective planes is with the Desarguesian planes.

Here is one way to construct the unique Desarguesian plane of order q , q being a prime power. Let V be a 3-dimensional vector space over the field of order q . Take the 1-dimensional subspaces of V as the points of the plane and the 2-dimensional subspaces as the lines: a point lies on a line if and only if it is a subset of the line. As each 2-dimensional subspace of V contains $q+1$ 1-dimensional subspaces, this plane has order q .

In V , no 2-dimensional subspace is different from any other: in fact any one can be mapped onto any other by a non-singular linear map of V . The geometrical consequence of this is that no line of a Desarguesian plane is different from any other and any one can be mapped onto any other by a collineation (i.e. isomorphism) of the plane onto itself.

These planes are called Desarguesian because Desargues' Theorem is true in them. This theorem is illustrated in Figure 2 where two triangles ABC and $A'B'C'$ are shown; the points L, M, N are intersections of sides of the triangles, as shown in the figure. Desargues' Theorem states that if the lines AA' , BB' , CC' all pass through a point, then the points L, M, N all lie on a line.

A proof of the theorem is easy in the present context. Let O be the point $\langle t \rangle$, the 1-dimensional subspace spanned by t . Let A be the point $\langle u_0 \rangle$. Because the points O, A, A' lie on a line, A' lies in the 2-dimensional subspace spanned by t and u_0 ; i.e. A' is the point $\langle \alpha t + \beta u_0 \rangle$ for some scalars α , β . As the three points are different, neither α nor β are zero. Now

$$\begin{aligned} \langle \alpha t + \beta u_0 \rangle &= \langle t + \frac{\beta}{\alpha} u_0 \rangle \\ &= \langle t + u \rangle \end{aligned}$$

where $u = \frac{\beta}{\alpha} u_0$. Also

$$\langle u_0 \rangle = \langle u \rangle .$$

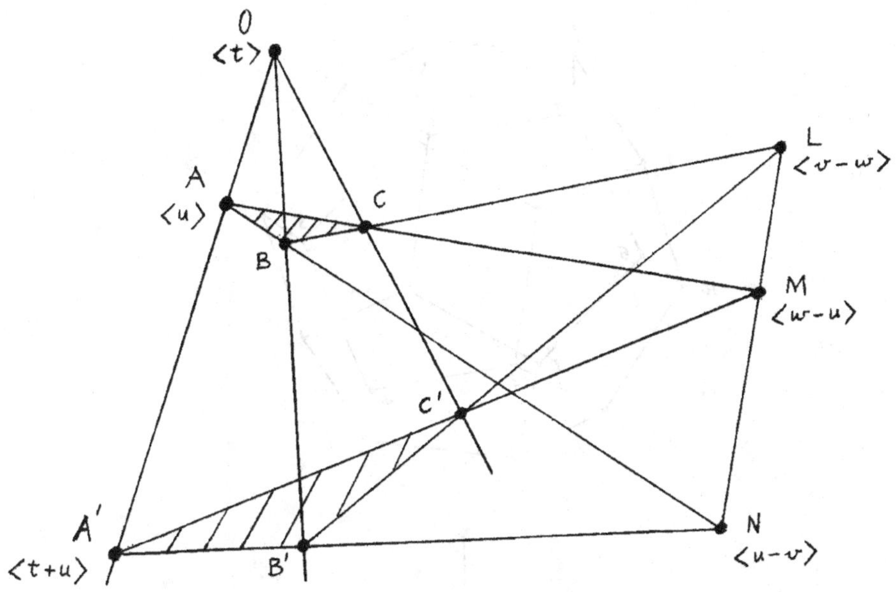

FIGURE 2

Hence A can be taken as $< u >$ with A′ as $< t + u >$. In a similar way, B can be taken as $< v >$ with B′ as $< t + v >$ and C as $< w >$ with C′ as $< t + w >$. Now to find L : it is the intersection of the subspace generated by v and w with the subspace spanned by $t + v$ and $t + w$; i.e. L is the point $< v - w >$. Similarly M and N are the points $< w - u >$ and $< u - v >$. As the sum of the three vectors $v - w$, $w - u$ and $u - v$ is zero, L, M, N are collinear.

The Desargues' configuration contains 10 points and 10 lines, with each point on 3 lines and each line through 3 points.

Actually, Desargues' Theorem is not characteristic of projective planes defined over fields in this way, but Pappus' Theorem is. Desargues' Theorem is characteristic of planes defined over skew fields, but as every finite skew field is a field, the distinction between the two disappears for finite planes.

Another characteristic of Desarguesian planes is that they can be imbedded in 3-dimensional projective spaces, and they are the only planes for which this is true. Any attempt to develop a theory of projective spaces of dimension higher than 2 leads necessarily to Desargues' Theorem.

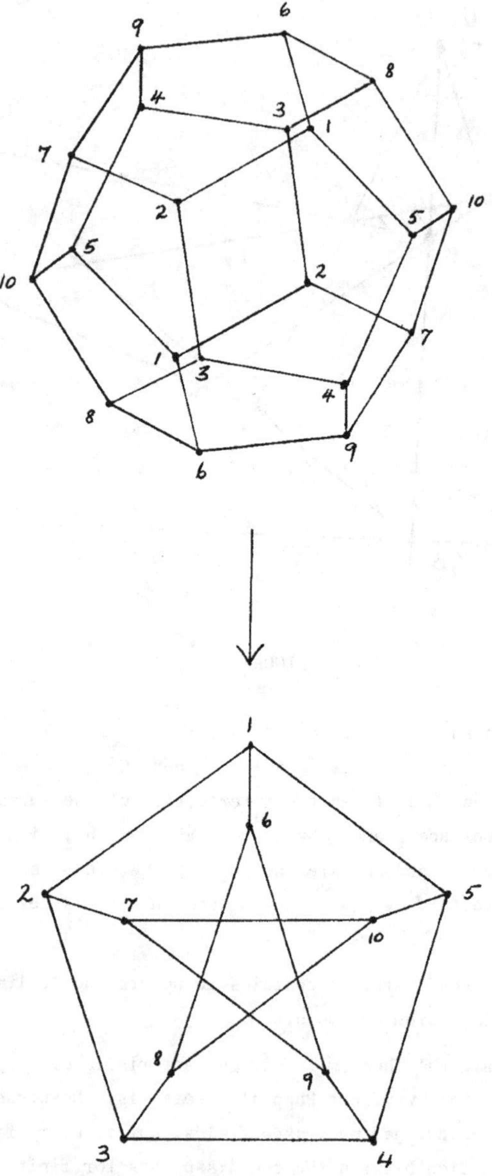

FIGURE 3

It is worth mentioning something about one infinite plane, the real projective plane which can be formed from the 3-dimensional real vector space $\mathbb{R}^3$. There is a 1-1 correspondence between the 1-dimensional subspaces of $\mathbb{R}^3$ and the lines through the origin in space; and there is a 1-1 correspondence between these lines and pairs of antipodal points on the unit sphere, a line corresponding to the two points where it cuts the sphere. There is also a 1-1 correspondence between the 2-dimensional subspaces of $\mathbb{R}^3$ and the planes through the origin in space, and hence with the great circles on the unit sphere. This gives the classical representation of the projective plane as the set of pairs of antipodal points on the unit sphere with lines being the great circles. Figure 3 shows a regular dodecahedron, the 20 vertices of which lie on a sphere as 10 pairs of antipodal points: in the figure antipodal points are given the same number. In the projective plane, pairs of antipodal points are coalesced into one point: if two of these points are joined by an edge if and only if their respective constituents are joined by at least one edge on the dodecahedron, the result is the Petersen graph, certainly the most prominent graph at this conference.

If one line and all the points on it are removed from a projective plane the result is called an affine plane: it is a set of points and lines satisfying the axioms.

<u>Axiom 1</u>. *Each pair of points is joined by one line.*

<u>Axiom 2</u>. *If a point P does not lie on a line ℓ, then there is one line through P which does not meet ℓ (i.e. is parallel to ℓ).*

<u>Axiom 3</u>. *There is a configuration of four points, no three of which are collinear.*

All lines have the same number of points and this number is called the order of the plane. The process of forming an affine plane can be reversed: from a given affine plane, a projective plane of the same order can be formed.

As no line in a Desarguesian plane is different from any other, all the affine planes obtained from a Desarguesian plane in this way are isomorphic to one another as affine planes.

Here is another way of constructing Desarguesian affine planes. Take the points as the members of a 2-dimensional vector space V over a field and take the lines as the 1-dimensional subspaces of V and all their cosets in V.

If the field involved has order $q = r^m$, then V is equally a 2m-dimensional vector space over the field of order r, and each 1-dimensional subspace is equally an m-dimensional subspace of this 2m-dimensional space. This leads directly into

2. TRANSLATION PLANES

These are the non-Desarguesian planes which are easiest to comprehend and most known planes are of this type.

Let V be a vector space of dimension 2m over a field of order r and let S be a *spread* of m-dimensional subspaces of V : for S to be a spread it is required that each non-zero member of V be a member of exactly one of the subspaces of S . If the members of V are taken as the points, and the subspaces of S and all their cosets are taken as the lines the result is an affine plane: these planes and the projective planes formed from them are called *translation planes*.

As far as the relationship between Desarguesian and translation planes is concerned two things are clear: each Desarguesian plane is a translation plane; every translation plane of prime order is Desarguesian. On the other hand, every "proper" prime power is the order of a non-Desarguesian translation plane.

The projective plane obtained from a non-Desarguesian translation affine plane has a very special property. The line which is added on, commonly called the line at infinity, is fixed by every collineation of the plane. Thus, in these planes, one line is very definitely different from all the others.

Here is another way to look at translation planes. In ordinary analytic geometry, the algebraic study of Desarguesian affine planes, the points are given co-ordinates (x,y) from a field and the lines are taken as the solution sets of the linear equations of the type

$$y = mx + c$$

$$x = d \ .$$

In the corresponding algebra of translation planes the points are given co-ordinates (x,y) from a vector space instead of a field, and the lines are taken as the solution sets of equations of the type

$$y = M(x) + c$$

$$x = d \ ,$$

where M is one of set M of suitably chosen linear maps. It is easy to see that two necessary conditions for M to satisfy are that there be a 1-1 correspondence between the maps of M and the members of V , and that if M_1 and M_2 are two (different) members of M , then $M_1 - M_2$ should be non-singular. These two conditions are also sufficient; the normalizing condition that M contain the zero matrix is usually added.

Here is an example of such a plane. Take the vector space as F_3^2, the two dimensional space of 9 column vectors $\begin{pmatrix} a \\ b \end{pmatrix}$ over the field of order 3 and the set M as the set of the following 9 matrices

$$\begin{pmatrix} 0 & 0 \\ 0 & 0 \end{pmatrix} \quad \begin{pmatrix} 1 & 0 \\ 0 & 1 \end{pmatrix} \quad \begin{pmatrix} 2 & 0 \\ 0 & 2 \end{pmatrix}$$

$$\begin{pmatrix} 0 & 1 \\ 2 & 0 \end{pmatrix} \quad \begin{pmatrix} 1 & 1 \\ 1 & 2 \end{pmatrix} \quad \begin{pmatrix} 2 & 1 \\ 1 & 1 \end{pmatrix}$$

$$\begin{pmatrix} 0 & 2 \\ 1 & 0 \end{pmatrix} \quad \begin{pmatrix} 2 & 2 \\ 2 & 1 \end{pmatrix} \quad \begin{pmatrix} 1 & 2 \\ 2 & 2 \end{pmatrix}$$

The crucial thing now is that the difference of any two of these matrices is non-singular.

Another way to look at this construction is to regard the matrices of M as defining an operation of multiplication on F_3^2 by the role

$$\begin{pmatrix} a \\ b \end{pmatrix} \begin{pmatrix} c \\ d \end{pmatrix} = \begin{pmatrix} a & * \\ b & * \end{pmatrix} \begin{pmatrix} c \\ d \end{pmatrix}$$

where the $*$'s are replaced by the two members of F_3 which make the matrix one of the given 9. This multiplication plays the role in translation planes that field multiplication plays in Desarguesian planes: lines are the solution sets of equations of the type

$$y = mx + c$$

$$x = d\ .$$

In constructing translation planes the trick is to fill in the rest of a matrix once the left hand column is given.

Here is an infinite class of planes, the Hall planes, defined by a set M of matrices over a field F : take all matrices of the type

$$\begin{pmatrix} a & 0 \\ 0 & a \end{pmatrix}$$

$$\text{or} \quad \begin{pmatrix} a & -\frac{1}{b}(a^2-\alpha a-\beta) \\ b & \alpha - a \end{pmatrix}, \quad b \neq 0$$

where $x^2-\alpha x-\beta$ is irreducible over F. As every possible column vector occurs as the left hand column of one of these matrices, a proof that a projective plane has been constructed consists in showing that the difference of any two of them is non-singular.

For each prime power q there is one Hall plane of order q^2, those constructed from different irreducible quadratics over the same field being isomorphic with one another. The 9 matrices mentioned earlier are the matrices of the Hall plane of order 9 defined by the polynomial $x^2 + 1$ which is irreducible over the field of order 3. The matrices of the second type mentioned above are the complete set of 2×2 matrices over the given field which have $x^2-\alpha x-\beta$ as their irreducible polynomial: for this reason $PGL(2,q)$ acts as a collineation group of the Hall plane of order q^2. The Hall involved here is Marshall Hall Jr who discovered these planes.

There is a construction principle due to T.G. Ostrom which is applicable in many situations but has its simplest use in the technique of derivation, particularly when applied to translation planes.

Consider a 4-dimensional vector space V over a field of order q and let S be a spread of 2-dimensional subspaces of V. A subset T of S containing $q + 1$ subspaces contains $(q + 1)(q^2 - 1)$ non-zero vectors. As $q^2-1 = (q-1)(q+1)$ it is conceivable that it might be possible to arrange these $q + 1$ subspaces into $q + 1$ others, each of which has $q - 1$ vectors from each of the original ones. If this is possible and the set of new subspaces form a set T', then T can be replaced by T' in S to form a new spread

$$(S - T) \cup T'$$

and hence another translation plane having V as its points. This new plane is said to be *derived* from the original one. (Anyone thinking projectively will recognize T as a regulus with T' the opposite regulus).

One problem with translation planes is that there are too many of them. T.G. Ostrom, at a conference at Washington State University in 1973, had this to say. "The number of known finite translation planes has become unmanageably large. Furthermore, there is an increasing amount of ambiguity in what we mean by a "known" plane. ... I suggest that when one constructs a plane the important question is not whether it is "new" or not, but does it have properties that differ in some interesting manner from the (known) properties of the known planes. ... I suspect that there are numerous cases of non-isomorphic planes that do not differ in a way that many people would find interesting."

And with these comments, let us move on to the next class of planes.

3. PLANES OF PRIME POWER ORDER WHICH ARE NOT TRANSLATION PLANES

As every known projective plane has prime power order, every known plane which is not a translation plane is included in this category. Infinite classes of them are known, but as none seem to be presentable in a brief meaningful way I will have to refrain from giving an example and stick to generalities.

In the section on translation planes I pointed out that a translation plane of prime order is necessarily Desarguesian. In fact, the only known planes of prime order are the Desarguesian ones. In the other direction, it is only for the small primes 2, 3, 5, 7 that the complete story is known: there is just one projective plane of each of these orders. As far as I can tell nothing is known about higher primes. Thus a major question is:

> Are there any projective planes of prime order apart from the Desarguesian ones?

A general question that might be asked is: why is every known plane of prime power order? The answer seems to be that every known plane has been constructed in one way or another from a finite field and in such a way that the construction carries over the prime power order of the field to the order of the plane. This is clear enough in the case of translation planes, but remains true, in ways more subtle, for other planes also.

As there is not much that can be said in a short time about these planes, let us move on.

4. PLANES OF ORDER NOT A PRIME POWER

Our knowledge of the possible orders of projective planes is easy to summarize.

1. Every known plane has prime power order .
2. There is just one theorem which excludes some natural numbers from being the order of a plane. This is the Bruck-Ryser Theorem which excludes n if $n \equiv 1$ or $2 \bmod 4$ and n is not the sum of two squares or equivalently n has a prime divisor $p \equiv 3 \bmod 4$ in its square free part.

Thus the Bruck-Ryser Theorem excludes

6, 14, 21, 22, 30, 33

but not

$$10,\ 12,\ 15,\ 18,\ 20,\ 24,\ 26,\ 28,\ 34\ .$$

The smallest number in doubt is 10, which explains the concentration on this particular case. I think it can be taken as one of the central problems of combinatorics today:

> Does there exist a projective plane not of prime power order; and, in particular, does there exist a plane of order 10?

The plane of order 10 has received a lot of attention. The latest news seems to be that if it exists it has a collineation group of order 1. This follows from recent work of Richard Anstee, Marshall Hall Jr, John Thompson, Zvonimir Janko, Sue Whiteside and earlier work of others. Marshall Hall believes that a plane of order 10 must contain what he calls a "primitive 20 point configuration M_{20}": this is a set of 20 points which every line of the plane would meet in 0, 2 or 4 points; it would contain 20 lines each containing 4 of the 20 points, and each of the points would lie on 4 of these lines. He has a computer program at Caltech which can check in a few minutes whether a configuration like this can be completed to a plane of order 10, but he believes that the number of distinct configurations like this is very large. (I am grateful to Marshall Hall for a personal communication containing the information in this paragraph).

On another tack, the existence of a projective plane of order 10 is equivalent to the existence of 9 mutually orthogonal latin squares of side 10. Two are easy to find, but three are not known. In the early sixties E.T. Parker found a square with about a million orthogonal mates, but no two of those were orthogonal to one another.

So, how big is the problem of the plane of order 10?

Another standard way of looking at projective planes is through sets of permutations. Figure 4 shows two fixed lines ℓ, m and a fixed point I : it also shows how a point P , not on ℓ or m , induces a permutation f of the members of the set Σ of points which lie on ℓ but not on m . The characteristic property of these permutations is that they are sharply 2-transitive: if $a_1, a_2, b_1, b_2 \in \Sigma$, $a_1 \neq a_2$, $b_1 \neq b_2$ then there exists a unique permutation f like this with $f(a_1) = b_1$ and $f(a_2) = b_2$. (A proof of this is easily composed from the plane). For a plane of order n , the sharply 2-transitive set contains $n(n-1)$ permutations acting on a set of n points. The combinatorial nature of the problem can be emphasized by pointing out that S is a set of permutations; it cannot be assumed, a priori, that S has any algebraic structure.

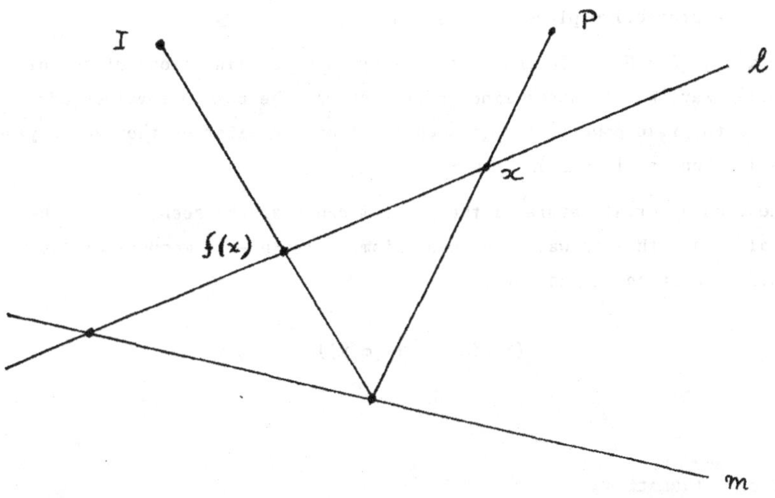

FIGURE 4

Thus the existence of a plane of order 10 is equivalent to the existence in the symmetric group S_{10} of a set of 90 permutations having a certain property. The most naive way to look for a plane would be to check these subsets in S_{10}. However, the number of subsets of order 90 in S_{10} is

$$\frac{(10!)!}{(10!-90)!90!}$$

a number which is easily proved to be greater than 10^{387}.

Is it likely that a plane of order not a prime power does exist? I don't know, but I want to suggest that it is not a hopeless task to look for one, because there are so many places to look and so little searching has been done. Let me finish by describing an algebraic condition which is sufficient for the existence of a plane of order n.

Let G be an abelian group of order $n - 1$, written additively, and suppose that there is defined on G a permutation ϕ with the properties

(i) $\phi(0) = 0$

(ii) each pair (a,b) of $G \times G$, $a \neq b$, $a \neq 0$, $b \neq 0$ can be written in a unique way in the form

$$(a,b) = (x,\phi(x)) - (y,\phi(y)) .$$

Then there is a projective plane of order n .

In fact, $G \times G$ would then act as a group of collineations of the plane in a very special way, but I cannot find any reason why the possible values of n should be restricted to prime powers, though such functions do exist on the cyclic groups of order $n - 1$ when n is a prime power.

The combinatorial nature of the problem can again be seen, for if the function ϕ was so algebraic that is was an automorphism, only $n - 1$ members of the whole of $G \times G$ could be written in the form

$$(x,\phi(x)) - (y,\phi(y)) .$$

Department of Mathematics,
University of Auckland,
Auckland, New Zealand.

A SURVEY OF GRAPH GENERATION TECHNIQUES

RONALD C. READ

This talk deals with various recently-developed methods for generating, by computer, catalogues of all graphs of some given kind. This includes discussion of the generation of graphs, digraphs, tournaments, self-complementary graphs, trees, and others. The present state of the art of graph generation is presented, together with some ideas on future prospects.

In this paper I shall give some information on recent advances in the generation of catalogues of graphs; but first it might be as well to say a little about why one should want to generate such catalogues at all - why, for example, one would wish to produce all the graphs on 8 vertices.

There are many uses to which such a list could be put. Scrutiny of the list, by hand or by computer, may suggest conjectures, or settle some question by turning up a counterexample. It may also enable one to get general ideas about graphs and their properties. Sometimes a list of graphs will supply numerical information for enumerative problems where a theoretical solution is absent, or provide a source from which specimen graphs can be taken for use in one of the real-life problems to which graph theory can be applied. All this is implicit in a succinct remark of Faradzhev [6] to the effect that graph theory is at present in a "botanical" stage of its development and that a "herbarium" of graphs is a useful thing to have around.

The situation regarding the existence of lists of graphs up to about 1976 is indicated in Table I (which is undoubtedly far from complete). This table is much the same as the one given in [21] at a time when it represented up-to-date information.

1. EARLY METHODS

How are comprehensive lists of graphs compiled? Small lists can easily be constructed by hand without much danger of graphs being overlooked, or counted twice. Figure 1 shows, for example, how the graphs on 4 vertices might be prepared. Each column (representing a given value of q, the number of edges) is generated from the graphs in the previous column by adding to each a single edge in all possible ways. This procedure will produce some graphs many times over, but for graphs this small there is no difficulty in eliminating the duplicates. Note that Figure 1 gives the results for $q \leq 3$; this is all that is necessary, since the graphs with $q > 3$ will be complements of graphs already listed, and therefore do not need to be generated.

Table 1. Some results up to c. 1976

Objects	p	Number		Year	Ref.
Graphs	6	156	I. Kagno	1946	[14]
	7	1044	B.R. Heap	1965	
	8	12346	"	1969	[12]
	9	274668	Dewdney et al.	1974	[1]
Digraphs	4	218			
	5	9608	R.C. Read	1966	[20]
Trees	≤ 13	1301	P.A. Morris	1972	[17]
	≤ 18	123867	R.J. Frazer	1973	[11]
Tournaments	≤ 7	456	P. McWha	1973	[16]
Self-complementary	8	10			
graphs	9	36	P.A. Morris	1971	[18]
Cubic graphs	≤ 14	509	Bussemaker, Cobeljic Cvetkovic & Seidel	1976	[3]

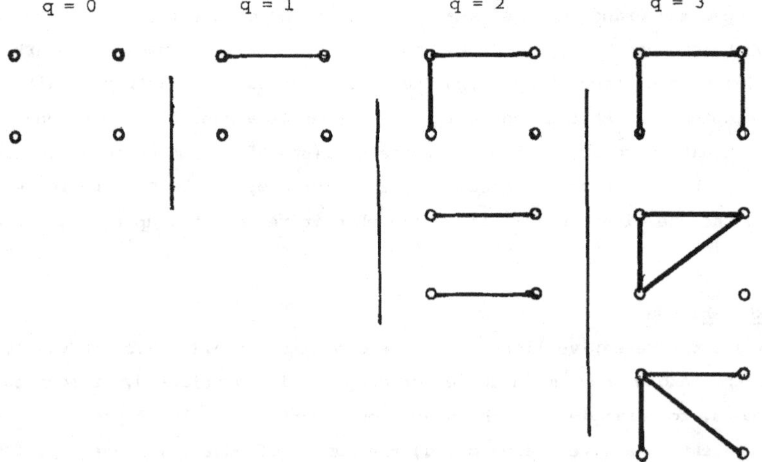

FIGURE 1

With larger graphs the elimination of duplicates becomes a problem; but putting that aside for the moment we see that Figure 1 illustrates a fairly general technique. It depends on producing graphs with $q+1$ edges from those on q edges, a procedure that can be set out as follows:

Step 1. Start with a list L_q of all graphs with q edges.

Step 2. Take each element of L_q in turn and from it generate candidates for the output list L_{q+1}.

(This is done by means of what we shall call an "augmenting operation". For the generation of graphs this could be the addition of a new edge in all possible ways. Clearly the augmenting operation must be chosen so that every graph in L_{q+1} must be generated at least once. It usually happens that these graphs are generated many times over, however, and it is for this reason that the next step is required.)

Step 3. As each candidate, G, for L_{q+1} is produced, determine whether it is already in L_{q+1}, that is to say, whether L_{q+1} contains some graph isomorphic to G. If so, reject G and continue processing candidates for L_{q+1}; if not, then add G to L_{q+1}.

A good method for storing the lists, and for testing for isomorphism, is to make use of a "code". A code for a set of graphs is a mapping of the set into the set of strings of symbols of some kind, in such a way that two graphs are isomorphic if, and only if, they have the same image (i.e. the same code). A typical code for graphs is that defined from the adjacency matrices, as in Figure 2. The upper triangular elements of an adjacency matrix are read off by rows to give a binary string, as shown in Figure 2a. This string will depend on how the vertices of the graph were numbered. We therefore consider all possible numberings, and use some criterion to pick out one of these strings to be "the" code for the graph. Commonly one chooses the string representing the largest integer. Figure 2b shows the derivation of another binary string from a differently labelled version of the same graph. Since this string is clearly maximal it would be the code for this graph (or, more strictly, for the isomorphism class to which this graph belongs).

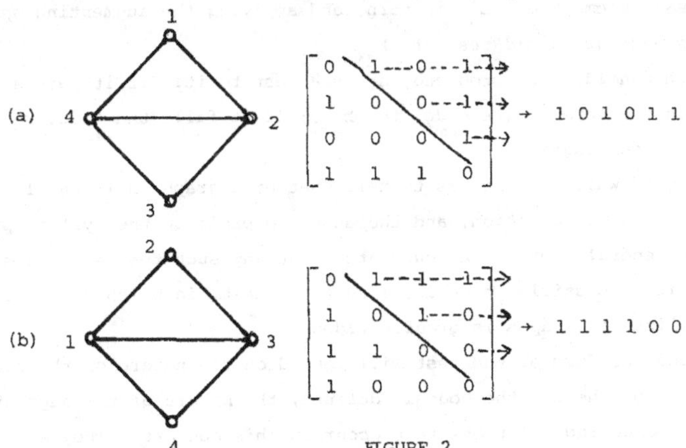

FIGURE 2

Thus to implement Step 3 above we code each candidate as it is produced, and search the current list L_{q+1} to see whether that code is already there. This can be a lengthy business; but at least the search is reduced to looking for the occurrence of a simple integer - the one having the code as its binary expression.

This procedure, or something very like it, lay behind the methods used to produce the lists in Table I. Its effectiveness is limited by two main considerations:

(a) The coding process, as described above, requires a number of operations proportional to p! - the number of labellings. This can be improved, but in any case the coding problem is as hard as the graph isomorphism problem, which is generally believed to be exponential, (see [22]).

(b) The lists are extremely long; usually to an extent that they cannot be contained in the immediate-access storage of a computer.

Of these two, (a) is not usually the main problem, since the graphs being labelled are not all that large. It is consideration (b) that mostly limits the extent to which lists of graphs can be produced. For large problems the list L_{q+1} must be stored in peripheral storage - on disk or (worse) on tape, and since this list must be searched for every new *candidate* for L_{q+1}, even the ones that turn out to be already in L_{q+1}, the whole procedure will clearly be very slow.

2. ORDERLY ALGORITHMS

It would be nice if there were some way of telling whether a newly-produced candidate was already in L_{q+1} without having to search L_{q+1} itself. This seems almost too much to hope for, but it is possible. In 1975 I stumbled upon a class of algorithms which do just that. These "orderly algorithms" can be applied to many problems of the type we are considering, i.e. in which the graphs are generated as a sequence of lists $L_0, L_1, \ldots, L_q$, each being produced from the one before. The general form of an orderly algorithm is as follows:

Step 1. Start with L_q.

Step 2. Take each element of L_q in turn, and applying the augmenting operation to produce a sequence of candidates for L_{q+1}.

Step 3. As each candidate is produced, apply a test to it. If it passes the test add this graph (in the form of its code) to L_{q+1}; if it fails the test, reject it, and process the next candidate.

What is this "test" which enables us to tell whether a graph is in the list or not? This is clearly the vital question, and the answer depends on the type of problem being tackled. In general there is no guarantee that any such test even exists; but for many problems it is possible to devise a suitable test, in which case the time taken to prepare lists of graphs is greatly reduced.

The existence and the form of the test will depend on the nature of the problem; in particular it depends on the way the code is defined, the nature of the augmenting operation, and on three ordering relations that occur in this context. They are

(a) the order in which the codes of the graphs appear in the lists L_q and L_{q+1} (the "list order");

(b) the order in which the augmenting operation produces candidates for L_{q+1} from elements of L_q;

(c) the order implicit in the definition of the code, e.g. the fact that we choose the *largest* binary string to be the code, rather than taking the first one in some other ordering.

I showed in [21] that if the augmenting operation, the code, and these orderings, satisfy three simple criteria, then an orderly algorithm exists; otherwise it will not.

As an example, consider the problem of generating graphs. An orderly algorithm exists under the following circumstances:

(a) The lists are in descending order of code.

(b) The augmenting operation adds edges by changing trailing 0's in the code to 1's, and does this from left to right.

(c) The code is defined as given earlier.

The appropriate test is then:

Test: If the candidate is "canonical" (i.e. its labelling is already the one that produces the (maximal) code) then it passes the test; otherwise it fails.

Note that since the test requires determining if the candidate is canonical, we are still committed to a large number, possibly $p!$, of operations. However, if the candidate is not canonical this will usually be discovered before all these operations have been performed; whereas if it is necessary to perform them all, to show canonicity, at least we have an addition to L_{q+1} for our pains. The important point is that the timeconsuming searching of lists has been avoided.

A slight modification of the definition of the code enables a similar orderly algorithm to be devised for generating digraphs, and in this way R. Cameron and I produced a catalogue of the 1,540,744 digraphs on 6 vertices in 1976.

There is also an orderly algorithm for generating tournaments. In these graphs the number of arcs is always $p(p-1)/2$, so p is the only parameter. Accordingly the problem is that of generating L_{p+1} from L_p. A method of doing this was described in [21], and the listing of all 6880 tournaments on 8 vertices was carried out a year or two ago [26].

Further applications of orderly algorithms will be described later.

3. TREES

Let us now turn to the problem of generating trees. Terry Beyer, of the University of Oregon, has developed elegant and fast algorithms for generating both rooted and unrooted trees; but since he has not yet published them I do not feel I can discuss them here. Instead I shall discuss algorithms for generating extended binary trees, as given in two papers that appeared in the same recent issue of J.A.C.M. [19,24].

An extended binary tree (EBT) is a rooted tree in which every vertex has either two upward branches or none. An (ordinary) binary tree (BT) is one where there is also the possibility of only one upward branch, in which case it is either a left branch or a right branch. Vertices with no upward branches are called "leaves"; the others are "internal vertices".

EBT's and BT's are closely related. In fact if we delete all the leaves from an EBT we obtain a BT, and it is easily verified that this correspondence is one-to-one.

See Figure 3. Hence if we generate EBT's we also generate BT's.

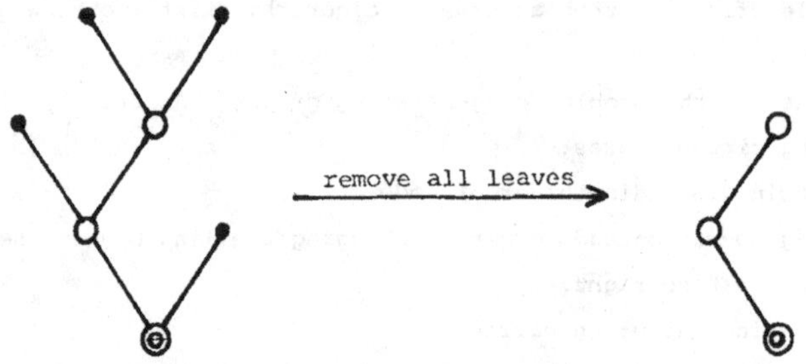

FIGURE 3

In [19] Proskurowski gives an algorithm for generating EBT's. It depends on a novel method of coding these trees, that is illustrated in Figure 4. We perform a walk around the tree in the familiar fashion, as indicated by the arrows, and write down '0' when we pass a leaf and '1' on meeting an internal vertex for the first time. In this way we get a string consisting of p 1's and $p+1$ 0's, where p is the number of internal vertices.

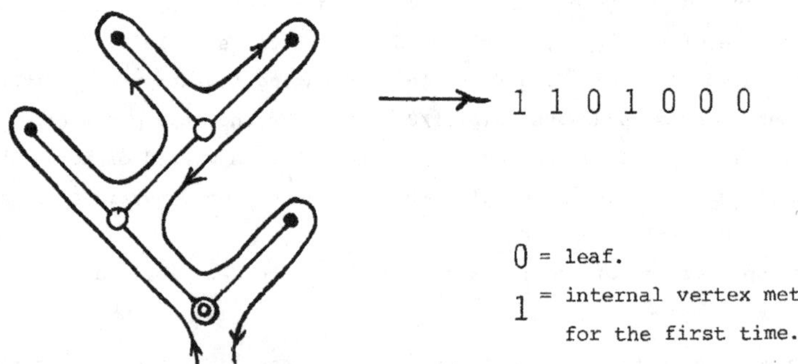

FIGURE 4

The generating algorithm produces the list L_{p+1} of EBT's on $p+1$ internal vertices from the list L_p. It is particularly simple since there is no test - every tree generated is added to L_{p+1}. Thus the whole algorithm depends on the augmenting operation, which is defined directly from the code. From left to right, replace each trailing zero by the string '100'. This is illustrated in Figure 5.

To generate L_{p+1}, start with L_p having the codes in descending order of magnitude. Apply this augmenting operation to each code in order, and append all the new EBT's produced to the growing list L_{p+1}.

In the second of the two papers mentioned [24], Solomon and Finkel give what is essentially the same algorithm, though the notation is different. In addition to this, however, they notice something new, namely, that it is possible to generate any given list, L_p, without needing to have the previous list L_{p-1} already available. Here is how it is done.

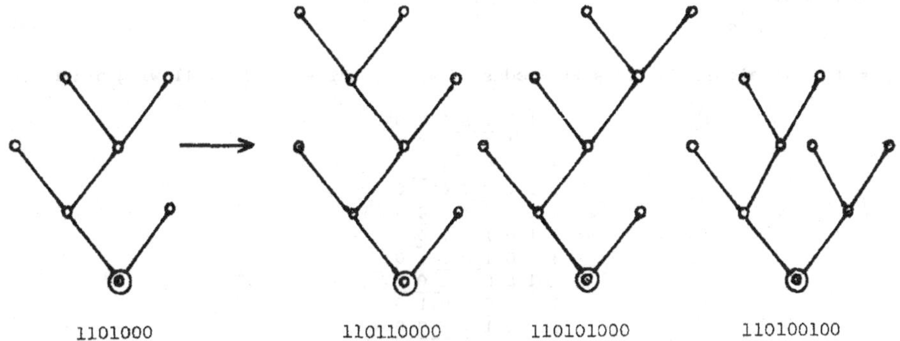

1101000 110110000 110101000 110100100

FIGURE 5

Suppose we have generated the list L_4, for example, as far as the EBT (call it T) having code 110110000. We can easily tell from which tree of L_3 the tree T would be generated by the algorithm above; we reverse the step by which T was formed, that is, we find the last occurrence of '100', and replace it by a '0'. This gives us the tree T_0 with code 1101000. The successor to T is then formed by replacing the *next* trailing zero in the code of T_0 by 100, and we obtain 110101000. Note that the final effect is to move the last '1' one place to the right.

A snag arises if the code of T ends with '100'. T is then the last tree produced from T_0, and to find its successor we must backtrack to find the successor of T_0 in L_{p-1}. This may require further backtracking to find the successor of a tree in L_{p-2}, and so on - possibly right back to L_1. Thus we have the prospect of having to juggle with all the lists simultaneously, something that is quite feasible (it was mentioned at the end of [21]) but which offers no great advantages.

For EBT's, however, there is no problem of this kind since the backtracking is so easy to perform. In fact, in what follows it may not even be very obvious that any backtracking is taking place, even though it is. I now give an algorithm for generating EBT's on p vertices directly, without any information on smaller trees. This algorithm is a sort of *pastiche* of the algorithms given by Proskurowski and by Solomon and Finkel in the papers mentioned above.

We shall modify the code for an EBT by omitting the final zero (which is redundant anyway). This prevents us from picking up the string '100' if it occurs at the end of the code as previously defined.

<u>Algorithm for generating EBT's</u>

1. Start with the code consisting of p 1's followed by p 0's.
2. Find the rightmost occurrence of the string '100'. If there is none, exit.

3. Delete this '100' and any bits to its right.
4. Append a '0' to what is left; then enough 1's to make up the total of p 1's; then enough 0's to make up the total of p 0's. This gives the next code in the list.
5. Repeat from 2.

For $p = 4$, this algorithm gives the codes of the 14 EBT's in the following order:

```
1 1 1 1 0 0 0 0
1 1 1 0 1 0 0 0
1 1 1 0 0 1 0 0
1 1 1 0 0 0 1 0
1 1 0 1 1 0 0 0
1 1 0 1 0 1 0 0
1 1 0 1 0 0 1 0
1 1 0 0 1 1 0 0
1 1 0 0 1 0 1 0
1 0 1 1 1 0 0 0
1 0 1 1 0 1 0 0
1 0 1 1 0 0 1 0
1 0 1 0 1 1 0 0
1 0 1 0 1 0 1 0
```

(The rightmost occurrences of '100' have been underlined.)

4. GRAPHS WITH GIVEN PARTITION

It is well-known (see [22]) that we can simplify the isomorphism problem for graphs by taking advantage of the fact that under any isomorphism vertices map onto vertices of the same degree. In much the same way we can simplify the coding of graphs. Instead of running through all the $p!$ permutations of the rows and columns of the adjacency matrix (corresponding to the $p!$ ways of labelling the vertices) we consider only those permutations that permute vertices of the same degree among themselves. In this way, if p_i is the number of vertices with degree i, we can find a code by considering only $p_1!p_2!p_3!\ldots$ permutations - possibly a much smaller number than $p!$. The sequence $\{p_i\}$ is called the partition of the graph.

This observation is not applicable directly to the problem of generating graphs in general, since it seems unlikely that an orderly algorithm would exist when the very definition of the coding process varies from one graph to another; but it suggests the possibility of an orderly algorithm for generating just those graphs with a given partition.

An algorithm for producing graphs with a given partition, albeit with some duplications, was given by Farrell [10] in 1971, and independently by James and Riha [13]. A similar algorithm (but more elaborate, for a reason that we shall come to) was given by Faradzhev [7].

Let us illustrate the algorithm by an example. Suppose the given partition is $1^3 2^3 3^3 4^3$, i.e. we have $p_1 = p_2 = p_3 = p_4 = 3$. Thus the vertices are classified by degree, into four classes. We provisionally letter the vertices and display their degrees as follows.

A B C D E F G H I J K L
4 4 4 3 3 3 2 2 2 1 1 1

In all possible ways we decide to how many vertices in each class the vertex A will be joined. We may decide, for example, that it will be adjacent to 1 vertex of degree 4, 2 of degree 3 and 1 of degree 2 (and hence none of degree 1). Since the graphs are unlabelled we lose no generality in assuming, for example, that A is joined to the *first* two vertices of degree 3, D and E. The same applies to the other classes.

We now remove vertex A from the graph, and look to the rest of the graph. The degree requirements can be written.

B	C	D	E	F	G	H	I	J	K	L
3'	4	2'	2'	3	1'	2	2	1	1	1

Note that D and E now have degree 2, as have H and I, but that we must distinguish these two pairs of vertices since D and E were originally of degree 3, unlike H and I. This is the reason for the primes on some of the degrees. We see that the number of classes tends to increase, since vertices with different "histories" must be distinguished. In our example, for instance, the adjacencies for B must be chosen from six different classes of vertices. In general, when a vertex is made adjacent to some, but not all, vertices in a given class, that class splits into two classes for the next iteration of this procedure.

This algorithm will certainly generate all the graphs with the given partition, but will usually produce some duplicates. We may well ask whether, in the spirit of orderly algorithms, there is some way to determine, each time a graph is produced, whether it is one that we have had already. This question was answered in [4] in which it is shown that each canonical graph is produced exactly once, so that by rejecting any graph that is not canonical we get exactly one representative of each isomorphism class. Canonicity here means having a code that is maximal over permutations that permute vertices of the same degree among themselves.

The test for canonicity may still require a large number of steps - indeed, for regular graphs it still needs the full $p!$ permutations - but for most partitions it is quite fast. R. Cameron, C. Colbourn and I tried for a long time to devise some efficient way of eliminating duplicates, but without success. (We suspect that there may be a theorem to the effect that this is not possible in less than factorial time). Accordingly, in implementing this algorithm for the application described in the next section we made some improvements to Farrell's algorithm so as to reduce the number of duplicates, but resigned ourselves to having to perform a possibly lengthy canonicity check after each graph was produced. In this our approach differed from that of Faradzhev, whose algorithm (see [7]) performs what is essentially a canonicity check *pari passu* with the process of generation.

5. GENERATING THE 10-VERTEX GRAPHS

The discovery of an orderly algorithm for graphs with a given partition paved the way to the successful completion of a project that had for long seemed to be on the borderline of feasibility, namely, the generation of all the graphs on 10 vertices. There are 12,005,168 such graphs, and the ability to generate them one partition at a time offered many advantages, not the least of which was the safeguard against the loss of great quantities of data in the event of a machine failure. Nevertheless we were still reluctant to embark on a long process of computing until we had some way of checking that we were getting the right answers. The number of graphs with given numbers of edges were known (see [25]) but what we really wanted was the number of graphs for each partition. These were unknown, but eventually N.C. Wormald and I found a way of computing them (it is described in [23]), and we then had all that was necessary to generate these graphs in convenient small batches, and check the accuracy of the computation.

As already mentioned, we saw no way of avoiding a canonicity check requiring the full $p_1!p_2!p_3!\ldots$ permutations, but this turned out to be a blessing in disguise. By counting how many permutations gave the maximal code, the program obtained the order of the automorphism group of each graph. This was included in the output. Moreover, this gave us the number of ways that each graph could be labelled, and in this way the results of the program were checked not only against the number of unlabelled graphs for each partition but also against the numbers of labelled graphs which, by then, were also known (see [23]). In view of this double checking of the output of the generating program it seems unlikely that the output contains any errors. The computation was performed by N.C. Wormald and me during April and May of 1980 and took about 16 hours CPU time on an IBM 370. The catalogue (for graphs up to 22 edges - the rest can be obtained by complementation) occupies two magnetic tapes.

6. SOME MISCELLANEOUS PROBLEMS

By tinkering with the definition of the code of a graph one can obtain orderly algorithms for a variety of different problems. Consider, for example, that of generating all Hamiltonian graphs. Figure 6 shows, for $p = 5$, a coding procedure which gives a code with the property that a graph is Hamiltonian if, and only if, its code begins with p 1's. It does not follow immediately that there is an orderly algorithm for this problem, but in fact there is, and by starting with the simplest Hamiltonian graph (the circuit) the complete set can be generated.

$$\begin{bmatrix} \cdot & a_1 & x & x & a_5 \\ & \cdot & a_2 & x & x \\ & & \cdot & a_3 & x \\ & & & \cdot & a_4 \end{bmatrix} \rightarrow a_1a_2a_3a_4a_5xxxxx$$

FIGURE 6

By generalizing this way of defining the code we can generate all graphs having some specified subgraph. For details see [5].

An interesting corollary of this result is the following. Suppose we use this method to generate all graphs having, as subgraph, the complement $\bar{G}$ of some given graph G. Each of these graphs has all the edges of $\bar{G}$ and some more. Each of the complements will thus have some subset of the edges of G. In other words we have generated all non-isomorphic subgraphs of G.

In conclusion I should like to mention, and pay tribute to, the extensive work of Faradzhev and his coworkers in the field of graph generation. It can be seen from his papers on the subject [2,6,7,8] that Faradzhev relies very heavily on a sophisticated algorithm for generating graphs with a given partition. Thus it is natural that a lot of his work is concerned with regular graphs, and we can note the following achievements.

Regular graphs:

Degree 3, up to 18 vertices. Number = 41301

Degree 4, up to 14 vertices. Number = 88168

Degree 5, up to 12 vertices. Number = 7848

Regular bipartite graphs:

Degree 3, up to 22 vertices. Number = 4132

Degree 4, up to 18 vertices. Number = 1980.

Faradzhev [9] has also generated all the self-complementary graphs on 12 vertices. It is not clear exactly when this was done, but it must have been before 1976, and therefore antedates the generation of these graphs by M. Kropar and me in 1977 [15]. The methods used were, incidentally, completely different.

Faradzhev has also carried out the enumeration of strongly regular graphs on up to 28 vertices, and of 3-vertex-connected graphs up to 12 vertices and 19 edges.

REFERENCES

[1] Baker, H.H., Dewdney, A.K., Szilard, A.L., Generating the nine-point graphs, *Math. Comp.* 28 (127), (1974) 833-838.

[2] Baraev, A.M., Faradzhev, I.A., The construction and computer investigation of homogeneous and inhomogeneous bipartite graphs, *Algorithmic Studies in Combinatorics* (Internat. Colloq. Combinatorics and Graph Theory, Orsay 1976) 25 - 60, 185.

[3] Bussemaker, F.C., Cobeljic, S., Cvetkovic, D.M., Seidel, J.J., Computer investigation of cubic graphs, T.H.-Report 76-WSK-01 Technological University, Eindhoven, Department of Mathematics, 1976.

[4] Colbourn, C.J., Read, R.C., Orderly algorithms for graph generation, *Intern. J. Computer Math,* 7A (1979) 167-172.

[5] Colbourn, C.J., Read, R.C., Orderly algorithms for generating restricted classes of graphs, *J. Graph Theory* 3 (1979) 187-195.

[6] Faradzhev, I.A., Constructive enumeration of combinatorial objects, *Internat. Colloq. CNRS* No.260, Combinatoire et Theorie des Graphes, Paris (1976), 131-135.

[7] Faradzhev, I.A., Generation of non-isomorphic graphs with a given distribution of the degree of vertices, *Algorithmic Studies in Combinatorics*, Internat. Colloq. Combinatorics and Graph Theory, Orsay (1976) 11-19, 185.

[8] Faradzhev, I.A., Constructive enumeration of homogeneous graphs, *Uspehi Mat. Nauk.* 31 (1976) 246.

[9] Faradzhev, I.A., The obtaining of a complete list of self-complementary graphs on up to 12 vertices, *Algorithmic studies in Combinatorics*, Internat. Colloq. Combinatorics and Graph Theory, Orsay (1976) 69-75, 186.

[10] Farrell, E.J., *Computer implementation of an algorithm for generating compositions and applications to problems in graph theory*, M.Math. Thesis, University of Waterloo, 1971.

[11] Frazer, R.J., Graduate course project, Department of Combinatorics and Optimization, University of Waterloo, unpublished (May 1973).

[12] Heap, B.R., The production of graphs by computer, *Graph Theory and Computing* (ed. R.C. Read), Academic Press (1972) 47-62.

[13] James, K.R., Riha, W., Algorithm 28: Algorithm for generating graphs of a given partition, *Computing* 16 (1976) 153-161.

[14] Kagno, I., Linear graphs of degree less than 7 and their groups, *Amer. J. Math.* 68 (1946) 505-529.

[15] Kropar, M., Read, R.C., On the construction of the self-complementary graphs on 12 nodes, *J. Graph Theory* 3 (1979) 111-125.

[16] McWha, P., Graduate course project, Department of Combinatorics and Optimization, University of Waterloo, unpublished (May 1973).

[17] Morris, P.A., A catalogue of trees on n nodes, n < 14, Mathematical observations, research and other notes, Paper No. 1 StA (mimeographed), Publications of the Department of Mathematics, University of the West Indies, 1971.

[18] Morris, P.A., Self-complementary graphs and digraphs, *Math. Comp.* 27 (1973) 216-217.

[19] Proskurowski, A., On the generation of binary trees, *J. ACM* 27 (1980) 1-2.

[20] Read, R.C., The production of a catalogue of digraphs on 5 nodes, Report UWI/CC1, Computing Centre, University of the West Indies.

[21] Read, R.C., Every one a winner, or how to avoid isomorphism search when cataloguing combinatorial configurations, *Annals of Discrete Math.* 2 (1978) 107-120.

[22] Read, R.C., Corneil, D.G., The graph isomorphism disease, *J. Graph Theory* 1 (1977) 339-363.

[23] Solomon, M., Finkel, R.A., A note on enumerating binary trees, *J. ACM* 27 (1980) 3-5.

[24] Stein, M.L., Stein, P.R., Enumeration of linear graphs and connected linear graphs up to P = 18 points, Report LA-3775 UC-32, Mathematics and Computers, TID-4500, Los Alamos Scientific Laboratory of the University of California.

[25] Thompson, T., Undergraduate course project, Department of Combinatorics and Optimization, University of Waterloo, unpublished (1978).

Department of Combinatorics and Optimization
University of Waterloo
Waterloo
Ontario N2L 3G1
CANADA.

GRAPHS AND TWO-DISTANCE SETS

J.J. SEIDEL

1. Introduction.

Sets of points whose mutual distances take only two values have an intrinsic interest. Upper bounds to their cardinality may depend on the specific metric space under discussion, on its dimension, and on the actual distances. Sets of optimal cardinality often have interesting combinatorial properties.

A two-distance set defines a graph in a natural way. Conversely, graphs give rise to two-distance sets. Thus, results on two-distance sets contribute to the theory of graphs. It is with this application in mind that the present paper surveys what is known about two-distance sets, both in Euclidean and in non-Euclidean spaces.

Section 2 explains and illustrates how graphs may be viewed as spherical two-distance sets in Euclidean space. For such sets we derive the absolute bound in section 3, and the special bound in section 4. In section 5 spherical two-distance sets in d-space are related to sets of equiangular lines in (d+1)-space, with either positive definite or indefinite inner product. Euclidean and non-Euclidean two-distance sets are discussed in sections 6 and 7; we mention recent work by Bannai [1] and by Neumaier [9]. Finally, section 8 describes the significance of root systems for certain classes of graphs.

2. Graphs.

Two-distance sets provide models for graphs. In explaining this, we restrict ourselves to regular graphs (for non-regular graphs cf.[10]). Let n denote the number of vertices, and k the valency of any regular graph. The (1,0)-adjacency matrix A of the graph has largest eigenvalue k, with the all-one vector as eigenvector. Let s denote the smallest eigenvalue of A, and let $n-d-1$ be its multiplicity. It is easy to check that

$$G := A - sI - \frac{k-s}{n} J$$

is a symmetric positive semidefinite matrix of size n and rank d with vanishing row sums. From linear algebra it follows that G is the Gram matrix (the matrix of the inner products) of n vectors in Euclidean space $\mathbb{R}^d$. Since G has constant diagonal, these vectors are on a sphere. Since the off-diagonal entries of G take only two values, these vectors form a spherical two-distance set. Normalizing to the unit sphere we denote by α and β the cosines of the angles between the vectors. The following examples are of special interest. They are obtained from well-known [5] graphs, or their complements, all of which are strongly regular.

	Petersen	Clebsch	Schläfli	T(8)	$L_2(5)$	T(10)	Suzuki	Fischer
n	10	16	27	28	25	45	416	31671
k	6	10	16	12	8	16	100	3510
-s	2	2	2	2	2	2	4	9
d	4	5	6	7	8	9	65	782
α	1/6	1/5	1/4	1/3	3/8	3/8	1/5	1/10
-β	2/3	3/5	1/2	1/3	1/4	1/4	1/15	1/80

More general, any graph may be viewed as a spherical two-distance set, cf.[10]. The strongly regular graphs are special, since for them the Gram matrix G constructed above has precisely 2 distinct eigenvalues. It is well-known [10] that this occurs if and only if the corresponding set of vectors forms a eutactic star, that is, the orthogonal projection into $\mathbb{R}^d$ of an orthonormal frame in $\mathbb{R}^n$.

3. The absolute bound for spherical two-distance sets.

Let X, of finite cardinality n, denote a subset of the unit sphere

$$\Omega_d = \{\xi \in \mathbb{R}^d \mid <\xi,\xi> = 1\}$$

in Euclidean space of d dimensions with inner product

$$<\xi,\eta> = \xi_1\eta_1 + \ldots + \xi_d\eta_d .$$

Assume that the vectors of X admit only two inner products $\neq 1$, say α and β. In $\mathbb{R}^2$ the maximum n equals 5, attained by the vertices of the regular pentagon. In $\mathbb{R}^3$ the maximum n equals 6, attained by the vertices of the octahedron, but also by any 6 of the 12 vertices of the icosahedron which do not contain an antipodal pair. Indeed, such sets have inner products $\pm 1/\sqrt{5}$. For general $\mathbb{R}^d$, at least $n = \frac{1}{2}d(d+1)$ may be achieved, viz. the $\binom{d+1}{2}$ points with coordinates $(1^2\, 0^{d-1})$, which in $\mathbb{R}^{d+1}$ lie on the linear manifold $\Sigma_{i=1}^{d} x_i = 2$. However, the following theorem [3] yields an upper bound for n in terms of d, called the absolute bound.

Theorem. $n \le \frac{1}{2}d(d+3)$, for the cardinality n of any spherical two-distance set in $\mathbb{R}^d$.

Proof. For each vector y in a two-distance set X of cardinality n with admissable inner products α and β we define the function

$$F_y(\xi) := (<y,\xi> - \alpha)(<y,\xi> - \beta), \quad \xi \in \Omega_d .$$

These are n polynomials of degree ≤ 2 in the variables $\xi_1,\ldots,\xi_d$ restricted to Ω_d. The linear space of all such polynomials has a basis consisting of the $\frac{1}{2}d(d+3)$ polynomials $\xi_1^2,\xi_1\xi_2,\ldots,\xi_d^2,\xi_1,\ldots,\xi_d$. The polynomials $F_y(\xi), y \in X$, are linearly independent since

$$F_y(x) = \delta_{x,y}(1-\alpha)(1-\beta) \text{ for } x,y \in X.$$

Therefore, their number n cannot exceed the dimension $\frac{1}{2}d(d+3)$.

Remark 1. If $\alpha + \beta = 0$ then the polynomials $F_y(\xi)$ are homogeneous of degree 2, and we arrive at the bound $\frac{1}{2}d(d+1)$ for the number of equiangular lines in $\mathbb{R}^d$, cf.[8].

Remark 2. The only known [3] cases for which the theorem holds with equality are

$$(n,d) = (5,2)\ ,\ (27,6)\ ,\ (275,22)$$

It is not difficult [11] to show that

$$n = \frac{1}{2}d(d+3) \text{ implies } \sum_{x,y\in X} \langle x,y\rangle^3 = 0.$$

Remark 3. The polynomials $F_y(\xi), y \in X$, in the above proof are "very" independent. The inequality of the theorem also holds for spherical sets X which are not quite two-distance sets, but like-two-distance sets, in the sense that the matrix $[F_y(x)]_{x,y\in X}$ is invertible. This applies for strictly diagonally dominant matrices, cf.[7].

4. The special bound for spherical two-distance sets.

Lemma. $\sum_{x,y\in X} \langle x,y\rangle^2 \geq n^2/d$, for any $X \subset \Omega_d$ spanning $\mathbb{R}^d$ with $|X| = n$. Equality holds iff X is a eutactic star.

Proof. Let $\gamma_1,\ldots,\gamma_d$ denote the nonzero eigenvalues of the Gram matrix G of the set X. Cauchy-Schwarz yields

$$(1+\ldots+1)(\gamma_1^2+\ldots+\gamma_d^2) \geq (\gamma_1+\ldots+\gamma_d)^2,$$

with equality iff $\gamma_1 = \ldots = \gamma_d$. Via

$$d.\ \text{trace}\,(G^2) \geq (\text{trace}\ G)^2$$

we arrive at the formula of the lemma. Equality holds iff G has at most two distinct eigenvalues, that is, iff X is a eutactic star, cf.[10].

<u>Theorem</u>. $n \le \frac{(1-\alpha)(1-\beta)d}{1+\alpha\beta d}$ for the cardinality n of any spherical two-distance set X with inner products α and β in $\mathbb{R}^d$, which satisfies

$\alpha + \beta = 0$ or $\sum_{x\in X} x = 0$.

Equality holds iff X is a eutactic star.

<u>Proof</u>. $(z-\alpha)(z-\beta) = \frac{1}{d} + \alpha\beta - (\alpha+\beta)z + z^2 - \frac{1}{d}$. Put $z = \langle x,y\rangle$ and sum over all $x,y \in X$, then

$$n(1-\alpha)(1-\beta) = n^2(\tfrac{1}{d} + \alpha\beta) - (\alpha+\beta)\sum_{x,y\in X}\langle x,y\rangle + \sum_{x,y\in X}\langle x,y\rangle^2 - \frac{n^2}{d}.$$

Now use the assumption and apply the lemma. For the case of equality we refer to the examples of section 2.

<u>Remark</u>. We may distinguish (cf.[3],[11]) between regular and various kinds of strongly regular graphs according as the two-distance set X is a spherical t-design for t = 1,2,3,4, following:

t = 1	t = 2	t = 3	t = 4
$\Sigma\langle x,y\rangle = 0$	$\Sigma\langle x,y\rangle^2 = \frac{n^2}{d}$	$\Sigma\langle x,y\rangle^3 = 0$	$\Sigma\langle x,y\rangle^4 = \frac{3n^2}{d(d+1)}$
	special bound	Krein bound	absolute bound
	n = 5,10,...	n = 5,16,27,100, 112,162,275,..	n = 5,27,275,..
regular graph	strongly regular graph	Smith graph	extremal graph

5. <u>Equiangular lines</u>.

Any set of n equiangular lines in $\mathbb{R}^{d+1}$ gives rise to a spherical two-distance set of n-1 points in $\mathbb{R}^d$. Indeed, for any unit vector u along any of the lines, consider the unit vectors at acute angle with u along the n-1 remaining lines, and project them into a hyperplane perpendicular to u. For instance, the 5 neighbours of any vertex of the isocohedron form a regular pentagon. For instance, the 2 × 28 vectors $\pm(3^2(-1)^6)$ define a set of 28 equiangular lines in $\mathbb{R}^7$, and our construction yields the two-distance set of 27 points in $\mathbb{R}^6$ mentioned earlier.

Conversely, given any two-distance set X on the unit sphere in $\mathbb{R}^d$, can we lift it up to a set of equiangular lines in $\mathbb{R}^{d+1}$? The answer is affirmative if we can find a radius $r \ge 1$ and an angle φ such that

$$1-\alpha = r^2(1-\cos\varphi),\quad 1-\beta = r^2(1+\cos\varphi).$$

This works if $\alpha + \beta \leq 0$. However, if $\alpha + \beta > 0$ then $r < 1$, and we have to proceed differently. In the case $\alpha + \beta > 0$ we define the space $\mathbb{R}^{d,1}$ of all vectors $x = (x_0; x_1, \ldots, x_d)$ provided with the indefinite inner product

$$\langle x, x' \rangle = -x_0 x_0' + x_1 x_1' + \ldots + x_d x_d' .$$

If x runs through the set $X \subset \mathbb{R}^d$ then

$$y := r^{-1}(\sqrt{1-r^2};\ x) \in \mathbb{R}^{d,1}$$

runs through a set Y of n vectors y satisfying

$$\langle y, y \rangle = 1 \quad , \quad \langle y, y' \rangle = \pm \cos \varphi .$$

The vectors of Y are outside of the isotropic cone in $\mathbb{R}^{d,1}$. The plane spanned by any 2 vectors y and y' of Y is Euclidean, and the lines spanned by y and y' are at angle φ. Thus we obtain a set of n equiangular lines in $\mathbb{R}^{d,1}$ at angle φ. For example, the two-distance sets mentioned in section 2 yield the following sets of equiangular lines:

$$10 \text{ in } \mathbb{R}^5,\ 16 \text{ in } \mathbb{R}^6,\ 28 \text{ in } \mathbb{R}^7 \text{ at } \cos\varphi = \frac{1}{3},$$

$$25 \text{ in } \mathbb{R}^{8,1},\ 45 \text{ in } \mathbb{R}^{9,1} \text{ at } \cos\varphi = \frac{1}{3},$$

$$416 \text{ lines in } \mathbb{R}^{65,1} \text{ at } \cos\varphi = \frac{1}{7},$$

$$31671 \text{ lines in } \mathbb{R}^{782,1} \text{ at } \cos\varphi = \frac{1}{17} .$$

6. Euclidean two-distance sets.

We now drop the assumption that the two-distance set is on a sphere.

Theorem. $n \leq \frac{1}{2}(d+1)(d+4)$, for the cardinality n of any two-distance set in Euclidean d-space.

We sketch two proofs.
The original proof, cf.[7], is similar to the proof of the theorem in section 3, and uses the polynomials

$$F_y(\xi) := (d^2_{\xi,y} - \alpha^2)(d^2_{\xi,y} - \beta^2),$$

where α and β are the admissable distances and

$$d^2_{x,y} = \langle x-y, x-y \rangle,$$

the Euclidean norm-square.

Bannai [1] gave a second proof, by use of the remark 3 of section 3. Indeed, let Y be any two-distance set in $\mathbb{R}^d$ which has sufficiently small diameter. By stereographic projection we project Y onto the set X on the unit sphere Ω_{d+1} in $\mathbb{R}^{d+1}$. It may be proved that X is a like-two-distance set on the sphere, for which the theorem of section 3 can be applied. Hence the cardinality of Y cannot exceed $\frac{1}{2}(d+1)(d+4)$ as well.

Bannai observed that equality cannot occur. In fact, it is well-known [4] that in $\mathbb{R}^2$ and in $\mathbb{R}^3$ the cardinality of two-distance sets cannot exceed the cardinalities 5 and 6, respectively, of spherical two-distance sets. Kristensen [6] shows that also for $\mathbb{R}^4$ and for $\mathbb{R}^5$ the spherical two-distance sets yield the two largest two-distance sets (of sizes 10 and 16, respectively). Maybe this is true*) for any dimension > 1.

7. Metric two-distance sets.

Neumaier [9] has observed that the first proof of the theorem of section 6 remains valid for arbitrary metric spaces, with an adapted notion of dimension. His argument runs as follows.

A distance matrix $D = [d^2_{x,y}]_{x,y\in X}$ is a symmetric matrix with zero diagonal and positive entries elsewhere. The rows and columns of D are indexed by the set X, of cardinality n. Define the matrix G by

$$G := -\frac{1}{2}(I - \frac{1}{n}J)\ D(I - \frac{1}{n}J),$$

and denote the entry $G_{x,y}$ by $\langle x,y\rangle$.

Lemma. $d^2_{x,y} = \langle x,x\rangle + \langle y,y\rangle - 2\langle x,y\rangle$.

Proof. By definition we have

$$2\langle x,y\rangle = -D_{x,y} + \frac{1}{n}\sum_i D_{i,x} + \frac{1}{n}\sum_i D_{i,y} - \frac{1}{n^2}\sum_{i,j} D_{i,j}.$$

In particular

$$2\langle x,x\rangle = \frac{2}{n}\sum_i D_{i,x} - \frac{1}{n^2}\sum_{i,j} D_{i,j}$$

The result follows by substitution.

Now consider G to be a Gram matrix, that is, the matrix of the inner products of a

*) Meanwhile, A. Blokhuis proved $n \le \frac{1}{2}(d+1)(d+2)$.

finite subset X of a linear space V over the reals provided with a nondegenerate real inner product. Then

$$\dim V = \operatorname{rank} G =: d.$$

By diagonalizing G, that is, by taking a suitable basis in V, we may write

$$\langle x,y\rangle = \varepsilon_1 x_1 y_1 + \ldots + \varepsilon_d x_d y_d \quad , \varepsilon_i \in \{1,-1\}.$$

From the lemma it follows that

$$d^2_{x,y} = \langle x-y, x-y\rangle .$$

Now let X be a two-distance set, that is, let the off-diagonal entries of the distance matrix D take only two values. Then we may repeat the first argument of the theorem of section 6, so as to arrive at the following.

Theorem. $n \le \frac{1}{2}(d+1)(d+4)$, for the size n of any two distance matrix of dimension d.

8. Elliptic two-distance sets.

Elliptic geometry of dimension $d-1$ over the reals is defined as follows. In Euclidean space $\mathbb{R}^d$ the lines, the planes, etc. through the origin are the elliptic points, the elliptic lines, etc. The angle between any two lines is interpreted as the elliptic distance between the corresponding elliptic points. Thus a set of equiangular lines corresponds to an equidistant elliptic set. An obvious elliptic two-distance set is a set of n lines through the origin in $\mathbb{R}^d$ having the angles 90° and 60° only. Such a set is called irreducible if it is not the union of two sets contained in proper orthogonal subspaces of $\mathbb{R}^d$. Three lines are said to form a star if they are mutually at 60° and lie in a plane. A set of lines at 60° and 90° is star-closed if with any two it contains the third line of a star.

Theorem [2]. The only irreducible star-closed sets of lines at 60° and 90° are the root systems A_d, D_d, E_8, E_7, E_6.

Hence the star-closed sets of lines at 60° and 90° are known. We indicate the significance of this result for graph theory. For any set of n lines at 60° and 90°, spanning $\mathbb{R}^d$, we consider 2n vectors of length $\sqrt{2}$ along the lines. The Gram matrix of the inner products of these vectors has size 2n, is positive semidefinite of rank d, has entries 2 on the diagonal and -2,0,1,-1 elsewhere. In this matrix we select a principal submatrix $2I-B$ having off-diagonal entries 0 and -1 only. The symmetric (1,0) matrix B may be interpreted as the adjacency matrix of a graph. Since $2I-B$ is positive semidefinite, this graph has largest eigenvalue ≤ 2. In particular, the graphs thus obtained from the root systems are the extended Dynkin diagrams, that is, the connected graphs having maximum eigenvalue 2 (cf.[2]).

Furthermore, in our matrix of size 2n we also select a principal submatrix $2I+A$ having off-diagonal entries 0 and 1 only. The symmetric (1,0) matrix A is interpreted as the adjacency matrix of a graph. Since $2I+A$ is positive semidefinite, this graph has smallest eigenvalue ≥ -2. Now for a long time the investigation of graphs with minimum eigenvalue -2 has been an important issue. In [2] this problem was attacked as follows. Let A be the adjacency matrix of a graph with n vertices and with smallest eigenvalue -2, of multiplicity $n-d$, say. The positive semidefinite matrix $2I+A$ is interpreted as the Gram matrix of n vectors at 90° and 60° in $\mathbb{R}^d$. According to our theorem, the star-closure of these vectors is one of the known root systems. Therefore, the study of the root systems serves to solve problems about graphs with smallest eigenvalue -2; we refer to [2] for the actual results.

What about graphs with smallest eigenvalue -3, say? This "next" case has been investigated by Shult and Yanushka [12], and has led to interesting results and new combinatorial notions, such as near n-gons.

Regular graphs with smallest eigenvalue -3 may be interpreted in terms of indefinite geometry as follows. For a regular graph on n vertices with valency k, let the adjacency matrix A have the smallest eigenvalue -3 of multiplicity $n-d-1$. Then $G := 3I+A-J$ has rank $d+1$, and precisely one negative eigenvalue $3+k-n$. Therefore, G is the matrix of the inner products of n vectors in $\mathbb{R}^{d+1}$, of equal length, outside of the light cone, at angles 90° or 120°. Thus, also two-distance sets in hyperbolic geometry may contribute to graph theory.

References

1. E. and E. Bannai, An upper bound for the cardinality of an s-distance subset in real Euclidean space, manuscript.

2. P.J. Cameron, J.M. Goethals, J.J. Seidel, E.E. Shult, Line graphs, root systems, and elliptic geometry, J.Algebra 43(1976), 305-327.

3. Ph. Delsarte, J.M. Goethals, J.J. Seidel, Spherical codes and designs, Geom. Dedic. 6 (1977), 363-388.

4. S.J. Einhorn, I.J. Schoenberg, On euclidean sets having only two distances between points, Indag. Math. 28 (1966), 479-504.

5. X.L. Hubaut, Strongly regular graphs, Discrete Math. 13 (1975), 357-381.

6. O. Kristensen, private communication via H. Tverberg.

7. D.G.Larman, C.A. Rogers, J.J. Seidel, On two-distance sets in Euclidean space, Bull. London Math. Soc. 9 (1977), 261-267.

8. P.W.H. Lemmens, J.J. Seidel, Equiangular lines, J. Algebra 24 (1973), 494-512.

9. A. Neumaier, Distance matrices, dimension, and conference graphs, Indag. Math., to be published.

10. J.J. Seidel, Eutactic stars, Coll. Math.Soc. Bolyai 18, Combinatorics, Keszthely (1976), 983-999.

11. J.J. Seidel, Strongly regular graphs, Surveys in Combinatorics (ed. B. Bollobás), London Math. Soc. Lecture Note Series 38 (1979), 157-180.

12. E. Shult, A. Yanushka, Near n-gons and line systems, Geom.Dedic. 9 (1980), 1-72.

Department of Mathematics
Eindhoven University of Technology
P.O. Box 513
5600 MB Eindhoven
THE NETHERLANDS.

FINITE RAMSEY THEORY IS HARD

JOHN SHEEHAN

The Ramsey number $r(G_1,G_2)$ of two finite simple graphs G_1 and G_2 is the least integer r such that in every partition (E_1,E_2) of the edges $E(K_r)$ of K_r either $G_1 \subseteq \langle E_1 \rangle$ or $G_2 \subseteq \langle E_2 \rangle$. Equivalently $r(G_1,G_2)$ is the least integer r such that for all graphs G with r vertices either G_1 is a subgraph of G or G_2 is a subgraph of its complement $\overline{G}$. The existence of $r(G_1,G_2)$ is guaranteed [16] by a theorem due to F.P. Ramsey. If G_1 and G_2 are isomorphic graphs - both isomorphic to a graph G say - we write r(G) for $r(G_1,G_2)$. The "classical" Ramsey numbers $r(K_m,K_n)$ were the first to be studied and only a few nontrivial values are known: $r(K_3) = 6$, $r(K_3,K_4) = 9$, $r(K_3,K_5) = 14$, $r(K_4) = 18$, $r(K_3,K_6) = 18$ and $r(K_3,K_7) = 23$. The first four numbers were computed by Greenwood and Gleason [14]. The difficulties of Graver and Yackel [13] in computing $r(K_3,K_6)$ and $r(K_3,K_7)$ suggest there is little hope for much further progress in this direction. On the other hand the study of generalized Ramsey numbers $r(G_1,G_2)$ for arbitrary finite graphs G_1 and G_2 has enjoyed surprising success in recent years. The first non-trivial result is due to Gerencsér and Gyárfas [10] who proved in 1967 that if $m \geq n \geq 2$ then $r(P_m,P_n) = m + [\frac{n}{2}] - 1$ (P_m is a path with m vertices). Since then numerous and in some cases quite surprising results have been obtained. For example if C_m and $K_{m,n}$ respectively denote a cycle of length m and the complete bipartite graph with colour classes m and n respectively then the numbers $r(P_m,K_n)$, $r(C_m,C_n)$, $r(P_m,C_n)$, $r(P_m,K_{1,n})$ are known. Burr [2] gives a beautiful survey account of these and many other results in this area.

We shall consider in this article (see [1], [15] for the relevant notation) the numbers $r(K_m + \overline{K}_n)$. Of course when $n = 1$ this is simply a classical Ramsey number $r(K_{m+1})$. However we hope to show that by generalizing what is apparently an intractable problem at least a little insight is provided into the inherent difficulties of the original problem.

§1 Goodman's Theorem

Let $k_m(G)$ be the number of K_m's contained in a graph G and let $T_m(n)$ be defined by

$$T_m(n) = \min_G \{k_m(G) + k_m(\overline{G})\}$$

where this minimum is taken over all graphs G on n vertices. Then

$$r(K_m) = \min\{n : T_m(n) > 0\}.$$

Now obviously the determination of $T_m(n)$ for all m and n is a more general problem than the determination of $r(K_m)$. Nevertheless when m = 3 Goodman [12] proved:

THEOREM 1

$$T_3(n) = \begin{cases} (n(n-2)(n-4))/24 & (n \equiv 0,2 \pmod 4) \\ ((n+1)(n-3)(n-4))/24 & (n \equiv 3 \pmod 4) \\ (n(n-1)(n-5))/24 & (n \equiv 1 \pmod 4). \end{cases}$$

Proof (an outline).

Let (E_1, E_2) be any 2-colouring of K_n and $M_3(n)$ the resulting number of monochromatic triangles. Let $\langle E_1\rangle$ have degree sequence $\{d_i\}$ and $\langle E_2\rangle$ degree sequence $\{\bar{d}_i\}$. Now consider the set of triangles which are not monochromatic i.e. they contain either "1 red and 2 blue edges" or "2 red and 1 blue edges". The number of such triangles is $\frac{1}{2}\sum_{i=1}^{n} d_i\bar{d}_i$. Hence, letting $e = |E(G)|$,

$$\begin{aligned} M_3(n) &= \binom{n}{3} - \frac{1}{2}\sum_{i=1}^{n} \bar{d}_i d_i \\ &= \binom{n}{3} - \frac{1}{2}\left[\sum_{i=1}^{n} ((n-2) - (d_i-1))d_i\right] \\ &= \binom{n}{3} + \sum_{i=1}^{n}\binom{d_i}{2} - (n-2)e \\ &\geq \binom{n}{3} + n\binom{(\sum d_i)/n}{2} - (n-2)e \\ &= \binom{n}{3} + n\binom{2e/n}{2} - (n-2)e \\ &= \binom{n}{3} - \frac{e(n(n-1) - 2e)}{n} \\ &\geq (n(n-1)(n-5))/24. \end{aligned} \qquad (1)$$

Now suppose $n \equiv 1 \pmod 4$. Then (1) becomes an equality if and only if $d_i = \bar{d}_i = (n-1)/2$ for each i. Hence $T_3(n) = (n(n-1)(n-5))/24$. The proof when $n \not\equiv 1 \pmod 4$ is obtained in exactly the same way except that in these cases the degrees d_i and $\bar{d}_i$ must be chosen as "equal as possible" □

REMARKS (1) We have

n	5	6	7	8	9	10	11	12
$T_3(n)$	0	2	4	8	12	20	28	40

So, in particular, $r(K_3) = \min\{n : T_3(n) > 0\} = 6$.

(2) Notice that always

$$T_3(n) \geq (n(n-1)(n-5))/24.$$

(3) The important thing to remember about Goodman's proof is that once the degree sequence $\{d_i\}$ is specified then $M_3(n)$ is determined. In otherwords if G is any graph on n vertices then the total number of triangles in G and $\overline{G}$ is simply

$$\binom{n}{3} - \frac{1}{2}\sum_{i=1}^{n} d_i\overline{d}_i$$

where G has degree sequence $\{d_i\}$. Unfortunately this no longer holds true for $T_4(n)$.

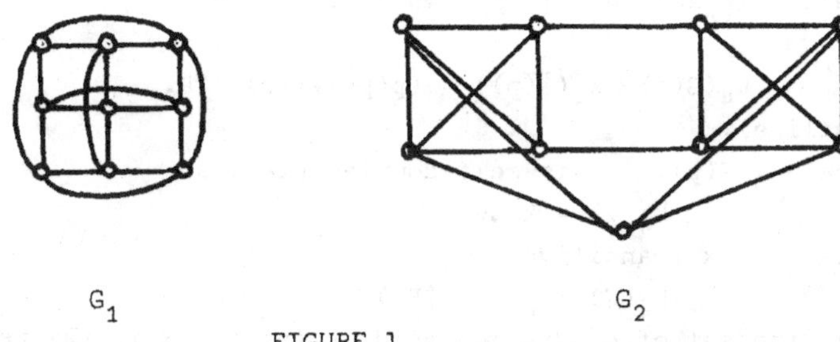

FIGURE 1

In Figure 1 both G_1 and G_2 are 4-valent graphs yet $k_4(G_i) + k_4(\overline{G}_i)$ equals 0 and 2 when i = 1 and 2 respectively. On the other hand $k_3(G_i) + k_3(\overline{G}_i)$ equals 12 (i = 1 and 2). Of course this is the underlying difficulty in any attempt to generalize Goodman's theorem.

The expected number of K_m's in a graph G and its complement is

$$\frac{2}{2^{\binom{m}{2}}}\binom{n}{m}$$

and Erdös conjectured [7] that

$$T_m(n) \sim \frac{2}{2^{\binom{m}{2}}}\binom{n}{m}.$$

In particular when m = 4 we have the conjecture that

$$T_4(n) \sim \frac{1}{32}\binom{n}{4}.$$

Giraud proved [11], using the same underlying idea as in the proof of Theorem 1, that for n large enough,

$$T_4(n) \geq \frac{1}{46}\binom{n}{4}.$$

In [9] we have counted the number of K_4's in a particular graph G and its complement $\overline{G}$. We obtain a beautiful formula which is expressible in terms of the classical Ramsey numbers. We consider this theorem in the next section.

§2 A theorem that counts.

Let $G(p)$ be the Paley graph where $p = 4k+1$ (p a prime power). Thus the vertices of $G(p)$ are the integers modulo p with vertex i joined to vertex j if and only if $i-j$ is a quadratic residue. Then $p = a^2 + b^2$ (a even, b odd) for some a and b. Write $a = 2n$, $b = 2m-1$. Then $k = n^2 + m(m-1)$. Let $f(p)$ be defined by

$$f(p) = ((k-2)^2 - n^2)/4.$$

Thus

$$f(p) \geq ((p-5)(p-17))/64$$

with equality if and only if $p = 4u^2+1$. Then

THEOREM 2

$$k_4(G(p) + k_4(\overline{G(p)}) = (p(p-1)f(p))/24.$$

Proof (Outline).

Write $G = G(p)$. G is self-complementary so

$$k_4(G) = k_4(\overline{G}).$$

Since G is vertex transitive

$$k_4(G) = \tfrac{1}{4}\, p\, k_3(\langle R\rangle)$$

where R is the set of quadratic residues (notice R is the neighbourhood $N(0)$ of the vertex 0). Let $H = N(0) \cap N(1)$. Since $\langle R\rangle$ is vertex transitive

$$k_3(\langle R\rangle) = \tfrac{1}{3}\left(\tfrac{p-1}{2}\right)k_2(\langle H\rangle).$$

Write $\hat{f}(p) = k_2(\langle H\rangle)$. The result is proved if we show $f = \hat{f}$. Now let $u,v \in H$ and suppose $uv \in E(\langle H\rangle)$. Then for some x and y (modulo p), $u = x^2$, $v = y^2$, $1-x^2 \in R$, $1-y^2 \in R$ and $x^2-y^2 \in R$. The problem eventually comes down to evaluating

$$\sum_0^{p-1}\sum_0^{p-1}\left(\frac{1-x^2}{p}\right)\left(\frac{1-y^2}{p}\right)\left(\frac{x^2-y^2}{p}\right)$$

where $(\frac{n}{p})$ is the Legendre symbol. In turn the key to this evaluation is the transformation

$$x = \frac{t+1}{ut-1}, \qquad y = \frac{u+1}{ut-1}.$$

The details will be published in [9] □

COROLLARY

$$k_4(G(p)) + k_4(\overline{G(p)}) \geq \frac{1}{32}\left[\frac{p(p-1)(p-5)(p-17)}{4!}\right]$$

with equality if and only if $p = 4u^2+1$ □

Now suppose p is prime and $p = 4u^2+1$. It is interesting to compare this corollary with Theorem 1. Write $r_i = r(K_i) - 1$ ($i \geq 1$). Since $G(p)$ is regular of degree $\frac{n-1}{2}$ we know from the proof of Theorem 1 that

$$T_3(p) = k_3(G(p)) + k_3(\overline{G(p)})$$
$$= ((p-r_1)(p-r_2)(p-r_3))/24.$$

From the corollary

$$k_4(G(p)) + k_4(\overline{G(p)}) = ((p-r_1)(p-r_2)(p-r_3)(p-r_4)/24)/32.$$

Unfortunately this pattern does not continue for $k_m(G(p)) + k_m(\overline{G(p)})$ $(m > 4)$ and in particular for $k_5(G(p)) + k_5(\overline{G(p)})$. Thomasson [19] using exactly the same techniques has evaluated $k_4(G) + k_4(\overline{G})$ for conference graphs. Of course the Paley graph is itself a conference graph.

§3 Back to Ramsey theory.

For the last thirty years little improvement has been made on the bounds given by Erdös [6] and by Erdös and Szekeres [5] viz.,

$$cm2^{\frac{m}{2}} < r(K_m) < 2^{2m}. \qquad (1)$$

The lower bound here is proved using a simple probabilistic argument and the upper bound using the usual Ramsey "nesting argument". Using similar techniques Erdös et al [8] proved that for large n

$$(2^m-\varepsilon)n < r(K_m+\overline{K}_n) < 2^{2m-1}(n+1) - 1. \qquad (2)$$

For the sake of argument we favour the lower bound against the upper bound in (2):

(a) If $m = 1$ these bounds are asymptotically the same.

(b) Suppose $m = 2$. Let (E_1,E_2) be a colouring of $E(K_p)$ such that $\langle E_i\rangle \not\geq K_2 + \overline{K}_n$. Let $M_3(p)$ be the number of monochromatic triangles. Then, from Theorem 1,

$$\begin{aligned}\frac{p(p-1)(p-5)}{24} &\le T_3(p)\\ &\le M_3(p)\\ &\le \frac{|E_1|(n-1)}{3} + \frac{|E_2|(n-1)}{3}\\ &= \frac{T_2(p)(n-1)}{3}. \end{aligned} \qquad (3)$$

Hence $p \le 4n+1$ i.e. $r(K_2+\overline{K}_n) \le 4n+2$. We show [17] that equality holds here when $p = 4n+1$ is a prime and in a number of other cases. All we want to observe here however is that in both cases, $m = 1$ and $m = 2$, the lower bound in (2) is asymptotically correct.

(c) Suppose $m \ge 3$. Well, truth to say, we can do no more than wildly speculate. It is hoped that the wildness of our speculations can be seen from the difficulties that we encountered in section (2). Suppose we assume that Erdös' conjecture

$$T_m(p) \sim \frac{2}{2^{\binom{m}{2}}} \binom{p}{m}$$

is true. Again assume that in any colouring (E_1,E_2) of $E(K_p)$ with $\langle E_i\rangle \not\supseteq K_m + \overline{K}_n$ (n large) we obtain the expected number of K_m's. Then equation (3) generalizes to

$$\frac{2}{2^{\binom{m+1}{2}}} \binom{p}{m+1} \sim T_{m+1}(p) \le M_{m+1}(p) \le T_m(p)\,\frac{n-1}{m+1} \sim \frac{2}{2^{\binom{m}{2}}} \binom{p}{m} \frac{n-1}{m+1}$$

which in turn yields

$$p \le 2^m(n-1) + O(1).$$

Thus once again the lower bound in (2) is asymptotically correct. However, as indicated in section (2), we are a long long way from being able to make either of the above not unreasonable assumptions.

§4 <u>The Ramsey numbers</u> $r(K_3+\overline{K}_n)$.

We know very little about the Ramsey numbers $r(K_3+\overline{K}_n)$. From (2) we know that

$$(8-\varepsilon)n \le r(K_3+\overline{K}_n) \le 32n. \qquad (4)$$

Clancy [4] has proved that $r(K_2+\overline{K}_2, K_3+\overline{K}_2) = 13$ which in turn implies that $r(K_3+\overline{K}_2) \le 26$. Hence:-

<u>QUESTION</u> Is $r(K_3+\overline{K}_2) = 26$? □

The answer is no. In fact $r(K_3 + \overline{K}_2) < 26$. A proof of this can be given using the following

LEMMA (CLANCY [4]). *Let G be a graph of order 12 such that G contains no $K_2 + \overline{K}_2$ and $\overline{G}$ contains no $K_3 + \overline{K}_2$. Then* (i) $q(G) \le 30$, and (ii) $q(G) = 30$ iff $G \cong H$ *where H is the graph shown on the next page.*

The proof of this lemma is straightforward. Now suppose that there exists a two-colouring (E_1,E_2) of $E(K_{25})$ in which neither $\langle E_1\rangle$ nor $\langle E_2\rangle$ contains $K_3 + K_2$. From the result of Clancy, $\langle E_1\rangle$ and $\langle E_2\rangle$ must be regular of degree 12. Let x be an arbitrary vertex and consider $G_1 = \langle N_1(x)\rangle_1$ and $G_2 = \langle N_2(x)\rangle_2$ where $N_1(x)$ and $N_2(x)$ denote the neighbourhoods of x in $\langle E_1\rangle$ and $\langle E_2\rangle$ respectively. By the regularity of $\langle E_1\rangle$ and $\langle E_2\rangle$ it follows that

$$q(G_1) + q(G_2) = 60.$$

Thus, by the Lemma, we must have $G_1 \cong G_2 \cong H$. This must be true for <u>every</u> vertex x. It is easily checked that the latter condition is impossible to fulfill. Consequently, $r(K_3 + \overline{K}_2) < 26$.

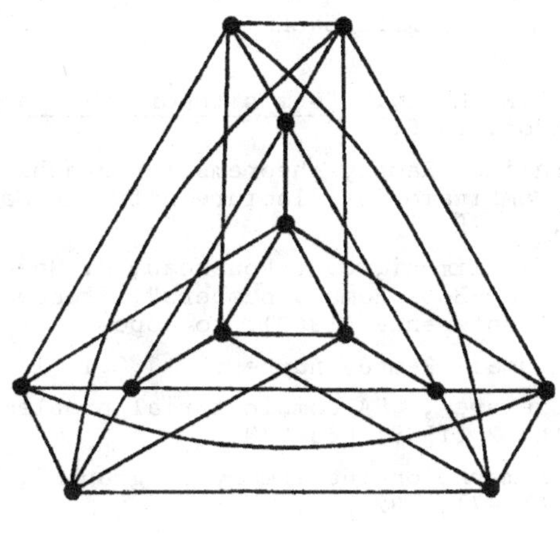

H

Since $r(K_2+\overline{K}_n) \le 4n+2$ we deduce that $r(K_3+\overline{K}_n) \le 16n+10$. Of course this improves on the upper bound in (4). We conjecture in fact that

<u>CONJECTURE 1</u> $r(K_3+\overline{K}_n) \le 8n+10$.

<u>Comment</u> Of course this is true when $n = 1$ and from the above when $n = 2$. It would also suggest that the lower bound in (4) is the correct one once again. Conjecture 1 would also follow if we could verify:-

<u>CONJECTURE 2</u> $r(K_2+\overline{K}_n, K_3+\overline{K}_n) \le 4n+5$.

<u>Comment</u> From Clancy's result this is true for $n = 2$. We can make little progress on this conjecture. Our final seemingly easier conjecture in this context is the following:-

<u>CONJECTURE 3</u> (The Monster Conjecture).

Set $k = [\frac{m-1}{n}]$. Then

$$r(K_1+\overline{K}_n, K_\ell+\overline{K}_m) \le \max\left\{(k+\ell)n+1, \left[\frac{(k+\ell+1)\{(k+\ell)(m-1)+\ell(k+1)n\}}{(k+1)(k+2\ell)}\right]+1\right\}.$$

<u>Comment</u> This conjecture is true (see [3],[18]) when $\ell = 2$. Provided a certain rectangular colouring exists (see [18]) it follows that $r(K_1+\overline{K}_n, K_\ell+\overline{K}_m)$ is at least as large as the upper bound given in Conjecture 3.

NOTES ADDED IN PROOF :

1. With reference to the classical Ramsey numbers $r(K_m,K_n)$, Grinstead has reportedly proved that $r(K_3,K_9) = 36$ and that $27 \le r(K_3,K_8) \le 29$.

2. I am indebted to the referee for the proof that $r(K_3+\overline{K}_2) < 26$.

REFERENCES

[1] J.A. Bondy and U.S.R. Murty, "Graph theory with applications", Macmillan, London, 1976.

[2] S.A. Burr, "Generalized Ramsey theorems for graphs - a survey" in Graphs and Combinatorics, lecture notes in Mathematics 406, Springer 1974, 52-75.

[3] G. Chartrand, A.D. Polimeni, C.C. Rousseau, J. Sheehan and M.J. Stewart, "On Star-Book Ramsey numbers", Proceedings of Kalamazoo Int. Conference (1980), to appear.

[4] M. Clancy, "Some small Ramsey numbers", J.G.T. 1 (1977), 89-91.

[5] P. Erdös and G. Szekeres, "A combinatorial problem in geometry", Compositio Math. 2 (1935) 463-470.

[6] P. Erdös, "Some remarks on the theory of graphs", Bull. Amer. Math. Soc. 53 (1947), 292-294.

[7] P. Erdös, "On the number of complete subgraphs contained in certain graphs", Magyar Tud. Akad. Mat. Kut. Int. Közl 7 (1962) 459-474.

[8] P. Erdös, R.J. Faudree, C.C. Rousseau and R.H. Schelp, "The size Ramsey number", Per. Math. Hungar. 9 (1978) 145-162.

[9] R.J. Evans, J.R. Pulham and J. Sheehan, "On the number of complete subgraphs contained in certain graphs", J.C.T. Ser B (to appear).

[10] L. Gerencsér and A. Gyárfas, "On Ramsey-type problems", Ann. Univ. Sci. Budapest Eötvös Sect. Math. 10 (1967), 167-170.

[11] G. Giraud, "Sur le problème de Goodman pour les quadrangles et la majoration des nombres de Ramsey", J.C.T. Ser. B 27 (1979), 237-253.

[12] A.W. Goodman, "On sets of acquaintances and strangers at any party", American Math. Monthly, 68 (1961), 107-111.

[13] J.E. Graver and J. Yackel, "Some graph theoretic results associated with Ramsey's theorem", J.C.T. 4 (1968), 125-175.

[14] R.E. Greenwood and A.M. Gleason, "Combinatorial relations and chromatic graphs", Canad. J. Math. 7 (1955), 1-7.

[15] F. Harary, Graph Theory, Addison-Wesley, Reading, Mass., 1969.

[16] F.P. Ramsey, "On a problem of formal logic", Proc. London Math. Soc. 30 (1930), 264-286.

[17] C.C. Rousseau and J. Sheehan, "On Ramsey numbers for books", J.G.T. 2 (1) (1978), 77-87.

[18] C.C. Rousseau and J. Sheehan, "A class of Ramsey problems involving trees", J. London Math. Soc. (2), 18 (1978), 392-396.

[19] A.G. Thomason, Ph.D. Thesis, Cambridge University, 1979.

Department of Mathematics

University of Aberdeen

Dunbar Street

Aberdeen AB9 2TY

SCOTLAND

FURTHER RESULTS ON COVERING INTEGERS OF THE FORM $1+k2^N$ BY PRIMES

R.G. STANTON

1. INTRODUCTION

In the preceding paper, we gave a history of the Polignac and Sierpinski problems; basically, both problems boil down to a discussion of the values of k for which the numbers $1+k2^n$ $(n \geq 0)$ are always composite for all n. For example, if k = 271,129, the numbers are always composite (and, indeed, are divisible by a member of the set of primes P = {3,5,7,13,17,241}).

Jacobi half-seriously claimed that in any mathematical problem "one must always invert". In this problem, the situation does seem clearer if we concentrate not on the numbers $1+k2^n$ but on the set of primes P. Let us ask the question in the inverse fashion: what sets P have the property that there exist integers k such that, for all $n \geq 0$,

$$1+k2^n \equiv 0 \bmod P.$$

The notation merely means "modulo at least one of the primes in the set P". The example in the last paragraph shows that such sets P do exist.

2. DISCUSSION OF SETS OF SMALL CARDINALITY

The fundamental tool in our discussions will be a very easy lemma.

Lemma 2.1 *If* $1+k2^n \equiv 0$ *mod* p *and* $1+k2^{n+a} \equiv 0$ *mod* p, *and if* a *is minimal, then* p *divides* 2^a-1, *and* p *does not divide* 2^b-1 *for* $b < a$.

Proof. We immediately have

$$k2^n \equiv -1 \equiv k2^{n+a} \bmod p.$$

Hence $1 \equiv 2^a \bmod p$, and p divides 2^a-1. Minimality is obvious.

We at once deduce several results.

Lemma 2.2 $|P| = 1$ *is impossible.*

Proof. $1+k2^n \equiv 0 \equiv 1+k2^{n+1}$ for $|P| = 1$.

But then $p|(2-1)$, and this is impossible.

It is useful at this stage to represent the various congruences by points on a linear graph and label each point by the modulus of the associated congruence.

Basically, Lemma 2.2 shows that adjacent points in the graph can not possess the same label; so the graph can not contain a section

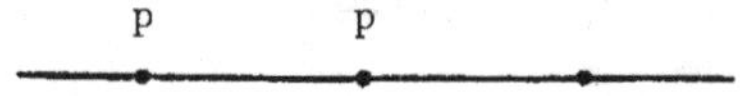

Now we prove

Lemma 2.3. $|P| = 2$ *is impossible.*

Proof. By Lemma 2.2, the system of congruences gives rise to an associated graph of the form

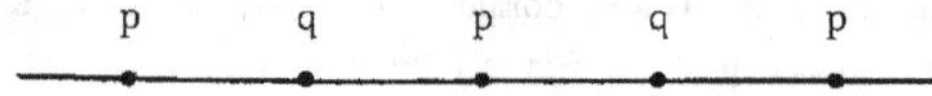

where $P = \{p,q\}$. Then Lemma 2.1 shows that $p \mid 2^2-1$, that is, $p = 3$. Similarly, $q = 3$, and we have a contradiction.

One can proceed onwards solely using graph-theoretic arguments. However, it is useful to employ density arguments as well. We note that the primes

$$p_2 = 3,\ p_3 = 7,\ p_4 = 5,\ p_5 = 31,$$

belong to exponents 2,3,4,5, respectively, where we say that p_i belongs to i when $p_i \mid 2^i-1$ (i being minimal). Clearly, there is at least one p_i for each $i > 1$, except for $i = 6$.

Now if $|P| = 3$, then $P = \{7,5,31\}$ can cover at most

$$\frac{1}{3} + \frac{1}{4} + \frac{1}{5} = \frac{47}{60}$$

of the points of the graph (actually, the number is less). So P must contain the prime 3. The graph thus has the form

p A p B p C p D p E

Points A and B must be associated with distinct primes; label A by q and B by r. Then C must be labelled q (and it follows that $q = 5$). But then any labelling of D leads to a contradiction. This proves

Lemma 2.4. *It is impossible to have* $|P| = 3$.

For $|P| = 4$, the density argument shortens the graph-theoretic argument quite a lot. Since

$$\frac{1}{3} + \frac{1}{4} + \frac{1}{5} + \frac{1}{7} = \frac{389}{420} < 1,$$

we see that $3 \in P$. Also,

$$\frac{1}{2}+\frac{1}{5}+\frac{1}{7}+\frac{1}{8} = \frac{271}{280} < 1.$$

Hence, either (or both) of 7 and 5 must be in P. The case of 3 and 5 leads to the graph

Now we may label A by r and B by s (neither p nor q is permitted). C must then be labelled r, and this forces the label s onto D; then $r = s = 17$, a contradiction.

If 3 and 5 are not in P, then 3 and 7 must be. But 3 belongs to exponent 2, 7 belongs to exponent 3. Between them, they label at most

$$\frac{1}{2}+\frac{1}{3}-\frac{1}{6} = \frac{4}{6}$$

of the points in the graph. This means that the only possibilities for exponents are 5 and 7. The figure

A p q p B q/p C p q p D q/p

leads to a contradiction, since C and A must both differ from B in label. Thus we have

Lemma 2.5. $|P| = 4$ *is impossible.*

3. THE COVERING APPROACH

The concept of labelling a linear graph that we employed in the last section is just a disguise for a combinatorial covering. Let us make a small table of the exponents to which any prime p belongs.

$2^2-1 = 3$	3 belongs to 2
$2^3-1 = 7$	7 belongs to 3
$2^4-1 = 3\cdot 5$	5 belongs to 4
$2^5-1 = 31$	31 belongs to 5
$2^6-1 = 7\cdot 3^2$	
$2^7-1 = 127$	127 belongs to 7
$2^8-1 = 3\cdot 5\cdot 17$	17 belongs to 8
$2^9-1 = 7\cdot 73$	73 belongs to 9
$2^{10}-1 = 3\cdot 31\cdot 11$	11 belongs to 10
$2^{11}-1 = 23\cdot 89$	23 belongs to 11
	89 belongs to 11

$2^{12}-1 = 3^2 \cdot 5 \cdot 7 \cdot 13$	13 belongs to 12
$2^{13}-1 = 8191$	8191 belongs to 13
$2^{14}-1 = 3 \cdot 43 \cdot 127$	43 belongs to 14
$2^{15}-1 = 31 \cdot 7 \cdot 151$	151 belongs to 15
$2^{16}-1 = 3 \cdot 5 \cdot 17 \cdot 257$	257 belongs to 16
$2^{17}-1 = 131071$	131071 belongs to 17
$2^{18}-1 = 7 \cdot 73 \cdot 3^3 \cdot 19$	19 belongs to 18
$2^{19}-1 = 524287$	524287 belongs to 19
$2^{20}-1 = 3 \cdot 11 \cdot 31 \cdot 5^2 \cdot 41$	41 belongs to 20
$2^{21}-1 = 127 \cdot 7^2 \cdot 337$	337 belongs to 21
$2^{22}-1 = 23 \cdot 89 \cdot 3 \cdot 683$	683 belongs to 22
$2^{23}-1 = 47 \cdot 178481$	47 belongs to 23
	178481 belongs to 23
$2^{24}-1 = 3^2 \cdot 5 \cdot 7 \cdot 13 \cdot 17 \cdot 241$	241 belongs to 24

Now let us consider the set of primes $P = \{3,5,7,13,17,241\}$ with its associated set of exponents

$$\exp P = \{2,4,3,12,8,24\}.$$

The LCM of the elements in exp P is a very important number (in this case, it is 24); let us denote it by

$$\mathrm{LCM}(\exp P) = 24.$$

If we think of the linear graph associated with P, we need only draw 24 nodes, since the set of labels is periodic with period 24 (in general, the set of labels is periodic with period LCM (exp P)).

Alternatively, we may replace the graph by a (periodic) set of boxes, and represent the situation by placing a prime in each compartment. The prime

3	5	3	7	3	5	3	13	3	5	3	17	3	5	3	7	3	5	3	13	3	5	3	241

3 goes into every second compartment; the prime 5 into every fourth compartment; the prime 7 into every third compartment (but do not put it in a compartment already occupied). With this agreement, we see that all compartments are filled; this is just another way of saying that every congruence is satisfied for some $p \in P$.

It is clear from this example of filling in the compartments that the primes can be inserted in various ways (for example, there are 2 ways of putting in the 3's; when the 3 and 5 have been inserted, the 241 could be placed in any of the 6 remaining places). Each allocation of primes leads to a different set of congruences and thence, by employing the Chinese Remainder Theorem, to a different k-value in the expressions $1+k2^n$.

Altogether, if we use a computer programme to apply the Chinese Remainder Theorem, we find 48 values of k in the range $0 < k < M = 3\cdot5\cdot7\cdot13\cdot17\cdot241$ (obviously, if a value k is found in this range, then $k+\alpha M$, where α is an arbitrary positive integer, will also work).

Of these 48 values of k, 21 values are odd (the other 27 values are obtained by multiplying odd k-values by powers of 2 that keep the result < M; clearly, such even k-values work since, for example, $1+(2k)2^n = 1+k(2^{n+1})$). We list these 21 odd values of k.

271129	1518781	2931991
271577	1624097	3083723
482719	1639459	3098059
575041	2131043	3555593
603713	2131099	3608251
903983	2541601	4067003
965431	2931767	4573999

Some of these k-values, such as the eleventh and twelfth, are remarkably near to one another.

We conclude this section by giving analogous lists. For $P = \{3,5,7,13,19,37,73\}$, we have

$$\mathrm{LCM}(\exp P) = \mathrm{LCM}\{2,4,3,12,18,36,36\} = 36.$$

In the range $0 < k < M$, there are 144 values of k; their are 75 basic odd values of k.

78557	19558853	44103533	60909197
2191531	20312899	44743523	61079749
2510177	20778931	45181667	61196987
2576089	21610427	45414683	62888633
7134623	22047647	45830431	63190223
7696009	27160741	46049041	63723707
8184977	29024869	50236847	63833243
10275229	30423259	51299477	63891497
10391933	31997717	51642601	65623711
11201161	32548519	51767959	66620329
12151397	33234767	52109063	66887071
12384413	33485483	52343539	66941839
12756019	34167691	53085709	67837073
13065289	34471877	55726831	68468753
13085029	34629797	56330011	68496137
15168739	34636643	57396979	
16391273	36120983	57616051	
18140153	38592529	57732559	
18156631	41403227	57940433	
19436611	42609587	60143641	

Finally, for $P = \{3,5,17,257,65537,641,6700417\}$, we have

$$\mathrm{LCM}(\exp P) = \mathrm{LCM}\,\{2,4,8,16,32,64,64\} = 64.$$

For this case, there are 64 values of k in $0 < k < M$, of which 33 are odd, namely,

201446503145165177
1007236913771681629
1697906240793858917
2331023822106839599
2935363331541925531
3367034409844073483
3914042604075779837
4863495246870308311
5036162578625852633
5590196669446332863
6705290764721718679
7284449444083822547
7338408328871591041
8374418985177323101
8489531205043036249
8719755778992191057
9122930260258969411

10388883947908195607
10691053702625738573
10906889241776812549
11655119660289929963
12018470371577942239
12576017419215635147
12865596758896687081
13784837166984260513
14417814010808873611
15050931866999740789
15232607222643746927
15511380746462593381
16432279042259212613
16839675648176649271
17643209860943100443
18044976967326326029

4. THE CASE OF FIVE PRIMES

First we use a density argument to show that $3 \in P$. Suppose this is not the case. Then the fractions of compartments covered by the various primes are

$$\frac{1}{3}, \frac{1}{4}, \frac{1}{5}, \frac{1}{7}, \frac{1}{8}, \frac{1}{9}, \frac{1}{10}, \frac{1}{11}, \frac{1}{12}, \ldots$$

Clearly we need $\frac{1}{3}$ (that is, $7 \in P$); then $\frac{1}{5}, \frac{1}{7}, \frac{1}{8}, \frac{1}{9}$, are insufficient. So we need $\frac{1}{3}$ and $\frac{1}{4}$; but, together, they can only account for $\frac{1}{3} + \frac{1}{4} - \frac{1}{12} = \frac{1}{2}$ of the compartments. Then $\frac{1}{2} + \frac{1}{5} + \frac{1}{7} + \frac{1}{8} < 1$, which will not do. Thus, we have

<u>Lemma 4.1</u>. *If* $|P| = 5$, *then* $3 \in P$.

Now the same density argument shows that densities of $\frac{1}{3}$ and $\frac{1}{4}$ can not both be missing. Hence, we get

<u>Lemma 4.2</u>. *If* $|P| = 5$, *either* $\{3,5\} \subset P$ *or* $\{3,7\} \subset P$.

If $\{3,5\} \subset P$, then we have the following figure.

3	A	3	5	3	B	3	5	3	C	3	5	3	D	3	5	3	E	3	5	3	F	3 ...

In Box A, we must place prime r; in box B, we must place prime s. If we place r in C, then t must be placed in D; then r is in E and F is forced to have prime s. This leaves nothing to go in place H (r is in place G). Hence A,B,C, must be filled by r,s,t, and this speedily produces a contradiction at place E.

Now consider $\{3,7\} \subset P$; the box figure looks as follows.

	r				s		t				r		s								
3	A	3	7	3	B	3	C	3	7	3	D	3	E	3	7	3	F	3		3	...

The primes in boxes C and A differ from that in box B, and from one another. Let them be r,s,t, respectively. Then box D must contain r and E must contain s. This leaves no prime for box F, and we have proved

Lemma 4.3. *It is not possible to have* $|P| = 5$.

5. THE CASE OF SIX PRIMES

From the example P = 3,5,7,13,17,241 , we know there is a solution for the case of 6 primes. It is remarkable that this solution is unique.

First, suppose that 3 is not an element of P; then the possible density fractions are

$$\frac{1}{3}, \frac{1}{4}, \frac{1}{5}, \frac{1}{7}, \frac{1}{8}, \frac{1}{9}, \frac{1}{10}, \ldots$$

It is not possible to have $\frac{1}{3}$ absent from the list; hence $7 \in P$. By the same argument, not both of $\frac{1}{4}$ and $\frac{1}{5}$ are missing. Indeed, even if both are present, they give a compound density of only

$$\frac{1}{3} + \frac{1}{4} + \frac{1}{5} - \frac{1}{12} - \frac{1}{20} - \frac{1}{15} + \frac{1}{60} = \frac{36}{60} .$$

This is insufficient, even if we add $\frac{1}{7} + \frac{1}{8} + \frac{1}{9}$. So we have proved

Lemma 5.1. *If* $|P| = 6$, *then* $3 \in P$.

The ensuing discussion is lengthier than in Section 4, but quite analogous. We omit it, and state only the final result as

Lemma 5.2. *If* $|P| = 6$, *then* $P = \{3,5,7,13,17,241\}$.

Lemma 5.2 illustrates an important fact, namely, that, if we fix the cardinality of P, then only a finite set of solutions occurs. We state this result as

Lemma 5.3. *The number of minimal s-sets, for s specified, is finite (this number is zero for* $s < 6$, *one for* $s = 6$*).*

<u>Proof</u>. Assume there are infinitely many minimal s-sets. Then density considerations show that some prime occurs in an infinitude of minimal s-sets. Let T be a set of t primes which occur together in an infinitude of minimal s-sets with t maximal (obviously $t < s$). Let C be the LCM of the gap lengths of the members of T; then there is a minimal s-set S in which all the members of S-T have gap length $> (s-t)C+1$. But then the graph corresponding to S has an interval of length C containing only members of T, implying that T is a T-set and so contradicting the minimality of S.

6. FURTHER REMARKS

The case of 7 primes in P has been studied by J. Selfridge and J. van Rees, and I am indebted to them for showing me their results. Among their interesting results, it is particularly worthwhile to mention the following.

(1) There are 20 solutions for 7 primes; of these 4 have LCM(exp P) = 36, 15 have LCM(exp P) = 48, one has LCM(exp P) = 64. The last one is, of course, the one derived from the Fermat numbers.

(2) The set P need not include the prime 3. The following set P with $|P| = 16$, does not contain 3.

$P = \{5,7,11,13,17,19,31,37,41,61,73,97,109,151,241,257\}$.

(3) It is impossible to have a set P with largest prime < 73. In this sense, the "smallest" set P is $\{3,5,7,13,19,37,73\}$.

(4) If the largest prime in a set P is specified to be p, then the minimal covers for the first 6 p's are as follows:

p = 73	P = {3,5,7,13,19,37,73}
p = 109	P = {3,5,7,13,19,37,109}
p = 151	P = {3,5,7,11,13,31,41,61,151}
p = 181	P = {3,5,7,11,13,19,31,37,41,61,181}
p = 241	P = {3,5,7,13,17,241}
p = 257	P = {3,5,7,13,17,97,257}.

Department of Computer Science
University of Manitoba
Winnipeg
Manitoba R3T 2N2
CANADA

DISTRIBUTIVE BLOCK STRUCTURES AND THEIR AUTOMORPHISMS

R.A. BAILEY

The experimental units in a statistical experiment are frequently grouped into blocks in one or more ways. When the different families of blocks fit together in a well-behaved way we have a distributive block structure. We show that the orbits of the automorphism group of a distributive block structure on pairs of experimental units are precisely the sets which the combinatorial structure leads one to expect. Possible generalizations of this result are discussed.

1. BLOCK STRUCTURES

Let Ω be a set. An equivalence relation ρ on Ω is *uniform* if all its equivalence classes have the same size. We shall refer to equivalence classes of ρ as ρ-*blocks* (or simply *blocks*) throughout this paper, and denote the ρ-block containing an element ω of Ω by $\rho(\omega)$.

Let ρ and σ be two equivalence relations on Ω. We define two further relations, $\rho \wedge \sigma$ and $\rho \vee \sigma$, on Ω as follows:

$$\alpha(\rho \wedge \sigma)\beta \text{ if and only if } \alpha \,\rho\, \beta \text{ and } \alpha \,\sigma\, \beta;$$

$$\alpha(\rho \vee \sigma)\beta \text{ if and only if } \exists\, \gamma \in \Omega \text{ such that } \alpha \,\rho\, \gamma \text{ and } \gamma \,\sigma\, \beta.$$

(The second relation is illustrated in Figure 1.)

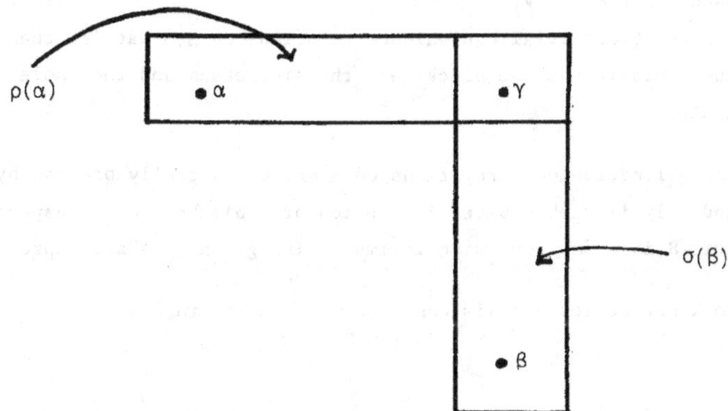

FIGURE 1.

Then $\rho \wedge \sigma = \sigma \wedge \rho$, and $\rho \wedge \sigma$ is also an equivalence relation. Its blocks are the non-empty intersections of ρ-blocks with σ-blocks. Even if ρ and σ are both uniform, $\rho \wedge \sigma$ is not necessarily uniform. On the other hand, $\rho \vee \sigma \neq \sigma \vee \rho$ in general: it is well-known, and easy to prove, that $\rho \vee \sigma$ is an equivalence relation if and only if $\rho \vee \sigma = \sigma \vee \rho$, which is the situation illustrated in Figure 2. It is

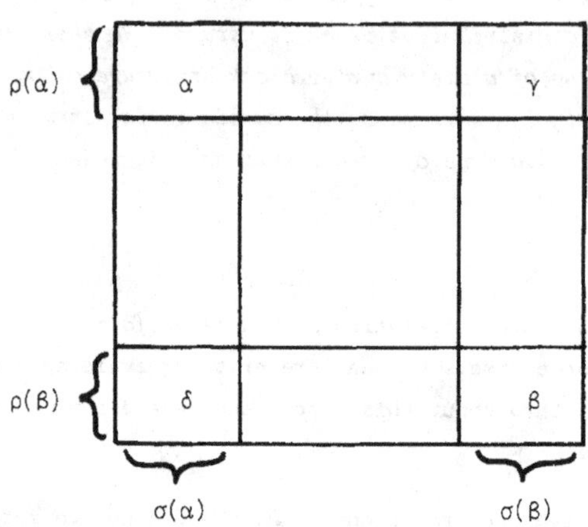

FIGURE 2.

shown in [7] that if ρ, σ and $\rho \wedge \sigma$ are uniform then so is $\rho \vee \sigma$.

Definition. A *block structure* is a pair (Ω, S), where Ω is a set and S is a finite set of uniform equivalence relations on Ω such that

(i) S is closed under $\wedge$ and $\vee$;

(ii) S contains the two trivial equivalence relations on Ω, that is, the equivalence relations whose blocks are the singletons and the whole of Ω respectively.

The set of all equivalence relations on Ω may be partially ordered by $\leq$, where $\rho \leq \sigma$ if and only if each σ-block is a union of ρ-blocks. With respect to this partial order, S is a *lattice*, with infimum being given by $\wedge$ and supremum by $\vee$.

These block structures are discussed in more detail in [7].

2. EXAMPLES

Sets of experimental units considered by statisticians often form block structures. Two of the most common of these are described below.

Example 1. In a row-and-column design (Figure 3(i)) $\Omega = \Delta_1 \times \Delta_2$ and the equivalence relations correspond to subsets of $\{1,2\}$. If $J \subseteq \{1,2\}$ then the corresponding equivalence relation ρ_J is defined by

$$(\alpha_1,\alpha_2)\rho_J(\beta_1,\beta_2) \text{ if and only if}$$

$$\alpha_i = \beta_i \text{ for all } i \in J.$$

Thus the trivial equivalence relations correspond to ϕ and $\{1,2\}$, while the subsets $\{1\}$ and $\{2\}$ correspond respectively to the row and column relations. The lattice diagram for S is shown in Figure 3(ii).

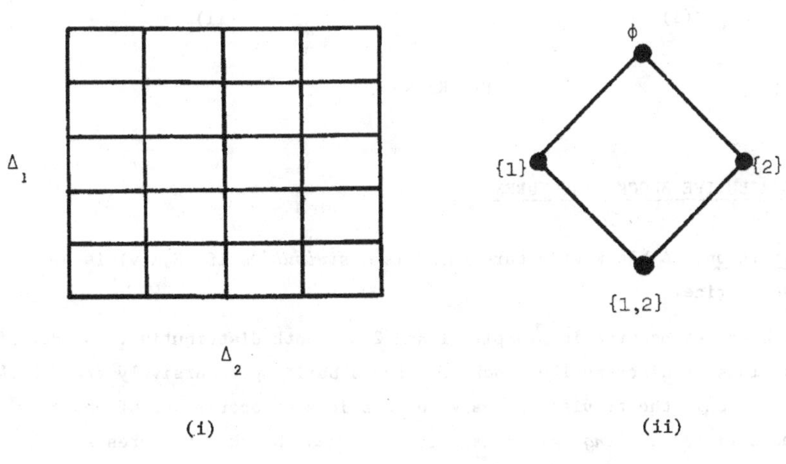

FIGURE 3.

Example 2. In a plots-within-blocks design (Figure 4(i)) we again have $\Omega = \Delta_1 \times \Delta_2$. However, in this case the only non-trivial equivalence relation in S is the column one; that is, equality of the first subscript has no significance unless the second subscript is also equal (see Figure 4(ii)). In statistical terminology (see [6], for example), blocks (subscript 2) *nest* plots (subscript 1). In Example 1 there is no such nesting.

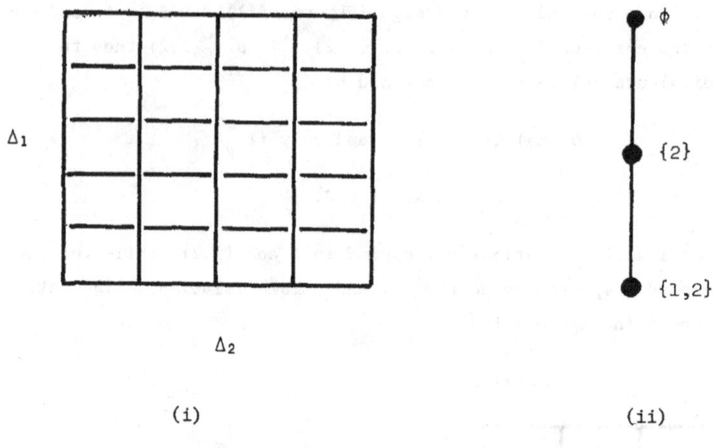

FIGURE 4.

3. DISTRIBUTIVE BLOCK STRUCTURES

Definition. A block structure (Ω,S) is *distributive* if $(S,\wedge,\vee)$ is a distributive lattice.

The block structures in Examples 1 and 2 are both distributive. Nelder [6] describes a class of distributive block structures built up recursively from block structures with only the trivial equivalence relations by operations of *crossing* (as in Example 1) and *nesting* (as in Example 2). These block structures also lie in the class of block structures constructed from partially ordered sets in the manner described below. The partial order here is the nesting relation illustrated in Example 2. Before we explain the construction we need one more definition.

Definition. Let $(I,\leq)$ be a partially ordered set. A subset J of I is *ancestral* if whenever $j \in J$ and $j \leq i$ then $i \in J$.

Construction. Let $(I,\leq)$ be a finite partially ordered set. For $i \in I$, let Δ_i be a set with $|\Delta_i| \geq 2$ unless $|I| = 1$ (to avoid degenerate cases). Let $\Omega = \prod_{i\in I} \Delta_i$. For $J \subseteq I$, let ρ_J be the equivalence relation on Ω defined by

$$\alpha \rho_J \beta \text{ if and only if } \alpha_i = \beta_i \text{ for all } i \in J.$$

Let A be the set of ancestral subsets of I, and let

$$S = \{\rho_J\colon J \in A\}.$$

Then (Ω,S) is a distributive block structure, because $(S,\wedge,\vee)$ is isomorphic to $(A,\cup,\cap)$. We call (Ω,S) a *poset block structure.*

This construction is given in [3] and [7]. Much literature on the design of experiments (see, for example [10] and [8], as well as standard texts such as [4] and [5]) is concerned, even if not explicitly, with the block structures obtained from this construction. Throckmorton [8] gives the following example of a poset block structure which is not obtainable by Nelder [6]'s construction.

Example 3. Take $(I,\leq)$ to be the partially ordered set shown in Figure 5(i). The dual of the lattice of ancestral subsets of I is shown in Figure 5(ii). Figure 6 shows a realization of the distributive block structure. Apart from Ω itself, only

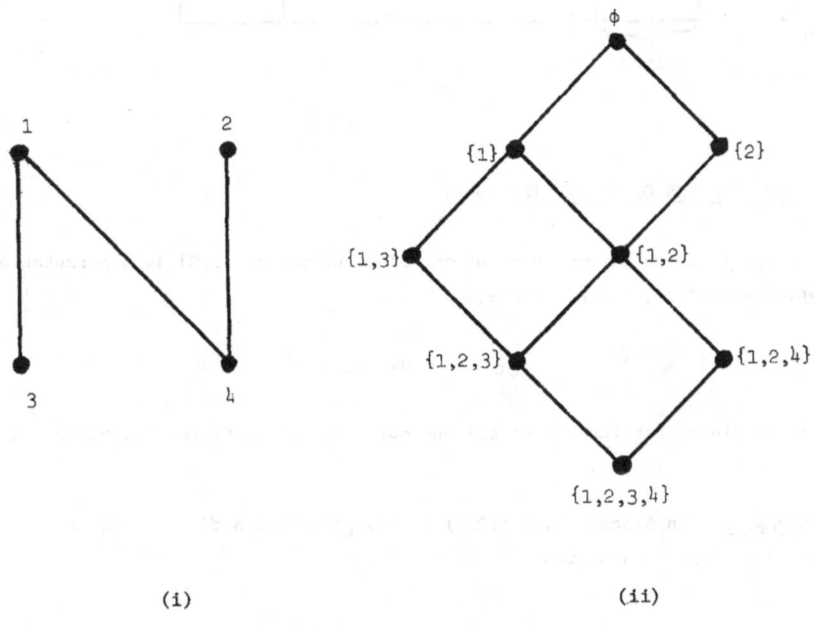

FIGURE 5.

four types of block are shown. The remaining types are intersections of the types shown: for example, a 1,2 -block is a *square*, the intersection of a row and a column.

It is proved in [7] that *every* distributive block structure is isomorphic to a poset block structure. In particular, a distributive block structure is determined, up to isomorphism, by its lattice of equivalence relations and the cardinalities $|\Delta_i|$.

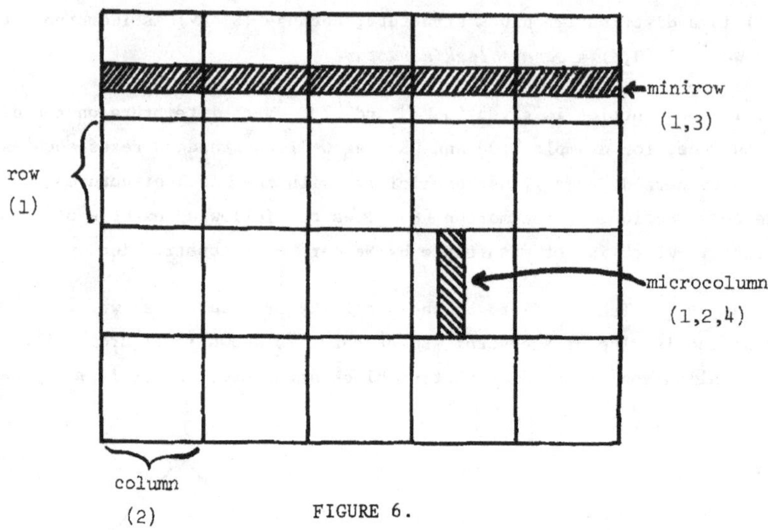

FIGURE 6.

4. AUTOMORPHISMS OF BLOCK STRUCTURES

Definition. An *automorphism* of the block structure (Ω,S) is a permutation g of Ω which satisfies, for all $\rho \in S$,

$$\alpha \rho \beta \iff \alpha g \rho \beta g .$$

It is clear that the set of all automorphisms of (Ω,S) forms a group, which we shall denote by $\mathrm{Aut}(\Omega,S)$.

Examples. In Example 1, $\mathrm{Aut}(\Omega,S)$ is the permutation *direct* product $\mathrm{Symm}(\Delta_1) \times \mathrm{Symm}(\Delta_2)$, with action defined by

$$(\delta_1,\delta_2)(g_1,g_2) = (\delta_1 g_1,\delta_2 g_2).$$

In Example 2, $\mathrm{Aut}(\Omega,S)$ is the permutation *wreath* product $\mathrm{Symm}(\Delta_1)$ wr $\mathrm{Symm}(\Delta_2)$, which as a *set* is equal to $(\mathrm{Symm}(\Delta_1))^{\Delta_2} \times \mathrm{Symm}\ \Delta_2$, and whose action is given by

$$(\delta_1,\delta_2)(f,g_2) = (\delta_1(\delta_2 f),\delta_2 g_2).$$

Holland [3] defines the automorphism group of a poset block structure to be a *generalized wreath product*. From the arguments in [3] we can give the explicit form of these groups as follows.

Theorem 1. Let $(I,\leq)$ be a partially ordered set, and let (Ω,S) be the poset block structure defined by $(I,\leq)$ and sets Δ_i for $i \in I$. Let $\Omega_i = \prod_{j>i} \Delta_j$, and let π^i be the natural projection from Ω onto Ω_i. Let $F_i = (\mathrm{Symm}(\Delta_i))^{\Omega_i}$. Then $\mathrm{Aut}(\Omega,S) = \prod_{i\in I} F_i$ and the action is defined by

$$\alpha f = \beta,$$

where $\alpha = (\alpha_i) \in \Omega$, $\beta = (\beta_i) \in \Omega$, $f = (f_i) \in F$, and

$$\beta_i = \alpha_i(\alpha\pi^i f_i).$$

5. ORBITS

Definition. Let (Ω,S) be a poset block structure based on the partially ordered set $(I,\leq)$. Let J be an ancestral subset of I. The *association set* A_J is the subset of $\Omega \times \Omega$:

$$\{(\alpha,\beta):\ J \text{ is the maximal ancestral subset of } I \text{ such that } \alpha\ \rho_J\ \beta\}.$$

If (α,β) are in A_J then α and β are J-*associates*.

Note that if α and β are J-associates then ρ_J is the *minimal* equivalence relation in S relating α and β. Moreover, the A_J form a partition of $\Delta \times \Delta$.

If Ω is a set of experimental units with associated random variables, the question arises as to what is an appropriate model for the covariance of those random variables. The covariance of the random variables associated with α and β is assumed by some authors, for example John [4], to be a function of the association set containing (α,β), and by other authors, for example, Nelder [6] and Bailey [2], to be a function of the orbit of $\mathrm{Aut}(\Omega,S)$ on $\Omega \times \Omega$ which contains (α,β). Thus an important question, recognized implicitly in [9] is: when do the association sets coincide with the orbits?

Theorem 2. If (Ω,S) is a distributive block structure, the association sets coincide with the orbits of $\mathrm{Aut}(\Omega,S)$ on $\Omega \times \Omega$.

Proof. Since $\mathrm{Aut}(\Omega,S)$ preserves each equivalence relation ρ_J for J ancestral, it is clear that each A_J is a union of orbits of $\mathrm{Aut}(\Omega,S)$.

To show that each A_J is a single orbit of $\mathrm{Aut}(\Omega,S)$, we first observe that α and β are J-associates if and only if

(i) $\alpha_i = \beta_i$ for all $i \in J$;

(ii) if $i \notin J$, there is some $k \geq i$ such that $\alpha_k \neq \beta_k$.

Now suppose that $(\alpha,\beta) \in A_J$ and $(\gamma,\delta) \in A_J$. For $i \in J$, choose $g_i \in \mathrm{Symm}(\Delta_i)$ such that $\alpha_i g_i = \gamma_i$, and define $f_i \in F_i$ by $\Omega_i f_i = \{g_i\}$. If $i \notin J$ there are two possibilities: either there is some $k > i$ such that $\alpha_k \neq \beta_k$ or $\alpha_i \neq \beta_i$ but $\alpha_k = \beta_k$ for all $k > i$. In the first case choose g_i and h_i in $\mathrm{Symm}(\Delta_i)$ such that $\alpha_i g_i = \gamma_i$ and $\beta_i h_i = \delta_i$. Since π^i does not forget the value of the k-th coordinate, we may define f_i by

$$\varepsilon\pi^i f_i = \begin{cases} g_i & \text{if } \varepsilon_k = \alpha_k \\ h_i & \text{otherwise} \end{cases} .$$

In the second case $J \supseteq \{k\colon k > i\}$. Since $\gamma\ \rho_J\ \delta$, we have $\gamma_k = \delta_k$ for all $k > i$. But γ and δ are J-associates, so γ and δ are not related by $\rho_{J\cup\{i\}}$, so $\gamma_i \neq \delta_i$. Now choose $g_i \in \mathrm{Symm}(\Delta_i)$ such that $\alpha_i g_i = \gamma_i$ and $\beta_i g_i = \delta_i$, and define $f_i \in F_i$ by $\Omega_i f_i = \{g_i\}$.

By construction, $\alpha f = \gamma$ and $\beta f = \delta$. Thus A_J is a single orbit of $\mathrm{Aut}(\Omega,S)$ on $\Omega \times \Omega$, and this completes the proof.

Note. Examination of the proof shows that the conclusion of Theorem 2 remains true if, for each i, $\mathrm{Symm}(\Delta_i)$ is replaced by any doubly transitive group on Δ_i.

6. OTHER BLOCK STRUCTURES

Non-distributive block structures are, in general, not determined up to isomorphism by their lattices of equivalence relations and suitable collections of cardinalities. The simplest non-distributive block structure has the lattice shown in Figure 7: the non-trivial equivalence relations correspond to the rows, columns and letters of a Latin square. In general there are many isomorphism classes of Latin squares of a given size, and hence many isomorphism classes of block structures.

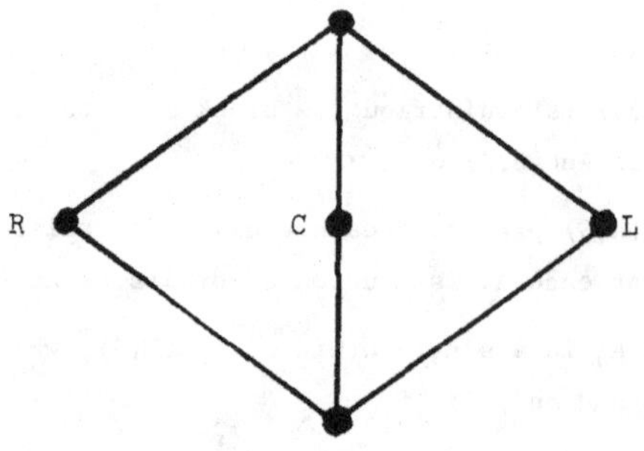

FIGURE 7.

Because of this lack of uniqueness, it seems unlikely that Theorem 2 can be extended to include a significantly larger class of block structures, although other examples are known where the association sets coincide with the orbits. For example, it is shown in [1] that if (Ω,S) is a block structure with $(S,\wedge,\vee)$ isomorphic to the lattice shown in Figure 7, so that (Ω,S) is specified by a Latin square Λ, then the association sets coincide with the orbits if and only if Λ is isomorphic to the composition table of an elementary abelian 2-group or the cyclic group of order 3.

Related to the "if" part of this result we have the following theorem.

Theorem 3. Let Ω be an n-dimensional vector space over GF(q), where $n \geq 2$. For each subspace V of Ω let ρ_V be the equivalence relation on Ω defined by

$$\alpha \, \rho_V \, \beta \text{ if and only if } \alpha - \beta \in V.$$

Let $S = \{\rho_V : V \text{ is a subspace of } \Omega\}$. Then

(i) (Ω,S) is a block structure;

(ii) $\mathrm{Aut}(\Omega,S) = \{\sigma_{\alpha,w} : \alpha \in GF(q), w \in \Omega\}$ where $\sigma_{\alpha,w} : v \mapsto \alpha v + w, v \in \Omega$.

(iii) the association sets of (Ω,S) are precisely the orbits of $\mathrm{Aut}(\Omega,S)$ on $\Omega \times \Omega$.

Because Latin squares or vector space structures can be imposed on distributive block structures with suitable cardinalities, giving lattices such as those shown in Figures 8 and 9, we can construct other block structures whose association sets are orbits of their automorphism groups. However, the problem of characterizing all such block structures does not seem close to solution.

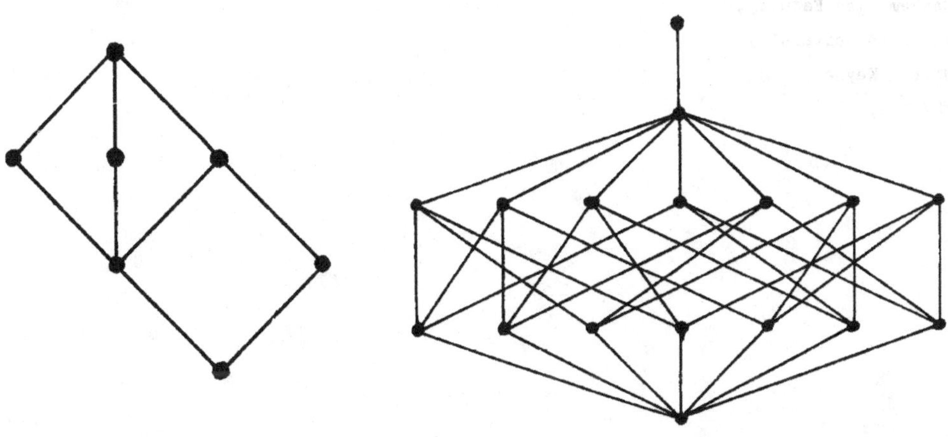

FIGURE 8. FIGURE 9.

REFERENCES

[1] R.A. Bailey, *Latin squares with highly transitive automorphism groups*, to appear.

[2] R.A. Bailey, *A unified approach to design of experiments*, J.R. Statist. Soc. A, to appear.

[3] W.C. Holland, *The characterization of generalized wreath products*, J. Algebra 13 (1969), 152-172.

[4] P.W.M. John, *Statistical design and analysis of experiments*. (Macmillan, 1971.)

[5] O. Kempthorne, *The design and analysis of experiments*. (Wiley, 1957.)

[6] J.A. Nelder, *The analysis of randomized experiments with orthogonal block structure*, Proc. Roy. Soc. A 283 (1965), 147-178.

[7] T.P. Speed and R.A. Bailey, *On a class of association schemes derived from lattices of equivalence relations*, Proceedings of the First Western Australian Algebra Conference (1980).

[8] T.N. Trockmorton, *Structures of classification data*. Ph.D. Thesis, Iowa State University, 1961.

[9] F. Yates, *The formation of Latin squares for use in field experiments*, Empire J. Exp. Agric. 1, (1933), 235-244.

[10] G. Zyskind, *Error structures in experimental designs*. Ph.D. Thesis, Iowa State University, 1958.

Mathematics Faculty,
The Open University,
Milton Keynes,
MK7 6AA,
U.K.

05 C 99

CONNECTED SUBGRAPHS OF THE GRAPH OF MULTIGRAPHIC REALISATIONS OF A DEGREE SEQUENCE

DAVID BILLINGTON

An m-*graph is a graph, without loops, but with multiple edges of any multiplicity less than or equal to* m. *An exact* m-*graph is an* m-*graph with at least one edge of multiplicity* m. *A new proof is given that the graph* $R(\underset{\sim}{d}, L(m))$, *of all* m-*graphic realisations of a degree sequence,* $\underset{\sim}{d}$, *is connected. This is done by taking any two vertices of* $R(\underset{\sim}{d}, L(m))$, *say* G *and* H, *and finding a path between them which preserves any previously chosen edge of multiplicity* m *that occurs in both* G *and* H. *The construction of this path also establishes best possible upper and lower bounds on the length of the shortest path between any two vertices of* $R(\underset{\sim}{d}, L(m))$.

1. DEFINITIONS

By an m-*graph* we mean a graph, without loops, but with multiple edges of any multiplicity less than or equal to m. Call an m-graph *exact* if and only if it has an edge of multiplicity m. Let G be an m-graph and V(G) be the vertex set of G. For all $v \in V(G)$, the *degree of* v *in* G, $\deg_G(v)$, is the number of edges of G which contain v. The *degree sequence* of G, Deg(G), is the non-increasing list of the degrees of all the vertices in G. If $u,v \in V(G)$ then the *multiplicity of* [u,v] *in* G, $\text{mult}_G[u,v]$, is the number of edges of G which contain both u and v. So for all $u,v \in V(G)$, $0 \leq \text{mult}[u,v] \leq m$. If H is also an m-graph and Deg(G) = Deg(H) then there is a degree preserving bijection from V(G) to V(H). Hence whenever Deg(G) = Deg(H) we shall suppose that V(G) = V = V(H) and $\deg_G(v) = \deg_H(v)$ for all $v \in V$.

Let [t,u] and [v,w] be two edges of G such that t,u,v,w are four different vertices. The notation

$$[t,u] + [v,w] \rightarrow [t,v] + [u,w]$$

denotes the operation of replacing the two edges [t,u] and [v,w] of G by the two edges [t,v] and [u,w]. Such an operation on G is called an *exchange* on G. If f is an exchange on G the result is denoted by f(G). It is clear that an exchange on G preserves the degree of every vertex of G. Let $\underset{\sim}{d}$ be the degree sequence of some multigraph. By a multigraphic realisation of $\underset{\sim}{d}$ we mean a multigraph whose degree sequence is $\underset{\sim}{d}$. We now define $R(\underset{\sim}{d})$, the *graph of multigraphic realisations of* $\underset{\sim}{d}$. The vertices of $R(\underset{\sim}{d})$ are all the non-isomorphic multigraphic realisations of $\underset{\sim}{d}$. Two vertices are adjacent if and only if one can be transformed into the other by an exchange. $R(\underset{\sim}{d},L(m))$ is the induced subgraph of $R(\underset{\sim}{d})$ generated by all the m-graphic vertices of $R(\underset{\sim}{d})$. $R(\underset{\sim}{d},E(m))$ is the induced subgraph of $R(\underset{\sim}{d})$ generated by all the exact m-graphic vertices of $R(\underset{\sim}{d})$. We shall write $G \in R(\underset{\sim}{d},L(m))$ as an abbreviation for G is an m-graph which realises $\underset{\sim}{d}$. If P_{n+1} is a path with n+1 vertices in $R(\underset{\sim}{d},L(m))$, then $G \in P_{n+1}$ means G is a vertex of P_{n+1}.

2. SHORTEST PATHS IN $R(\underset{\sim}{d},L(m))$

Chungphaisan, in [1], showed that $R(\underset{\sim}{d},L(m))$ is connected, by showing that there is a path in $R(\underset{\sim}{d},L(m))$ from any vertex to a particular realisation of $\underset{\sim}{d}$. In this section we shall show that $R(\underset{\sim}{d},L(m))$ is connected by finding upper and lower bounds for the length of the shortest path between any two vertices of $R(\underset{\sim}{d},L(m))$.

If $G,H \in R(\underset{\sim}{d},L(m))$ we can define two new m-graphs, G - H and H - G, as follows. Put G' = G - H and H' = H - G. Define V(G') = V = V(H') and for all $u,v \in V$,

$$\text{mult}_{G'}[u,v] = \text{mult}_G[u,v] - \lambda(u,v), \text{ and}$$

$$\text{mult}_{H'}[u,v] = \text{mult}_H[u,v] - \lambda(u,v), \text{ where}$$

$\lambda(u,v) = \min(\text{mult}_G[u,v], \text{mult}_H[u,v])$.

We summarise some of the properties of G - H in the form of a lemma. Since G,H and G - H all have the same vertex set we shall denote the number of edges of G - H by $|G-H|$.

Lemma 1. *If* $G,H,N \in R(\underset{\sim}{d},L(m))$ *then the following eight properties hold.*

D1. G - H *is a submultigraph of* G *and an* m-*graph.*

D2. *If* $\text{mult}_{G-H}[u,v] > 0$ *then* $\text{mult}_{H-G}[u,v] = 0$, *for all* $u,v \in V$.

D3. $|G-H| = |H-G|$.

D4. $|G-H| = 0$ *if and only if* $G = H$.

D5. *If* f *is an exchange on* G *then* $|G-f(G)| = 2$.

D6. $|G-N| + |N-H| \geq |G-H|$.

D7. *For all* $v \in V$, $deg_{G-H}(v) = deg_{H-G}(v)$.

D8. $|G-H| \neq 1$.

Proof. Properties D1, D2, D3 and D5 follow from the definitions. Property D4 follows from D3 and the definitions.

To show D6 take any $u,v \in V$ and let $mult_G[u,v] = g$, $mult_N[u,v] = n$ and $mult_H[u,v] = h$. We first show that $mult_{G-N}[u,v] + mult_{N-H}[u,v] \geq mult_{G-H}[u,v]$. If $h \geq g$ then $mult_{G-H}[u,v] = 0$ and so the inequality holds. So suppose $g > h$, then $mult_{G-H}[u,v] = g - h$. If $g \geq h \geq n$ then $mult_{G-N}[u,v] = g - n \geq g - h$. If $n \geq g \geq h$ then $mult_{N-H}[u,v] = n - h \geq g - h$. Finally if $g \geq n \geq h$ then $mult_{G-N}[u,v] = g - n$ and $mult_{N-H}[u,v] = n - h$, so again the inequality holds. Thus $mult_{G-N}[u,v] + mult_{N-H}[u,v] \geq mult_{G-H}[u,v]$, for all $u,v \in V$. Now $|G-N| = \frac{1}{2}\sum_{v \in V} deg_{G-N}(v) = \frac{1}{2}\sum_{v \in V}\sum_{u \in V} mult_{G-N}[u,v]$, and similarly $|N-H| = \frac{1}{2}\sum_{v \in V}\sum_{u \in V} mult_{N-H}[u,v]$. So $|G-N| + |N-H| = \frac{1}{2}\sum_{v \in V}\sum_{u \in V}(mult_{G-N}[u,v] + mult_{N-H}[u,v]) \geq \frac{1}{2}\sum_{v \in V}\sum_{u \in V} mult_{G-H}[u,v] = \frac{1}{2}\sum_{v \in V} deg_{G-H}(v) = |G-H|$.

To show D7 take any $v \in V$ and let $\lambda(v) = \sum_{w \in V} \lambda(v,w)$.

Then $deg_{G-H}(v) = \sum_{w \in V} mult_{G-H}[v,w] = \sum_{w \in V} (mult_G[v,w] - \lambda(v,w))$

$$= \sum_{w \in V} mult_G[v,w] - \sum_{w \in V} \lambda(v,w) = deg_G(v) - \lambda(v).$$

Similarly $deg_{H-G}(v) = deg_H(v) - \lambda(v)$, and since $deg_G(v) = deg_H(v)$, D7 holds.

Property D8 can be proved as follows. Suppose $|G-H| \geq 1$. Then there is an edge $[x,x']$ in $G - H$. By D7 there is an edge in $H - G$ containing x, and an edge in $H - G$ containing x'. There is no edge in $H - G$ containing both x and x', by D2, and so $|H-G| \geq 2$. From D3 we have $|G-H| = |H-G| \geq 2$. ∎

Lemma 2. *If* $G,H \in R(\underset{\sim}{d},L(m))$ *with* $G \neq H$ *then there are four different vertices* $t,u,v,w \in V$ *such that* $[t,u]$ *and* $[v,w]$ *are edges of* $G - H$ *and* $[t,v]$ *is an edge of* $H - G$.

Proof. Since $G \neq H$, $|G-H| \geq 1$ and so there is an edge $[x,x']$ in $G - H$. Hence there is an edge $[x,y]$ in $H - G$, for some $y \in V \setminus \{x,x'\}$. Now either $[x',y]$ is an edge in $H - G$ or $[x',y']$ is an edge in $H - G$, for some $y' \in V\setminus\{x,x',y\}$. In the former case there is an edge $[y,z]$ in $G - H$ for some $z \in V \setminus \{x,x',y\}$, and so we have the desired configuration. So suppose the latter case. If there is a $z \in V \setminus \{x,x'\}$ such that either $[y,z]$ or $[y',z]$ is an edge in $G - H$, then again we have the desired configuration. If there is no such z, then both [y,x'] and [y',x] are edges in G - H and so the desired configuration has been established. ∎

Since an exchange can only alter two edges, $|G_1-G_2|$ is a measure of how "close" G_1 and G_2 are to each other. In seeking a shortest path from G_1 to G_2 we only admit exchanges on G_1 which bring us "closer" to G_2. The following definition makes this precise. Let $G_1,G_2 \in R(\underset{\sim}{d},L(m))$, $\{v',v''\} \subseteq V$ and $\{i,j\} = \{1,2\}$. An exchange, f, on G_i, is $(G_1,G_2;v',v'')$ - *admissible* if and only if A1, A2 and A3 all hold.

A1. $f(G_i) \in R(\underset{\sim}{d},L(m))$.

A2. $|f(G_i) - G_j| < |G_1-G_2|$.

A3. If $\text{mult}_{G_1}[v',v''] = m = \text{mult}_{G_2}[v',v'']$, then $\text{mult}_{f(G_i)}[v',v''] = m$.

Condition A3 ensures that a previously chosen "matched" edge of multiplicity m is unaltered by the exchange.

The diagrams in the proof of the following lemma are captioned with either G - H or H - G. Vertices are labelled with their names. The diagrams captioned with G - H satisfy the following four rules. An unbroken, or solid, line in G - H joining two vertices indicates that there is an edge in G - H containing those two vertices. A broken line in G - H joining two vertices indicates that these two vertices are not adjacent in G - H. The label m on a solid line in G - H indicates that the edge containing the joined vertices has multiplicity m in G - H. The label m on a broken line in G - H indicates that the multiplicity in G of the edge containing the joined vertices is m, even though there is no edge in G - H containing both of these vertices. The diagrams captioned with H - G satisfy the last four sentences provided G and H are interchanged.

Lemma 3. *If* $G,H \in R(\underset{\sim}{d},L(m))$ *with* $G \neq H$ *then, for all* $v',v'' \in V$, *there is a* $(G,H;v',v'')$ - *admissible exchange on either* G *or* H.

Proof. Take any $v',v'' \in V$ and keep them fixed. Lemma 2 ensures that the following two diagrams exist.

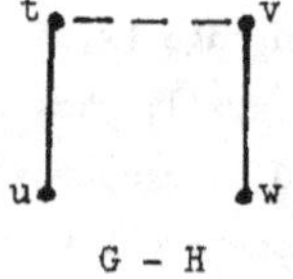

G - H

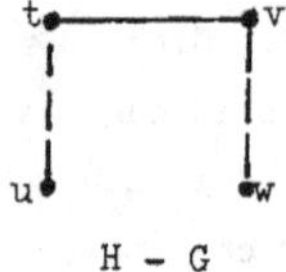

H - G

If $mult_G[u,w] < m$ then the exchange $[t,u] + [v,w] \rightarrow [t,v] + [u,w]$ on G is $(G,H;v',v'')$ - admissible.

If either $0 < mult_H[u,w] < m$ or $mult_H[u,w] = m = mult_G[u,w]$ and $\{u,w\} \neq \{v',v''\}$, then the exchange $[t,v] + [u,w] \rightarrow [t,u] + [v,w]$ on H is $(G,H;v',v'')$ - admissible.

If both $mult_{H-G}[t,w] > 0$ and $mult_{H-G}[u,v] > 0$ then the exchange $[t,w] + [u,v] \rightarrow [t,u] + [v,w]$ on H is $(G,H;v',v'')$ - admissible. Without loss of generality we shall suppose $mult_{H-G}[u,v] = 0$.

The remaining possibilities are $mult_G[u,w] = m$ and either $mult_H[u,w] = m$ and $\{u,w\} = \{v',v''\}$, or $mult_H[u,w] = 0$. We split the former possibility into two cases depending on whether $mult_{G-H}[u,v]$ is zero or positive. Thus there are three cases to consider. For the remainder of this proof we shall abbreviate $(G,H;v',v'')$ - admissible to admissible.

Case 1. This possibility is specified by $\{u,w\} = \{v',v''\}$ and the following two diagrams.

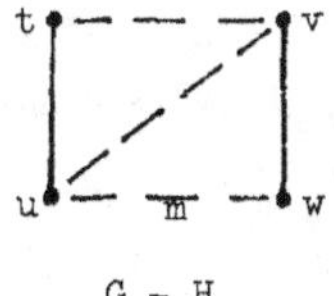

G - H

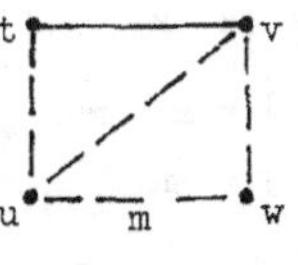

H - G

Since $deg_{G-H}(u) = deg_{H-G}(u)$, there exists $x_1 \in V \setminus \{t,u,v,w\}$ such that $mult_{H-G}[u,x_1] > 0$, and hence $mult_{G-H}[u,x_1] = 0$. The exchange $[t,v] + [u,x_1] \rightarrow [t,u] + [v,x_1]$ on H is not admissible if and only if $mult_H[v,x_1] = m$. So suppose $mult_H[v,x_1] = m$. If $mult_G[v,x_1] > 0$ then the exchange $[t,u] + [v,x_1] \rightarrow [t,v] + [u,x_1]$ on G is admissible. So suppose $mult_G[v,x_1] = 0$ and hence $mult_{H-G}[v,x_1] = m$.

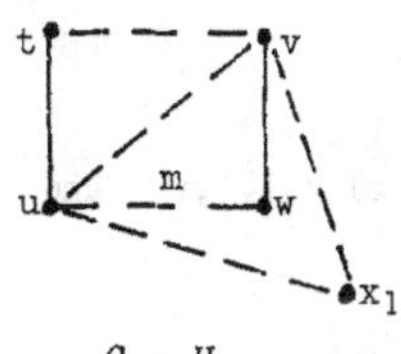

G - H

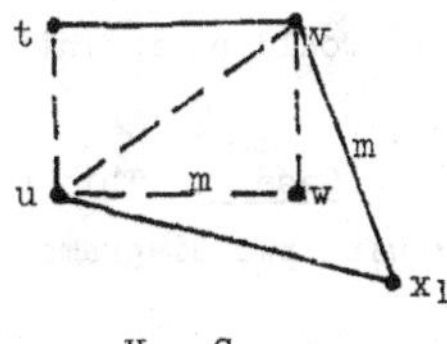

H - G

Since $\deg_{G-H}(v) = \deg_{H-G}(v)$, there exists $y_1 \in V \setminus \{t,u,v,w,x_1\}$ such that $\text{mult}_{G-H}[v,y_1] > 0$, and hence $\text{mult}_{H-G}[v,y_1] = 0$. The exchange $[t,u] + [v,y_1] \rightarrow [t,v] + [u,y_1]$ on G is not admissible if and only if $\text{mult}_G[u,y_1] = m$. So suppose $\text{mult}_G[u,y_1] = m$. If $\text{mult}_H[u,y_1] > 0$ then the exchange $[t,v] + [u,y_1] \rightarrow [t,u] + [v,y_1]$ on H is admissible. So suppose $\text{mult}_H[u,y_1] = 0$ and hence $\text{mult}_{G-H}[u,y_1] = m$.

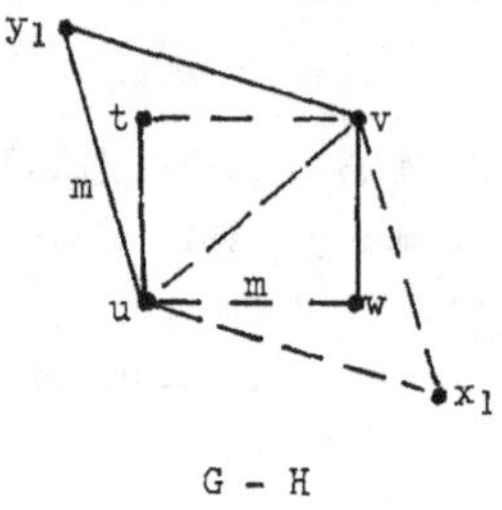

G - H

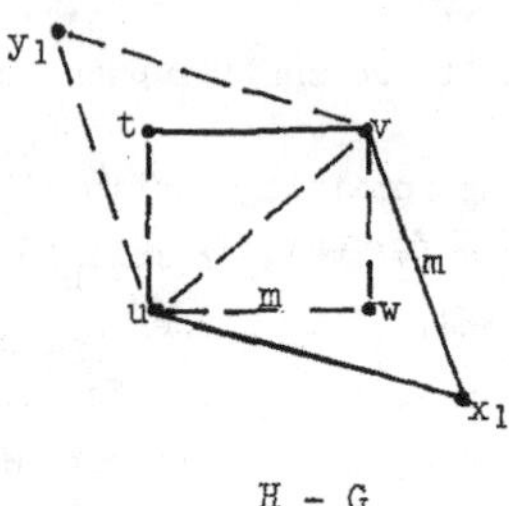

H - G

Similarly by alternately comparing the degrees of u and v in G - H and H - G we can establish the existence of distinct vertices $x_2,y_2,x_3,y_3,\ldots$ such that if $x \in \{x_1,x_2,\ldots\}$ and $y \in \{y_1,y_2,\ldots\}$ then

$\text{mult}_{G-H}[u,y] = m = \text{mult}_{H-G}[v,x]$, $\text{mult}_{G-H}[v,y] > 0$, $\text{mult}_{H-G}[u,x] > 0$ and $\text{mult}_{G-H}[u,x] = \text{mult}_{G-H}[v,x] = \text{mult}_{H-G}[u,y] = \text{mult}_{H-G}[v,y] = 0$.

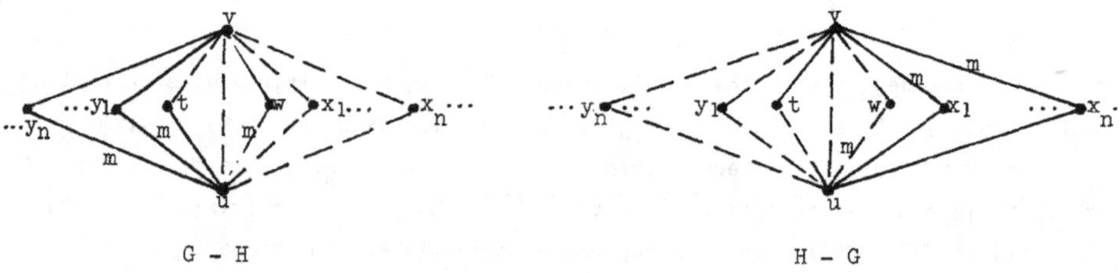

G - H

H - G

Eventually there must be an admissible exchange, as if not then V, deg(u) and deg(v) would be infinite.

Case 2. This possibility is specified by $\{u,w\} = \{v',v''\}$ and the following two diagrams.

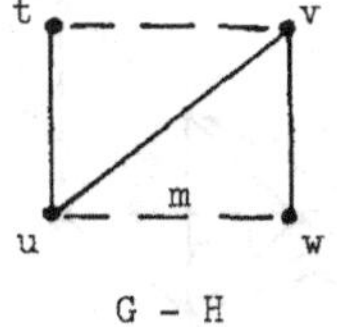

G − H

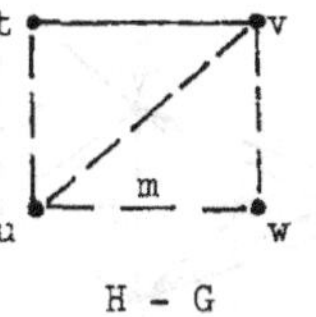

H − G

Since $\deg_{G-H}(u) = \deg_{H-G}(u)$, there exists $z \in V \setminus \{t,u,v,w\}$ such that $\mathrm{mult}_{H-G}[u,z] > 0$, and hence $\mathrm{mult}_{G-H}[u,z] = 0$. The exchange $[t,v] + [u,z] \to [u,v] + [t,z]$ on H is not admissible if and only if $\mathrm{mult}_H[t,z] = m$. The exchange $[t,v] + [u,z] \to [t,u] + [v,z]$ on H is not admissible if and only if $\mathrm{mult}_H[v,z] = m$. So suppose $\mathrm{mult}_H[t,z] = m = \mathrm{mult}_H[v,z]$. If $\mathrm{mult}_G[t,z] > 0$ then the exchange $[u,v] + [t,z] \to [t,v] + [u,z]$ on G is admissible. If $\mathrm{mult}_G[v,z] > 0$ then the exchange $[t,u] + [v,z] \to [t,v] + [u,z]$ on G is admissible. So suppose $\mathrm{mult}_G[t,z] = 0 = \mathrm{mult}_G[v,z]$ and hence $\mathrm{mult}_{H-G}[t,z] = m = \mathrm{mult}_{H-G}[v,z]$.

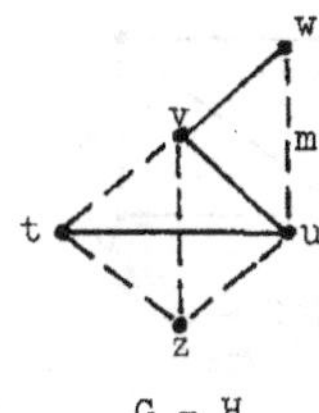

G − H

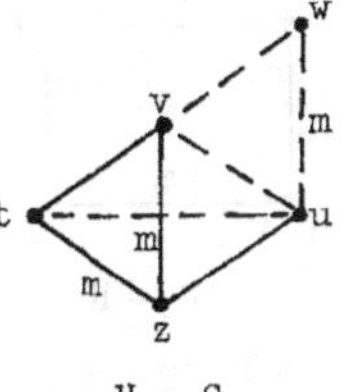

H − G

As in case 1 we alternately compare the degrees of v and z in G − H and H − G, always remembering that $\mathrm{mult}_{G-H}[w,z]$ could be as large as m. Since $\deg_{G-H}(z) = \deg_{H-G}(z)$, there are $x_0, x_1 \in V \setminus \{t,u,v,w,z\}$ such that $\mathrm{mult}_{G-H}[z,x_0] > 0$ and $\mathrm{mult}_{G-H}\{z,x_1\} > 0$ and hence $\mathrm{mult}_{H-G}[z,x_0] = 0 = \mathrm{mult}_{H-G}[z,x_1]$ If $i \in \{0,1\}$ then the exchange $[u,v] + [z,x_i] \to [u,z] + [v,x_i]$ on G is not admissible if and only if $\mathrm{mult}_G[v,x_i] = m$. So suppose $\mathrm{mult}_G[v,x_0] = m = \mathrm{mult}_G[v,x_1]$. If $i \in \{0,1\}$ and $\mathrm{mult}_H[v,x_i] > 0$ then the exchange $[u,z] + [v,x_i] \to [u,v] + [z,x_i]$ on H is admissible. So suppose $\mathrm{mult}_H[v,x_0] = 0 = \mathrm{mult}_H[v,x_1]$ and hence $\mathrm{mult}_{G-H}[v,x_0] = m = \mathrm{mult}_{G-H}[v,x_1]$.

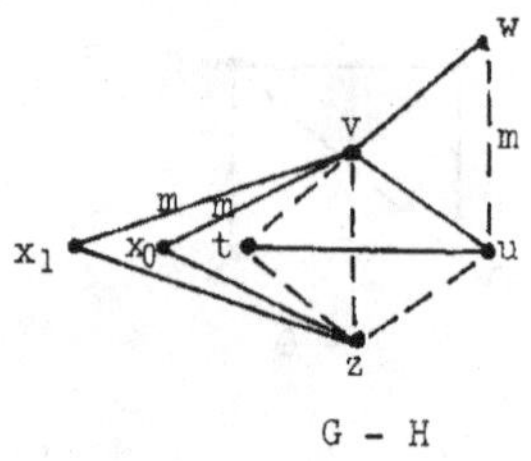

G - H

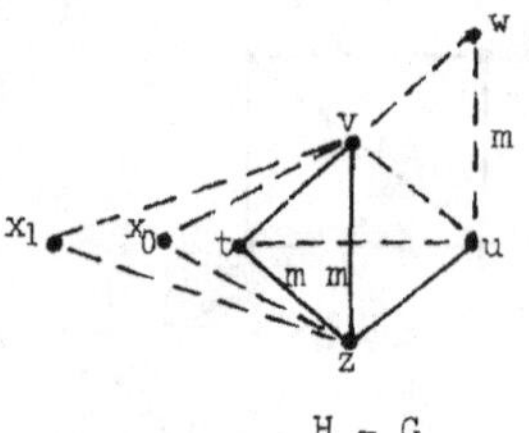

H - G

From now on the alternate comparison of the degrees of v and z in G - H and H - G yields distinct vertices $y_1, x_2, y_2, x_3, y_3, \ldots$, such that if $x \in \{x_0, x_1, \ldots\}$ and $y \in \{y_1, y_2, \ldots\}$ then $\mathrm{mult}_{G-H}[v,x] = m = \mathrm{mult}_{H-G}[z,y]$, $\mathrm{mult}_{G-H}[z,x] > 0$, $\mathrm{mult}_{H-G}[v,y] > 0$, and $\mathrm{mult}_{G-H}[v,y] = \mathrm{mult}_{G-H}[z,y] = \mathrm{mult}_{H-G}[v,x] = \mathrm{mult}_{H-G}[z,x] = 0$. Eventually there must be an admissible exchange, as if not then V, deg(v) and deg(z) would be infinite.

Case 3. This possibility is specified by the following two diagrams.

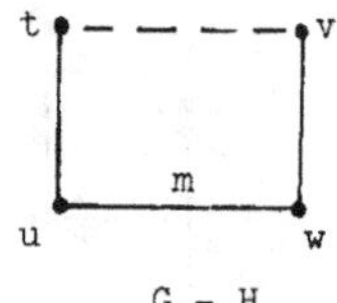

G - H

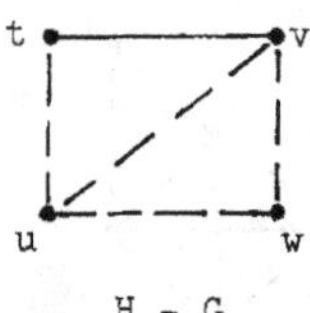

H - G

Since $\deg_{G-H}(u) = \deg_{H-G}(u)$, there are $x_0, x_1 \in V \setminus \{t,u,v,w\}$ such that $\mathrm{mult}_{H-G}[u,x_0] > 0$ and $\mathrm{mult}_{H-G}[u,x_1] > 0$, and hence $\mathrm{mult}_{G-H}[u,x_0] = 0 = \mathrm{mult}_{G-H}[u,x_1]$. If $i \in \{0,1\}$ then the exchange $[t,v] + [u,x_i] \to [t,u] + [v,x_i]$ on H is not admissible if and only if $\mathrm{mult}_H[v,x_i] = m$. So suppose $\mathrm{mult}_H[v,x_0] = m = \mathrm{mult}_H[v,x_1]$. If $i \in \{0,1\}$ and either $0 < \mathrm{mult}_G[v,x_i] < m$ or $\mathrm{mult}_G[v,x_i] = m$ and $\{v,x_i\} \neq \{v',v''\}$ then the exchange $[t,u] + [v,x_i] \to [t,v] + [u,x_i]$ on G is admissible. Let $\{i,j\} = \{0,1\}$.

If $\mathrm{mult}_G[v,x_i] = m$, $\{v,x_i\} = \{v',v''\}$ and $\mathrm{mult}_G[v,x_j] = 0$ we have the following two diagrams.

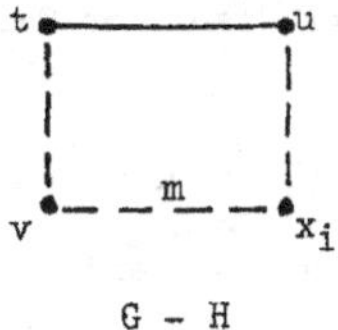

G - H

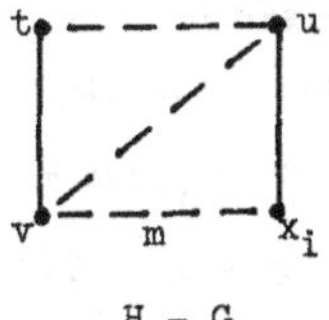

H - G

Since the exchange $[u,v] + [t,x_i] \to [t,v] + [u,x_i]$ on G is admissible, we may suppose that either $mult_{G-H}[u,v] = 0$ or $mult_{G-H}[u,v] > 0$ and $mult_{G-H}[t,x_i] = 0$.

If $mult_{G-H}[u,v] = 0$ we have the same configuration as Case 1, and hence an admissible exchange can be found. So suppose $mult_{G-H}[u,v] > 0$ and $mult_{G-H}[t,x_i] = 0$. If $mult_{H-G}[t,x_i] > 0$ then we have the same configuration as Case 2, and so an admissible exchange can be found. If $mult_{H-G}[t,x_i] = 0$ then the exchange $[t,v] + [u,x_i] \to [u,v] + [t,x_i]$ on H is admissible.

So suppose $mult_G[v,x_0] = 0 = mult_G[v,x_1]$, and hence $mult_{H-G}[v,x_0] = m = mult_{H-G}[v,x_1]$.

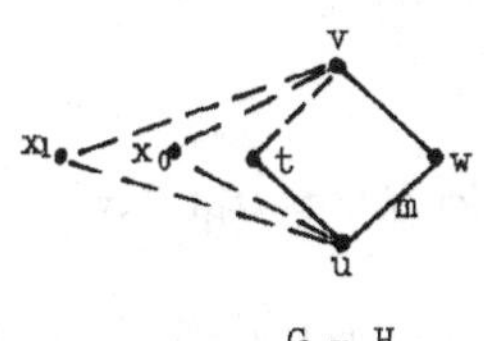

G - H

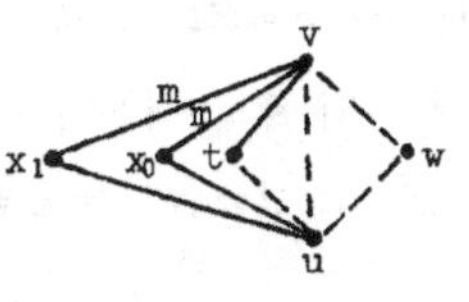

H - G

Since $deg_{G-H}(v) = deg_{H-G}(v)$, there exists $y_1 \in V \setminus \{t,u,v,w,x_0,x_1\}$ such that $mult_{G-H}[v,y_1] > 0$ and hence $mult_{H-G}[v,y_1] = 0$. The exchange $[t,u] + [v,y_1] \to [t,v] + [u,y_1]$ on G is not admissible if and only if $mult_G[u,y_1] = m$. So suppose $mult_G[u,y_1] = m$. If either $0 < mult_H[u,y_1] < m$ or $mult_H[u,y_1] = m$ and $\{u,y_1\} \neq \{v',v''\}$ then the exchange $[t,v] + [u,y_1] \to [t,u] + [v,y_1]$ on H is admissible.

If $mult_H[u,y_1] = m$ and $\{u,y_1\} = \{v',v''\}$ we have the following two diagrams.

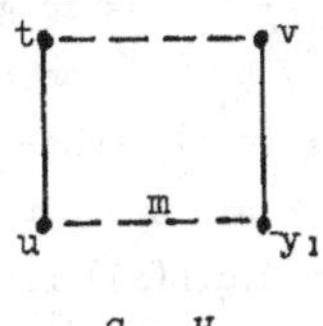

G - H

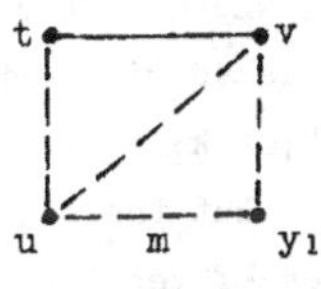

H - G

It is clear that the configuration is the same as either Case 1 or Case 2, and hence an admissible exchange can be found.

So suppose $\text{mult}_H[u,y_1] = 0$ and hence $\text{mult}_{G-H}[u,y_1] = m$.

By similarly comparing the degrees of u and v in G − H and H − G we obtain distinct vertices $x_2, y_2, x_3, y_3, \ldots$. Eventually there must be an admissible exchange, as if not V, deg(u) and deg(v) would be infinite. ■

The following theorem, which generalises Theorem 5.2 of [2], shows that $R(\underset{\sim}{d}, L(m))$ is connected. The proof constructs a path between any two vertices of $R(\underset{\sim}{d}, L(m))$ which preserves any previously chosen edge of multiplicity m that occurs in both of the vertices. This path gives an upper bound for the length of the shortest path between any two vertices of $R(\underset{\sim}{d}, L(m))$. Since an exchange can alter only two edges, a lower bound for the length of any path in $R(\underset{\sim}{d})$ from G to H is $\frac{1}{2}|G - H|$.

Theorem 4. *Suppose* $G, H \in R(\underset{\sim}{d}, L(m))$ *and* $G \neq H$. *Then for all* $v', v'' \in V$ *there is a path,* P_{n+1}, *in* $R(\underset{\sim}{d}, L(m))$ *from* G *to* H, *of length* n *such that the following two conditions hold.*

(i) $\frac{1}{2}|G-H| \leqslant n \leqslant |G-H| - 1$.

(ii) *If* $\text{mult}_G[v',v''] = m = \text{mult}_H[v',v'']$ *then* $M \in P_{n+1}$ *implies* $\text{mult}_M[v',v''] = m$.

Proof. Let $G_0 = G$ and $H_0 = H$. Using Lemma 3 we construct a sequence of m-graphs, $(G_0, G_1, \ldots, G_i, H_j, \ldots, H_1, H_0)$, from which we shall obtain the path P_{n+1}. Suppose the sequence $(G_0, \ldots, G_i, H_j, \ldots, H_0)$, where $G_i \neq H_j$ and $i, j \in \{0, 1, \ldots\}$, has been defined. By Lemma 3 there is a $(G_i, H_j; v', v'')$ – admissible exchange, f, on either G_i or H_j. If f acts on G_i, let $f(G_i) = G_{i+1}$; and if f acts on H_j, let $f(H_j) = H_{j+1}$. Hence either $(G_0, \ldots, G_i, G_{i+1}, H_j, \ldots, H_0)$ is defined and $|G_i - H_j| > |G_{i+1} - H_j|$; or $(G_0, \ldots, G_i, H_{j+1}, H_j, \ldots, H_0)$ is defined and $|G_i - H_j| > |G_i - H_{j+1}|$.

To each sequence $(G_0, \ldots, G_i, H_j, \ldots, H_0)$ associate the integer $k_{i+j+1} = |G_i - H_j|$, which is never 1, by Lemma 1 (D8). Then $|G-H| = |G_0 - H_0| = k_1 > k_2 \ldots$, where either $k_i = k_{i+1} + 1$ or $k_i = k_{i+1} + 2$. As long as $k_i > 0$ a sequence associated with either $k_{i+1} = k_i - 1$ or $k_{i+1} = k_i - 2$ can be constructed. Thus $|G-H| = k_1 > k_2 > \ldots > k_n = 2$ and so $|G-H| \geqslant 2 + (n-1)$, hence $n \leqslant |G-H| - 1$. But the sequence associated with $k_n = 2$ has $n + 1$ terms and is the sequence of vertices in a path, of length n, in $R(\underset{\sim}{d}, L(m))$ from G to H. Let $P_{n+1} = (M_0, M_1, \ldots, M_n)$ be the sequence associated with 2. Then $M_0 = G_0 = G$ and $M_n = H_0 = H$. Since $|M_{i-1} - M_i| = 2$, for all $i \in \{1, \ldots, n\}$, $|G-H| \leqslant |M_0 - M_1| + |M_1 - M_2| + \ldots + |M_{n-1} - M_n| = 2n$. Hence $\frac{1}{2}|G-H| \leqslant n$.

Finally each exchange was $(M_i, M_j; v', v'')$ - admissible for some $i,j \in \{0,1,\ldots,n\}$. So if $\text{mult}_G[v',v''] = m = \text{mult}_H[v',v'']$ then $\text{mult}_{M_i}[v',v''] = m$, for all $i \in \{0,1,\ldots,n\}$. ∎

The following example shows that the bounds on n are, in general, the best possible.

Let $G \in R(\underset{\sim}{d}, L(m))$ and let f be an exchange on G such that $f(G) \not\cong G$. Put $H = f(G)$. Then $|G-H| = 2$ by Lemma 1 (D5), and so $\frac{1}{2}|G-H| = 1 = |G-H| - 1$. Thus both bounds are attained.

Another example shows that it is not always possible to leave every "matched" edge of multiplicity m unaltered when finding a path between two vertices of $R(\underset{\sim}{d}, L(m))$. Let $\underset{\sim}{d} = (2m+3, 2m+3, 2m+2, 2m+1, 3)$ and $m \geqslant 3$. The following two diagrams specify G and H.

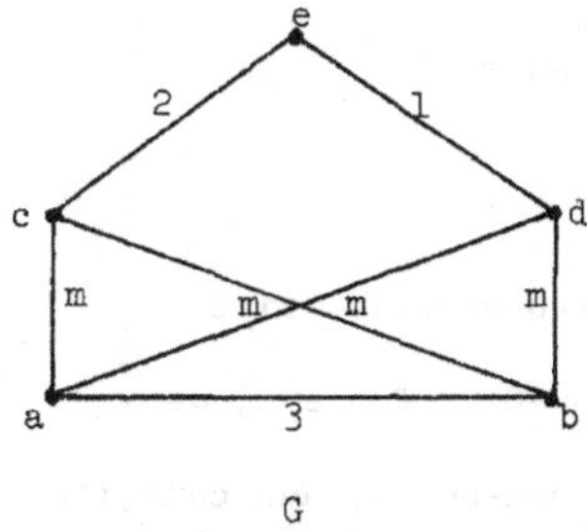

G

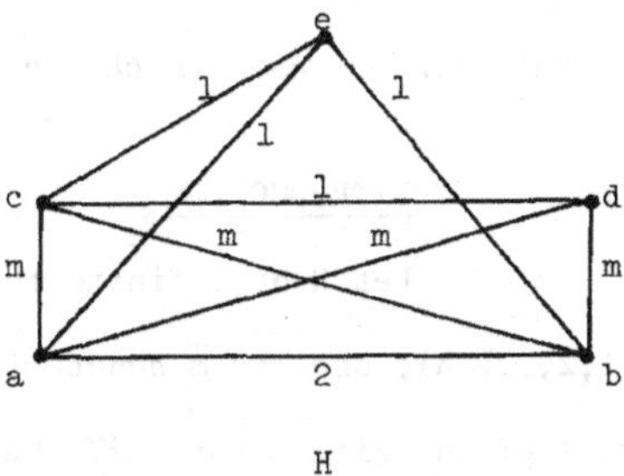

H

Then $G, H \in R(\underset{\sim}{d}, L(m))$ because $\deg(a) = 2m + 3 = \deg(b)$, $\deg(c) = 2m + 2$, $\deg(d) = 2m + 1$ and $\deg(e) = 3$. Now any exchange on G or H either alters an edge of multiplicity m or creates an edge of multiplicity m + 1.

In a subsequent paper we shall show that $R(\underset{\sim}{d}, E(m))$ is connected if there is a realisation of $\underset{\sim}{d}$ which has at least 3 edges of multiplicity m.

REFERENCES

[1] V. Chungphaisan, Conditions for sequences to be r-graphic, *Discrete Math.* 7 (1974), 31-39.

[2] D.R. Fulkerson, A.J. Hoffman, and M.H. McAndrew, Some properties of graphs with multiple edges, *Canad. J. Math.* 17 (1965), 166-177.

Department of Mathematics,
University of Melbourne,
Parkville, Victoria,
Australia, 3052.

A CONSTRUCTION FOR A FAMILY OF SETS AND ITS APPLICATION TO MATROIDS

J.E. DAWSON

Given a family $\mathcal{A}$ *of subsets of an ordered set* E, *we define a construction giving a family of sets* $\mathcal{B}$ *in 1-1 correspondence with* $\mathcal{A}$; *the same construction applied to* $\mathcal{B}$ *then gives* $\mathcal{A}$. *For each subset* X *of* E, $A\cap B \subseteq X \subseteq A\cup B$ *for exactly one pair of* $A\in\mathcal{A}$ *and corresponding* $B\in\mathcal{B}$. *When the family* $\mathcal{B}$ *is the basis collection of a matroid on* E, $\mathcal{A}$ *can be described simply in terms of the matroid structure. A polynomial is defined which, in this latter case, is the Tutte polynomial of the matroid.*

1. INTRODUCTION

Let E be a finite totally ordered set which we will write as $\{1,2,\ldots,n\}$, and let $\bar{m}$ denote $\{1,2,\ldots,m\}$ (so that $\bar{n} = E$). For $e\in E$, we will often write e for $\{e\}$. Let $\mathcal{A}\subseteq\mathcal{P}(E)$, where $\mathcal{A}$ is non-empty, and call its members $\mathcal{A}$- *sets*. We first associate a member of $\mathcal{A}$ with each subset of E, by defining a map $\alpha:\mathcal{P}(E)\to\mathcal{A}$.

Definition 1.1. *For* $I\subseteq E$, *let* $w(I) = \sum_{i\in I} 2^i$, *and for* $X,Y\subseteq E$, *let* $d(X,Y) = w((X\setminus Y)\cup(Y\setminus X))$. *For* $G\subseteq E$, *let* $\alpha(G)$ *be that member of* $\mathcal{A}$ *which minimizes* $d(G,\alpha(G))$.

It is trivial that $w:\mathcal{P}(E)\to\mathbb{N}$ is an injective map, and so the above definition defines $\alpha(G)$ uniquely. Also, if $X,Y\subseteq E$ with $z = \max((X\setminus Y)\cup(Y\setminus X))$, then $w(X) > w(Y)$ if $z\in X\setminus Y$ and $w(Y) > w(X)$ if $z\in Y\setminus X$. Thus we have the following result characterizing whether or not a particular element m is in $\alpha(G)$.

Lemma 1.2.

(a) *if every* $\mathcal{A}$*-set* A *such that* $A\setminus\bar{m} = \alpha(G)\setminus\bar{m}$ *(does, does not) contain* m, *then* m *(is, is not) in* $\alpha(G)$, *and*

(b) *if there are* $\mathcal{A}$*-sets* A', A" *such that* $A'\setminus\bar{m} = A''\setminus\bar{m} = \alpha(G)\setminus\bar{m}$ *and* $m\in A'$ *but* $m\notin A''$, *then* m *(is, is not) in* $\alpha(G)$ *if* m *(is, is not) in* G.

Proof Since $\alpha(G)\in\mathcal{A}$, (a) is trivial. Suppose the condition of clause (b) holds but that $m\in G\setminus\alpha(G)$. Then,

$$d(G,\alpha(G)) \geq 2^m + \sum\{2^i: \ i\in((G\setminus\alpha(G)) \cup(\alpha(G)\setminus G))\setminus\bar{m}\}$$

whereas

$$d(G,A') \leq 2^1+2^2 +\ldots+ 2^{m-1} + \sum\{2^i: \ i\in((G\setminus\alpha(G)) \cup (\alpha(G)\setminus G))\setminus\bar{m}\}$$

Since $2^1+2^2 +\ldots+ 2^{m-1} < 2^m$, this contradicts the definition of $\alpha(G)$. A similar proof applies if $m\in\alpha(G)\setminus G$. Thus (b) holds. □

We note that Lemma 1.2 determines $\alpha(G)\setminus\overline{m-1}$ in terms only of $\alpha(G)\setminus\bar{m}$, $G\setminus\overline{m-1}$ and $\{A\setminus\overline{m-1} : A\in\mathcal{A}\}$. Thus we can use it to construct $\alpha(G)$ from G and $\mathcal{A}$, by applying it successively to the cases m=n, m=n-1, ... , m=1. Also, it could have been used as the definition of the map $\alpha:\mathcal{P}(E)\to\mathcal{A}$. To do so we would need to see that $\alpha(G)$ is indeed in $\mathcal{A}$; however it is quite clear that if $A\setminus\bar{m} = \alpha(G)\setminus\bar{m}$ for some $A\in\mathcal{A}$, and $\alpha(G)\setminus\overline{m-1}$ is determined using Lemma 1.2, then $A\setminus\overline{m-1} = \alpha(G)\setminus\overline{m-1}$ for some $A\in\mathcal{A}$. We can describe the construction thus: "to change G into $\alpha(G)$, we work from the highest element downwards, making changes only where necessary to get $\alpha(G)\in\mathcal{A}$".

Example Writing {1,2} as 12, let n=3 and $\mathcal{A}$ = {12,13,23}. Then $\alpha^{-1}(12) = \{\emptyset,1,2,12\}$, $\alpha^{-1}(13) = \{3,13\}$ and $\alpha^{-1}(23) = \{23,123\}$.

From the example we notice that each $\alpha^{-1}(A)$ is of the form $\{X\subseteq E: H\subseteq X\subseteq F\}$, an interval of the lattice of subsets of E (a *lattice interval*), which we write $<H,F>$. So, for $A\in\mathcal{A}$, we now determine $\alpha^{-1}(A)$, showing that it is always of this form.

Let $H = \{m\in A\colon$ there exists $A'\in\mathcal{A}$ such that $A'\setminus\bar{m} = A\setminus\bar{m}$ and $m\notin A'\}$, and $F = E\setminus\{m\in E\setminus A\colon$ there exists $A'\in\mathcal{A}$ such that $A'\setminus\bar{m} = A\setminus\bar{m}$ and $m\in A'\}$.

Lemma 1.3. *For* $A\in\mathcal{A}$, $\alpha^{-1}(A) = <H,F>$.

Proof It is clear from Lemma 1.2(b) that $\alpha^{-1}(A) \subseteq <H,F>$. Suppose $G\notin\alpha^{-1}(A)$; say $d(G,A) > d(G,A')$ for some $A'\in\mathcal{A}$ and let $m = \max((A\setminus A') \cup (A'\setminus A))$. Suppose $m\in A'\setminus A$. Then $m\in E\setminus F$, but clearly $m\in G$. If, on the other hand, $m\in A\setminus A'$, then $m\in H$, and clearly $m\notin G$. Thus $G \not\subseteq <H,F>$, as required. □

Since $H\subseteq A\subseteq F$, writing $H = A\cap B$ and $F = A\cup B$ uniquely defines a set B ($B = H\cup(F\setminus A)$), which we describe next.

Lemma 1.4. *For* $A\in\mathcal{A}$, $\alpha^{-1}(A) = <A\cap\beta'(A), A\cup\beta'(A)>$ *where* $\beta'\colon\mathcal{A}\to\mathcal{P}(E)$ *is defined by:*

$m\in\beta'(A)$ *if and only if either*

(a) *for every* $\mathcal{A}$*-set* A' *such that* $A'\setminus\bar{m} = A\setminus\bar{m}$, $m\notin A'$, *or*

(b) $m\in A$, *and there is* $A'\in\mathcal{A}$ such that $A'\setminus\bar{m} = A\setminus\bar{m}$ and $m\notin A'$.□

This result gives a partitioning of $\mathcal{P}(E)$ into lattice intervals: $\mathcal{P}(E) = \cup\{\alpha^{-1}(A)\colon A\in\mathcal{A}\}$. Let us define $\mathcal{B} = \mathrm{im}(\beta')$. Then $\alpha|\mathcal{B}$ and β' are mutually inverse bijections between $\mathcal{A}$ and $\mathcal{B}$, with the property that $\beta'(A)\setminus\bar{m} = \beta'(A')\setminus\bar{m}$ if and only if $A\setminus\bar{m} = A'\setminus\bar{m}$.

In the example above, $\beta'(12) = \emptyset$, $\beta'(13) = 3$, $\beta'(23) = 123$, and $\mathcal{B} = \{\emptyset,3,123\}$.

Theorem 1.5. *Let* $A\in\mathcal{A}$, $B = \beta'(A)$ *and* $m\in E$. *Then* 1a, 2a *and* 3a *are equivalent, as are* 1b, 2b *and* 3b.

(1a,1b) ($m \in B \setminus A$, $m \in A \setminus B$)

(2a,2b) *All A-sets* A' *such that* $A' \setminus \bar{m} = A \setminus \bar{m}$ *(do not, do) contain* m.

(3a,3b) *All B-sets* B" *such that* $B'' \setminus \bar{m} = B \setminus \bar{m}$ *(do, do not) contain* m.

Proof 1a ⇔ 2a and 1b ⇔ 2b follow from the definition of β' (also from Lemma 1.2.) 2a ⇒ 3a: assume 2a holds, and suppose $B'' \setminus \bar{m} = B \setminus \bar{m}$ and $A'' = \alpha(B'')$. Thus $A'' \setminus \bar{m} = A \setminus \bar{m}$. Now condition 2a remains true for A" as it is for A, and so, by the definition of β', $B'' = \beta'(A'')$ contains m. Thus 3a follows. 2b ⇒ 3b is shown similarly.

3a ⇒ 2a: assume 3a. Thus $m \in B$, 3b does not hold, and so 2b does not hold. Let $A'' \setminus \bar{m} = A \setminus \bar{m}$ with $m \notin A''$. Hence $\beta'(A'') \setminus \bar{m} = B \setminus \bar{m}$, and by the assumption 3a, $m \in \beta'(A'')$. Thus, by the definition of β', and since $A'' \setminus \bar{m} = A \setminus \bar{m}$, 2a holds.

3b ⇒ 2b is shown similarly. □

There is an obvious symmetry between (2a,2b) and (3b,3a) of Theorem 1.5. We define $\alpha' : \mathcal{B} \to P(E)$ in terms of $\mathcal{B}$ just as β' is defined in terms of $\mathcal{A}$ in Lemma 1.4; likewise we define $\beta : P(E) \to \mathcal{B}$ in terms of $\mathcal{B}$ in a manner analogous to Definition 1.1.

Corollary 1.6. $\mathrm{Im}(\alpha') = \mathcal{A}$, *and* $\alpha' = \alpha|\mathcal{B} = (\beta')^{-1} = (\beta|\mathcal{A})^{-1}$.

Proof As with $\alpha|\mathcal{B}$ and β', α' and $\beta|\mathrm{im}(\alpha')$ are mutually inverse bijections. Just as Theorem 1.5, (1a,1b) ⇔ (2a,2b), contains the definition of β', (1a,1b) ⇔ (3a,3b) asserts that $A = \alpha'(B)$ for A,B as in the Theorem (equivalently, for $B \in \mathcal{B}$ and $A = \alpha(B)$). Thus $\alpha' = \alpha|\mathcal{B}$, so $\mathrm{im}(\alpha') = \mathcal{A}$; just as $(\beta')^{-1} = \alpha|\mathcal{B}$, $(\alpha')^{-1} = \beta|\mathcal{A}$, and the result follows. □

We can now refer to α' as α; we simply have one description of α in terms of $\mathcal{A}$, and one of $\alpha|\mathcal{B}$ in terms of $\mathcal{B}$. Similarly for β. The following

Corollary is useful.

Corollary 1.7.

(i) $\alpha\beta = \alpha$ *and* $\alpha^2 = \alpha$.

(ii) $\alpha(G) = \alpha(H) \Leftrightarrow \beta(G) = \beta(H)$, *and, in this case, if* $G\cap H\subseteq F\subseteq G\cup H$ *then* $\alpha(F) = \alpha(G)$.

(iii) *For* $G\subseteq E$, $\alpha(G)\cap\beta(G)\subseteq G\subseteq\alpha(G)\cup\beta(G)$

(iv) *Let* $g\in G$. *Then* $g\notin\beta(G) \Leftrightarrow g\in\alpha(G\setminus g)$, *and in this case* $\alpha(G) = \alpha(G\setminus g)$.

In each of the above results we may interchange α *and* β.

Proof (i) Let $A\in\mathcal{A}$ and $B = \beta(A)$. It was shown earlier that $\alpha^{-1}(A) = \langle A\cap B, A\cup B\rangle$, and similarly $\beta^{-1}(B) = \langle A\cap B, A\cup B\rangle$. As $\alpha|\mathcal{B}$ is a bijection, $(\alpha\beta)^{-1}(A) = \beta^{-1}(B)$, and the first result follows. Clearly α is idempotent.

(ii) The equivalence holds as $\alpha\beta = \alpha$ and similarly $\beta\alpha = \beta$. Also $\alpha^{-1}(\alpha(G)) = \langle\alpha(G) \cap \beta(G), \alpha(G) \cup \beta(G)\rangle$, which is a lattice interval; since it contains G and H, it also contains F.

(iii) follows as $G\in \alpha^{-1}(\alpha(G)) = \langle\alpha(G) \cap \beta(G), \alpha(G) \cup \beta(G)\rangle$.

(iv) Let $g\notin \beta(G)$. Then $G\setminus g \in \alpha^{-1}(\alpha(G))$ and so $\alpha(G\setminus g) = \alpha(G)$. Now $G \subseteq \alpha(G) \cup \beta(G)$, so $g\in \alpha(G) = \alpha(G\setminus g)$ as required. The converse proof is similar. □

Let us call $\mathcal{A}$ and $\mathcal{B}$, as above, a *complementary pair* of subsets of $\mathcal{P}(E)$, or a *complementary pair on* E. We can form from $\mathcal{A}$ and $\mathcal{B}$ some other complementary pairs.

Theorem 1.8. *If* $\mathcal{A}$ *and* $\mathcal{B}$ *are a complementary pair of subsets of* $\mathcal{P}(E)$, *then each of the following also forms a complementary pair:*

(i) *on* E: $\mathcal{A}^* = \{E\setminus A: A\in\mathcal{A}\}$, *where* $\alpha^*(E\setminus G) = E\setminus\alpha(G)$, *with* $\mathcal{B}^*$ *and* β^* *defined similarly,*

(ii) *on* $E\setminus\bar{m}$ *for* $m\in E$: $\mathcal{A}' = \{A\setminus\bar{m}: A\in\mathcal{A}\}$ *where* $\alpha'(G\setminus\bar{m}) = \alpha(G)\setminus\bar{m}$, *with* $\mathcal{B}'$ *and* β' *defined similarly.*

(iii) *For* $m\in E$, *let* $A'\in\mathcal{A}'$ *of* (ii), *and* $B' = \beta'(A')$. *Let* $\mathcal{A}'' = \{A\cap\bar{m}: A\in\mathcal{A}, A\setminus\bar{m} = A'\}$, *and for* G *such that* $G\setminus\bar{m} = A'$, *let* $\alpha''(G\cap\bar{m}) = \alpha(G)\cap\bar{m}$. *Then with* $\mathcal{B}''$ *and* β'' *defined similarly,* $\mathcal{A}''$ *and* $\mathcal{B}''$ *form a complementary pair on* $E\cap\bar{m}$.

Proof (i) is easy. For (ii), note that by the comment following Lemma 1.2, α is well-defined. (iii) is now straightforward. □

2. SOME ASSOCIATED OPERATORS

We define some further operators related simply to α and β. Let $\gamma(G) = \alpha(G)\cup\beta(G)$ and $\delta(G) = \alpha(G)\cap\beta(G)$. Then by Corollary 1.7 (i) and (ii) we have that α, β, γ and δ have the property that for $\{\zeta,\eta\}\subseteq\{\alpha,\beta,\gamma,\delta\}$, $\zeta\eta = \zeta$. Thus

$$\alpha^{-1}(\alpha(G)) = \beta^{-1}(\beta(G)) = \gamma^{-1}(\gamma(G)) = \delta^{-1}(\delta(G)) = <\delta(G), (G)>.$$

We can also define

$$\phi(G) = \beta(G)\cup G,\ \theta(G) = \beta(G)\cap G,\ \omega(G) = \alpha(G)\cup G \text{ and } \psi(G) = \alpha(G)\cap G$$

(which are all contained in $<\delta(G), \gamma(G)>$).

Proofs in this section are brief or absent; they use Lemma 1.4 and Corollary 1.7 (i) to (iii), and are easily seen with the help of a Venn diagram displaying G, α(G) and β(G).

Lemma 2.1 *The operators* ϕ,ψ,θ,ω *are idempotent, and* $\phi\psi = \phi$, $\psi\phi = \psi$, $\theta\omega = \theta$ *and* $\omega\theta = \omega$; *otherwise they commute, with* $\phi\omega = \gamma$, $\theta\psi = \delta$, $\phi\theta = \beta$ *and* $\psi\omega = \alpha$. □

Lemma 2.2 (i) $\phi^{-1}(\phi(G)) = \psi^{-1}(\psi(G)) = <\psi(G),\phi(G)>$ *and* $\theta^{-1}(\theta(G)) = \omega^{-1}(\omega(G)) = <\theta(G),\omega(G)>$.

(ii) *For* $g\in G$, $g\notin\psi(G) \Leftrightarrow g\in\phi(G\setminus g)$; *similarly* θ *and* ω.

Proof (i) The first equality follows as $\phi\psi = \phi$ and $\psi\phi = \psi$. Likewise, $\beta\phi = \beta$ and so $\phi(H) = \phi(G)$ implies $\beta(H) = \beta(G)$. As $\phi(H) = H \cup \beta(H)$, it is now easy to check that

$$\phi^{-1}(\phi(G)) = \{H\in<\delta(G),\ \gamma(G)>:\ H\setminus\beta(H) = G\setminus\beta(G)\}$$
$$= <\psi(G),\phi(G)>.$$

The second result is shown similarly.

(ii) This is immediate from Corollary 1.7 (iv). □

The last Lemma gives two partitions of $P(E)$ into lattice intervals, which are subintervals of $<\delta(G),\gamma(G)>$, with the following property.

Lemma 2.3 $\phi(H) = \phi(G)$ *and* $\theta(H) = \theta(G) \Rightarrow G = H$, *i.e.*, $<\psi(G),\phi(G)> \cap <\theta(G),\ \omega(G)> = \{G\}$.

Proof $G\setminus\theta(G) = G\setminus\beta(G) = \phi(G)\setminus\beta\phi(G)$, since $\beta\phi = \beta$. Thus $\theta(G)$ and $\phi(G)$ determine G. □

Although $\phi(G)$ and $\psi(G)$ may vary even for G within a fixed $<\delta(H),\gamma(H)>$, $\phi(G)\setminus\psi(G)$ may not, for $\phi(G)\setminus\psi(G) = \beta(G)\setminus\alpha(G)$; similarly $\theta(G)\setminus\omega(G) = \alpha(G)\setminus\beta(G)$. Finally we pair these operators in another way.

Lemma 2.4 $G\in<\theta(H),\ \phi(H)> \Leftrightarrow H\in<\psi(G),\omega(G)>$, *and in this case,*

$$\theta(H) \subseteq \theta(G) \subseteq G \subseteq \phi(G) \subseteq \phi(H) \quad ,$$
$$\phi(G)\setminus G \subseteq \phi(H)\setminus H,\ G\setminus\theta(G) \subseteq H\setminus\theta(H) \quad ,$$

and also the similar statements got by interchanging G *with* H, ϕ *with* ω *and* θ *with* ψ.

Proof   Either of the first conditions gives $\alpha(G) = \alpha(H)$ etcetera. The proof is then easy with a Venn diagram of $G, H, \alpha(G)$ and $\beta(G)$. □

Although these results are quite simply proved with the suggested Venn diagrams, they are of interest when $\mathcal{B}$ is the basis collection of a matroid on E.

3. APPLICATION TO MATROIDS

Suppose we let $\mathcal{B}$ be the *basis* collection of a matroid $\mathcal{E}$ on E. That is, $\mathcal{B}$ is a non-empty clutter on E (i.e. no $\mathcal{B}$-set contains another) and $\mathcal{E}$ is the collection of subsets of sets in $\mathcal{B}$, called the *independent* sets, satisfying the axiom: for $F \in \mathcal{E}$ and $B \in \mathcal{B}$, there exists $B' \in \mathcal{B}$ such that $F \subseteq B' \subseteq F \cup B$. Let $\mathcal{C}$ be the collection of the minimal dependent sets, called *circuits*. Let $\mathcal{B}^* = \{E \setminus B : B \in \mathcal{B}\}$, as in Theorem 1.8. Then $\mathcal{B}^*$ is the basis collection of another matroid $\mathcal{E}^*$ (the *dual* of $\mathcal{E}$), whose circuit collection $\mathcal{C}^*$ we also call $\mathcal{D}$, the *cocircuit* collection of the original matroid $\mathcal{E}$. We say that e *depends* on F, $e|F$ or $e \in [F]$, if $e \in F$ or there exists $C \in \mathcal{C}$ such that $e \in C \subseteq F \cup e$. Similarly $|^*$ and $[\ ]^*$ are defined relative to $\mathcal{C}^*$ $(=\mathcal{D})$. (We note that the matroid property of $\mathcal{B}$ is preserved also by construction (iii), but not by (ii), of Theorem 1.8.) It can be shown that $[\ ]$ has the properties of a closure operator (i.e. $F \subseteq [F]$ and $F \subseteq [G] \Rightarrow [F] \subseteq [G]$), that for $e \notin F$, $e \in [F] \Leftrightarrow e \notin [E \setminus F \setminus e]^*$, and that bases have the same number of elements. If $G \subseteq E$ we define the matroid $\mathcal{E}|G = \{F \subseteq G : F \in \mathcal{E}\}$ on G. For further details see [6], [4] or [3]. For such $\mathcal{B}$, the maps α and β have a particularly simple description in terms of the matroid structure.

Theorem 3.1   *Let $\mathcal{B}$ be the collection of bases of a matroid on E. Then, for $G \subseteq E$,*

(i) $\alpha(G) = G \cup \{e \notin G : e|G \setminus \bar{e}\} \setminus \{e \in G : e|^* E \setminus G \setminus \bar{e}\}$

(ii) $\beta(G) = G \cup \{e \notin G: e|^*E\setminus G\setminus \bar{e}\} \setminus \{e \in G: e|G\setminus \bar{e}\}$,

which we can write as

$$\beta(G) = G \cup \{\min(D): D \in \mathcal{D},\ D \cap G = \emptyset\} \setminus \{\min(C): C \in \mathcal{C},\ C \subseteq G\}$$

or $$\beta(G) = (E\setminus G) \cup \{\max(D \cap G): D \in \mathcal{D},\ D \cap G \neq \emptyset\} \setminus \{\max(C\setminus G): C \in \mathcal{C},\ C\setminus G \neq \emptyset\}.$$

Proof Note that we could equally write these results with "$\in G$" and "$\notin G$" replaced by "$\in E$" since it is not possible that $e|G\setminus \bar{e}$ and $e|^*E\setminus G\setminus \bar{e}$. It is easy to check the equivalence of the three descriptions of $\beta(G)$. Firstly we show that (i) is true for $G = B \in \mathcal{B}$, using the definition of α in terms of $\mathcal{B}$ (like that of β in terms of A in Lemma 1.4). Let $e \notin B$. If $e|B\setminus \bar{e}$, then clearly every $B' \in \mathcal{B}$ such that $B'\setminus \bar{e} = B\setminus \bar{e}$ does not contain e, and $e \in \alpha(B)$. However, if $e \nmid B\setminus \bar{e}$, then $B\setminus \bar{e} \cup \{e\} \in \mathcal{E}$, and so there is a basis B' such that $B\setminus \bar{e} \cup \{e\} \subseteq B' \subseteq B \cup \{e\}$, and so $B'\setminus \bar{e} = B\setminus \bar{e}$, with $e \in B'$. Therefore $e \notin \alpha(B)$. Similarly we can verify (i) with respect to the elements e in $B(=G)$. Further, since $B \in \mathcal{E}$ and $E\setminus B \in \mathcal{E}^*$, we can write $\alpha(B)\setminus B = \{e \in E: e|B\setminus \bar{e}\}$ and $B\setminus \alpha(B) = \{e \in E: e|^*E\setminus B\setminus \bar{e}\}$. To complete the proof, we need the next Lemma.

Lemma 3.2 *For* $G \in \langle B \cap \alpha(B), B \cup \alpha(B)\rangle$, $e \in [G\setminus \bar{e}] \Leftrightarrow e \in [B\setminus \bar{e}]$ *and* $e \in [E\setminus G\setminus \bar{e}]^* \Leftrightarrow e \in [E\setminus B\setminus \bar{e}]^*$.

Proof First we show by induction on $|G\setminus B|$ that $[G\setminus \bar{e}] \subseteq [B\setminus \bar{e}]$. Let $g \in G\setminus B$ $(\subseteq \alpha(B)\setminus B)$, so $g \in [B\setminus \bar{g}]$, and assume by induction that this result is true for $G\setminus g$ (for all e). If $g \leq e$ then $[G\setminus \bar{e}] = [G\setminus g\setminus \bar{e}] \subseteq [B\setminus \bar{e}]$. If $g > e$, then $g \in [B\setminus \bar{g}] \subseteq [B\setminus \bar{e}]$. As $[G\setminus g\setminus \bar{e}] \subseteq [B\setminus \bar{e}]$, we have $[G\setminus \bar{e}] \subseteq [B\setminus \bar{e}]$.

Now suppose $e \in [B\setminus \bar{e}] \setminus [G\setminus \bar{e}]$. Then for some $b \in B\setminus G\setminus \bar{e}$, some circuit $C \subseteq B\setminus \bar{e} \cup \{e\}$ contains e and b, whereby $b|B\setminus b \cup e$. As $b \in B\setminus G$, $b|^*E\setminus B\setminus \bar{b}$; since $e \in \bar{b}$, this contradicts $b|B\setminus b \cup e$. Thus the first result follows. The second part of the Lemma is shown similarly. □

Proof of Theorem (continued). Let $G \subseteq E$, and recall that $G \in \langle B \cap \alpha(B), B \cup \alpha(B)\rangle$ for $B = \beta(G)$. Then

$$\begin{aligned}\alpha(G) &= \alpha(B) = G\cup(\alpha(B)\setminus B)\setminus (B\setminus\alpha(B)) \\ &= G\cup\{e\in E : e|B\setminus\bar{e}\} \setminus \{e\in E : e|{*E}\setminus B\setminus\bar{e}\} \\ &= G\cup\{e\in E : e|G\setminus\bar{e}\} \setminus \{e\in E : e|{*E}\setminus G\setminus\bar{e}\}\ ,\end{aligned}$$

as required. For (ii), we have, for $e\notin G$,

$$\begin{aligned}e\in\beta(G) &\Leftrightarrow e\notin\alpha(G\cup e) \quad \text{(symmetrical with Corollary 1.7 (iv))} \\ &\Leftrightarrow e|{*E}\setminus(G\cup e)\setminus\bar{e} \\ &\Leftrightarrow e|{*E}\setminus G\setminus\bar{e}\ .\end{aligned}$$

We complete the proof of (ii) by a similar argument. □

Corollary 3.3 $A\in\mathcal{A}$ *if and only if* $\alpha(A) = A$, *i.e.*,

(i) *if* $C\in\mathcal{C}$ *and* $C\setminus\{\min(C)\} \subseteq A$, *then* $C \subseteq A$, *and*

(ii) *if* $D\in\mathcal{D}$ *and* $D\setminus\{\min(D)\} \subseteq E\setminus A$, *then* $D \subseteq E\setminus A$. □

The corresponding result for $\mathcal{B}$ turns out to be that $B\in\mathcal{B}$ if and only if B contains no circuit and $E\setminus B$ contains no cocircuit. It is shown in [5] that the (unique) lexicographic maximum basis $B = \{b_1,\ldots,b_n\}$ is "maximum" in the stronger sense that for any other basis $F = \{f_1,\ldots,f_n\}$, $b_i \geq f_i$ for each i when B and F are each written with elements in ascending order. The next result then follows easily from [5] or [2].

Proposition 3.4 *For* $G \subseteq E$, $\theta(G)$ *is the maximum basis of* $E|G$ and $E\setminus\phi(G)$ *is the maximum basis of* $E^*|(E\setminus G)$. □

We now see that some of the results of [1], §4 are applications to matroids of results in Sections 1 and 2. (Note that to get our results into the form of those in [1], we need to reverse the ordering of the set.) Thus $|G\setminus\psi(G)| = |G\setminus\alpha(G)|$ is the *internal activity* i(G) of G, and $|\alpha(G)\setminus G|$ is its *external activity* e(G). It follows from Proposition 3.4 that $|G\setminus\beta(G)|$ is its *dependence*, $d(G) = |G| - r(G)$ (where r(G) is the rank of G), and that

$|\beta(G)\backslash G|$ is r(E)-r(G), which we may call the *shortage* of G, s(G). Let us retain these definitions of i,e,d and s based on α and β, but drop the assumption that $\mathcal{B}$ is a matroid basis collection.

Now if $G \subseteq E$, $A = \alpha(G)$ and $B = \beta(G)$, then $A\cap B \subseteq G \subseteq A\cup B$, and so

$$e(B) = d(A) = |A\backslash B| = |A\backslash G| + |G\backslash B| = e(G) + d(G).$$

Similarly $i(B) = s(A) = i(G) + s(G)$. Also, the intervals $<A\cap B, A\cup B>$ each contain one $\mathcal{A}$-set and one $\mathcal{B}$-set. (In the matroid case, these are the structureless intervals of [1], Proposition 12.) Such an interval contains $\binom{s(A)}{i}\binom{d(A)}{j}$ sets G such that $s(G) = i$ and $d(G) = j$, since such G is of the form

$$B \cup \{j \text{ elements of } A\backslash B\} \backslash \{i \text{ elements of } B\backslash A\} .$$

This gives the result, shown by Crapo for the matroid case ([1], Theorem 1), that if $T(\mathcal{B};x,y) = \Sigma\{x^{i(B)}\, y^{e(B)} : B\in\mathcal{B}\}$, which is also equal to $\Sigma\{x^{s(A)}\, y^{d(A)} : A\in\mathcal{A}\}$, and $R(\mathcal{B};x,y) = \Sigma\{x^{s(G)}\, y^{d(G)} : G \subseteq E\}$, then $R(\mathcal{B};x,y) = T(\mathcal{B};\, x+1,y+1)$.

<u>Proposition 3.5</u> $R(\mathcal{B};x,y) = \Sigma\{x^{i(G)}\, y^{e(G)} : G \subseteq E\}$.

<u>Proof</u> A similar argument shows that the right-hand side is equal to $T(\mathcal{B};x+1,y+1)$. □

It is now useful to return to the assumption that $\mathcal{B}$ is a matroid basis collection, for then (see [1]) $R(\mathcal{B};x,y)$ does not depend on the ordering of E, and hence neither does T. This tells us that the number of $\mathcal{A}$-sets of given dependence and shortage is independent of the ordering of E.

The 1-1 correspondence between $\mathcal{A}$, $\mathcal{B}$, im(γ) and im(δ), established in Section 2, has some interesting consequences for matroids. In particular, Theorem 3.6 (ii) relates to the fact that matroids seem usually to have more bases than circuits ([6], p.74).

<u>Theorem 3.6</u> *A matroid has at least as many bases as* (i) *fully dependent flats* (ii) *proper upper segments of circuits (i.e.,* $C\backslash \min(C)\backslash \bar{m}$, *for* $C\in\mathcal{C}$, $m\in E$).

<u>Proof</u> (i) Fully dependent flats are in A.

(ii) Such a set F satisfies $\alpha(F) \supseteq F$ and $\beta(F) \supseteq F$. As $F \supseteq \delta(F)$, so $F = \delta(F)$ and F is in $im(\delta)$, which is in 1-1 correspondence with B. □

I wish to thank Dr. C.C. Heyde for arousing my interest in this subject.

REFERENCES

[1] H.H. Crapo, The Tutte polynomial, *Aeq. Math.* 3 (1969), 211-229.

[2] J.E. Dawson, Optimal matroid bases: an algorithm based on cocircuits, *Quart. J. Math. (Oxford)* (2) 31 (1980), 65-69.

[3] J.E. Dawson, A simple approach to some basic results in matroid theory. *J. Math. Anal. Appl.*, 75 (1980), 611-615.

[4] R. von Randow, *Introduction to the Theory of Matroids*, Lecture Notes in Economics and Mathematical Systems 109, Springer-Verlag, 1975.

[5] D.J.A. Welsh, Kruskal's theorem for matroids, *Proc. Camb. Phil. Soc.* 64 (1968), 3-4.

[6] D.J.A. Welsh, *Matroid Theory*, Academic Press, New York, 1976.

Division of Mathematics and Statistics,
CSIRO,
P.O. Box 218,
LINDFIELD, N.S.W. 2070,
AUSTRALIA.

REGULARITY AND OPTIMALITY FOR TREES

PETER EADES

A cost function for trees is introduced; roughly speaking, the cost of a tree is the time needed to traverse all of its branches. A tree of least cost for a given number of leaves is called optimal. Bounds for the cost of an optimal tree are proved, and these bounds are used to restrict the possible outdegrees in optimal trees. Finally, a method for proving regularity in optimal trees is presented.

1. INTRODUCTION

In this paper a *tree* is a directed rooted tree with the arcs directed away from the root. Such trees are commonly used to structure data in computers; they are also used as patterns for merging disc files. For details on such applications we refer the reader to Knuth [4].

The efficiency of the algorithms used to process such trees depends on the shape of the tree. In this paper we consider the problem of choosing trees of optimal shape. Our main result is that for a certain class of cost functions, regular trees are nearly optimal. We also give a method for proving the complementary result that optimal trees are nearly regular.

The *leaf* of a tree is a node of zero outdegree. A *branch* of a tree is a path from root to leaf. For other terminology in this paper we refer the reader to Knuth [4].

The cost function relates to the cost of traversing each branch of the tree, where the cost of visiting a node depends on its outdegree. The cost function is described precisely as follows. An increasing non-negative function $f: N \to R$ is given, where N denotes the nonnegative integers. The *cost of a node* with outdegree c is f(c); the *cost of a branch* is the sum of the costs of the nodes on the branch; and the *cost of a tree* is the sum of the costs of the branches of the tree. For example, if $f(c) = c^2$ then the cost of the branches in the tree in figure 1, from left to right, are $3^2 + 2^2 = 13$, $3^2 + 2^2 + 2^2 = 17$, $3^2 + 2^2 + 2^2 = 17$, $3^2 = 9$, $3^2 + 2^2 = 13$, $3^2 + 2^2 = 13$. Thus the cost of the tree is $13 + 17 + 17 + 9 + 13 + 13 = 82$.

This class of cost functions arises in the problem of searching for data stored at the leaves of a tree. In the case where f is linear, the cost of a tree T coincides with the cost of using T as a merge pattern for equal length disk files (see [4]).

We denote the cost of a tree T by G(T). A tree T_1 with n leaves is *optimal* if $G(T_1) \le G(T_2)$ for each tree T_2 with n leaves. The cost of an optimal tree with n leaves is denoted by $\Gamma(n)$.

If the principal subtrees of a tree T are $T_1, T_2, \ldots, T_c$, as in figure 2, where c is the outdegree of the root, then

$$G(T) = n\,f(c) + \sum_{i=1}^{c} G(T_i).$$

It follows that the

(1.1) $$\Gamma(n) = \min_{2 \le c \le n} \; n\,f(c) + \min_{\substack{\text{partitions}\\ n=(n_1+n_2+..+n_c)}} \sum_{i=1}^{c} \Gamma(n_i)$$

For convenience we denote

$$n\,f(c) + \min_{\substack{\text{partitions}\\ n=n_1+\ldots+n_c}} \sum_{i=1}^{c} \Gamma(n_i)$$

by $\Gamma_c(n)$; this quantity represents the cost of the best tree whose root has outdegree c.

Thus (1.1) becomes

(1.2) $$\Gamma(n) = \min_{2\le c\le n} \Gamma_c(n).$$

Algorithms using (1.2) to compute optimal trees are given in [4] and [5].

The case f(c) = c has been studied by Knuth [4] and Goebel and Hoede [2]. In this case the optimal trees can be characterized as trees in which all nodes except the fathers of leaves and the leaves have outdegree 3. Goebel and Hoede conjecture that this sort of regularity property can be generalized; for a few simple functions f they show that a binary tree is optimal. However, apart from these cases, little is known about Γ or the shape of optimal trees.

In this paper we give bounds on Γ which show that regular trees are optimal in an asymptotic sense. We use these bounds to prove that the possible outdegrees in an optimal tree is bounded. Finally, we give a method for determining whether

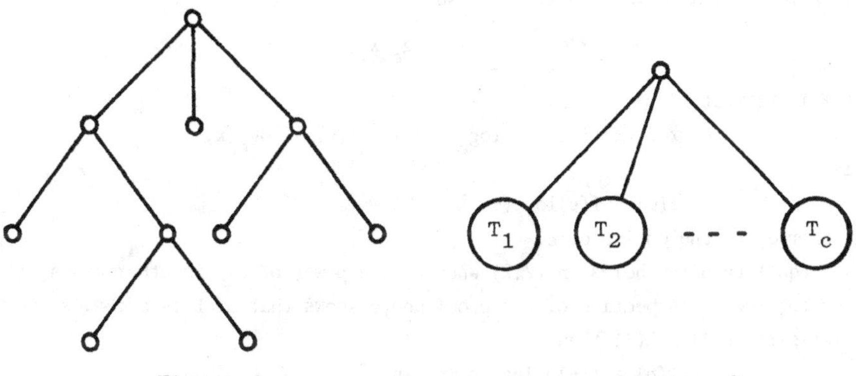

Figure 1　　　　Figure 2

optimal trees are regular.

2. COST BOUNDS

The cost of a c-regular tree is about

$$n\ f\ (c)\ *\ \text{depth of tree}$$
$$= f(c)\ n \log_c n + o(n),$$
$$= (f(c)/\log_2(c))\ n \log_2(n) + o(n)$$

and so the best regular trees have outdegree ε, where ε is an integer such that

$$\frac{f(\varepsilon)}{\log_2(\varepsilon)} \le \frac{f(c)}{\log_2(c)}$$

for all integers $c \ge 2$. The next theorem shows that there are no trees which have cost significantly lower than the cost of an ε-regular tree.

(2.1) Theorem

Suppose that there is an integer $\varepsilon \ge 2$ *such that* $f(c) - f(\varepsilon) \log_\varepsilon(c) \ge 0$ *for each integer* $c \ge 2$. *Then*

(2.2) $$\Gamma(n) \ge f(\varepsilon)\ n \log_\varepsilon n$$

for all $n \ge 1$.

Proof. Note that (2.2) holds for $n = 1$. Assume that (2.2) fails for the first time at $n = x$, i.e.

$$\Gamma(x) < f(\varepsilon)\ x \log_\varepsilon x.$$

Since $\Gamma(x) = \Gamma_c(x)$ for some $c \ge 2$, we have

$$x f(c) + \sum_{i=1}^{c} \Gamma(x_i) < f(\varepsilon) x \log_\varepsilon x$$

for some partition $x = x_1 + x_2 + \ldots + x_c$. Now $x_i < x$ for each x and so by induction,

$$\Gamma(x_i) \ge f(\varepsilon)\ x_i \log_\varepsilon x_i$$

for each i. Hence

$$x\ f(c) + f(\varepsilon) \sum_{i=1}^{c} x_i \log_\varepsilon x_i < f(\varepsilon)\ x \log_\varepsilon x.$$

Now $Z \to Z \log Z$ is a convex function and so

$$x \log_\varepsilon (x/c) \le \sum_{i=1}^{c} x_i \log_\varepsilon x_i,$$

and it follows that

$$x\ f\ (c) + f(\varepsilon) x \log_\varepsilon (x/c) < f(\varepsilon)\ x \log_\varepsilon x,$$

that is.

$$f(c) - f(\varepsilon) \log_\varepsilon(c) < 0,$$

which contradicts the choice of ε.

Remark. Equality often holds in (2.2) when n is a power of ε; in other cases, (2.2) is often too low. Inspection of the proof above shows that if L is a convex function which satisfies $\Gamma(1) \ge L(1)$ then

$$\Gamma(n) \ge f(\varepsilon) n \log_\varepsilon n + L\ (n).$$

Next we present a class of trees to bound Γ from above.

(2.3) Theorem

If $0 \le \alpha \le \lfloor \log_\varepsilon n \rfloor$, $x = \lfloor n/\varepsilon^\alpha \rfloor$ *and* $y = n = x\,\varepsilon^\alpha$,

then

(2.4) $\Gamma(n) \le f(\varepsilon)\, n\, \alpha + y\, \Gamma(x+1) + (\varepsilon^\alpha - y)\, \Gamma(x)$.

Proof If $\alpha=0$ then the claim is trivial.

Let T_1 denote the tree with a single node and if $\alpha > 1$, let T_α denote the tree

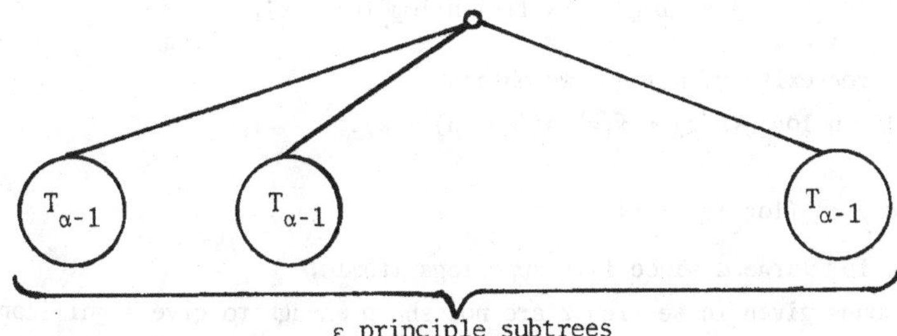

It is clear that the cost of T_α is $f(\varepsilon)\alpha\varepsilon^\alpha$.

If $n = x\,\varepsilon^\alpha + y$, then form a tree T by attaching an optimal tree with (x+1) leaves to y of the leaves of T_α and attach an optimal tree with x leaves to the other $(\varepsilon^\alpha - y)$ leaves of T_α. Clearly T has n leaves and $G(T) = f(\varepsilon) n\, \alpha + y\, \Gamma(x+1) + (\varepsilon^\alpha - y)\Gamma(x)$, and the result follows.

(2.5) Corollary

$$\Gamma(n) \le n\, f(\varepsilon)\ \ (\log_\varepsilon (n) + \varepsilon)$$

Proof. Consider (2.4) with $\alpha = \lfloor \log_\varepsilon n \rfloor$. Note that in this case, $x \le \varepsilon - 1$.

Hence

$$\begin{aligned}\Gamma(n) &\le \alpha\, f(\varepsilon)\, n + y\, \Gamma(x+1) + (\varepsilon^\alpha - y)\, \Gamma(x)\\ &\le f(\varepsilon)\, n \log_\varepsilon n + \varepsilon^\alpha\, \Gamma(x) + y\, (\Gamma(x+1) - \Gamma(x))\\ &\le f(\varepsilon)\, n \log_\varepsilon n + n\, \Gamma(x) + n\, (\Gamma(x+1) - \Gamma(x))\\ &= f(\varepsilon)\, n \log_\varepsilon n + n\, \Gamma(x+1)\\ &\le f(\varepsilon)\, n \log_\varepsilon n + n\, \varepsilon\, f(\varepsilon),\end{aligned}$$

since $\Gamma(x+1) \le \Gamma(\varepsilon) \le \varepsilon\, f(\varepsilon)$.

Remark. Although the bound (2.5) is asymptotically good, the o(n) term is too large. A smaller o(n) term may be obtained by considering (2.4) with small values of α. In many cases (2.4) holds with equality when $\alpha = 1$.

3. OUTDEGREE BOUNDS

An increasing function f is *superlogarithmic* if $\frac{f(c)}{\log(c)}$ is increasing for large enough c.

(3.1) Theorem

If f *is superlogarithmic then there is an integer* C *such that*

(3.2) $\Gamma(n) = \min_{2 \le c \le C} \Gamma_c(n)$ *for each* $n \ge 2$.

Proof. The theorem can be deduced from Lemma 2 of [2]; a sharp bound C is given in

[5] for linear functions f. Here we outline a general proof using the bounds in section 2. Suppose that $\Gamma_c(n) < \Gamma_\varepsilon(n)$;
then

$$nf(c) + \sum_{i=1}^{c} \Gamma(n_i) < f(\varepsilon)\, n(\log_\varepsilon n + \varepsilon)$$

for some partition $n = n_1 + n_2 + \ldots + n_c$, using Theorem (2.3). Hence

$$n\, f(c) + \sum_{i=1}^{c} f(\varepsilon)\, n_i \log_\varepsilon n_i < f(\varepsilon)\, n(\log_\varepsilon(n) + \varepsilon),$$

and, using the convexity of n log n we obtain

$$n\, f(c) + n \log_\varepsilon (n/c) < f(\varepsilon)\, n(\log_\varepsilon(n) + \varepsilon).$$

Hence

$$f(c) < f(\varepsilon)\, (\log_\varepsilon(c) + \varepsilon),$$

which is false for large c since f is superlogarithmic.

Remark The bounds given in section 2 are not sharp enough to give significant restrictions on c, except where f is increasing rapidly.

Remark Conversely, if $f(c)/\log_2(c)$ is a decreasing function, then it is clear that a node with n leaves is an optimal tree for large enough n.

4. GENERALIZED CONVEXITY

The proof of the regularity result $\Gamma(n) = \Gamma_3(n)$ for $n \geq 3$ in the case $f(c) = c$ (see [4] or [2]) depends on the convexity of Γ. Knuth [4] points out that in other cases (e.g. $f(c) = c+1$), Γ is demonstrably not convex and so a proof along the same lines will fail. Here we show that a relaxed convexity restraint, together with some computation, can prove $\Gamma(n) = \Gamma_\varepsilon(n)$ for large n.

A function H is k-*convex* if

$$H(a) + H(b) \geq H(a+y) + H(b-y)$$

whenever $\frac{1}{2}(b-a) \geq y \geq k$. It is clear that if Γ is k-convex then the minimum value of $\Gamma(n_1) + \Gamma(n_2) + \ldots + \Gamma(n_c)$ is obtained when the parts n_i of n differ by at most 2k. Hence

$$\Gamma_c(n) = nf(c) + \min \sum_{i=1}^{c} \Gamma(n_i)$$

where the minimum is over all partitions $n = n_1 + n_2 + \ldots + n_c$ with each part n_i at least $(n/c) - 2k$. Now suppose that $\Gamma(x) = \Gamma_\varepsilon(x)$ for $x_0 \leq x \leq n-1$, and $n > c(x_0+k)$, then

$$\Gamma_c(n) = n\, f(c) + \sum_{i=1}^{c} \Gamma(n_i)$$

where each part n_i of n is at least $(n/c) - 2k$. Now $n > c\,(x_0 + k)$ ensures that each n_i is at least x_0. Hence there are partitions $n_i = n_{i1} + n_{i2} + \ldots + n_{i\varepsilon}$ $(1 \leq i \leq c)$ such that

$$\Gamma(n_i) = \Gamma_\varepsilon(n_i) = f(\varepsilon)\, n_i + \sum_{j=1}^{\varepsilon} (n_{ij}).$$

It follows that

$$\Gamma_c(n) = nf(c) + nf(\varepsilon) + \sum_{i,j} \Gamma(n_{ij}).$$

But if we denote $n_{1j} + n_{2j} + \ldots + n_{\varepsilon j}$ by n_j',

then it follows that

$$\Gamma_c(n) = n\, f\,(\varepsilon) + \sum_{j=1}^{\varepsilon} (n_j'\; f(c) + \sum_{i=1}^{c} \Gamma(n_{ij}))$$

$$\geq n\, f\,(\varepsilon) + \sum_{j=1}^{\varepsilon} \Gamma_c\,(n_j')$$

$$\geq \Gamma_\varepsilon\,(n).$$

The following theorem summarizes this discussion.

(4.1) Theorem.

Suppose that Γ *is* k-*convex and* $\Gamma(n) = \Gamma_\varepsilon(n)$ *for* $x_0 \leq n \leq c\,(x_0 + 2k)$. *Then* $\Gamma_\varepsilon(n) \leq \Gamma_c(n)$ *for all* $n \geq x_0$.

Combining Theorem (4.1) with the results of section 3 gives a method for deciding whether the regularity result $\Gamma = \Gamma_\varepsilon$ holds.

(4.2) Theorem.

Suppose that Γ *is* k-*convex and there are integers* C *and* x_0 *such that for all* $n \geq x_0$, *there is an integer* $c \leq C$ *such that* $\Gamma(n) = \Gamma_c(n)$, *and further, if* $n \leq C(x_0 + k)$, $\Gamma(n) = \Gamma_\varepsilon(n)$. *Then* $\Gamma(n) = \Gamma_\varepsilon(n)$ *for all* $n \geq x_0$.

Remark The difficulty in using Theorem (4.2) is proving that Γ is k-convex a priori. In particular cases it may be possible to prove that if $\Gamma\,(n) = \Gamma_\varepsilon(n)$ and Γ is k-convex over a large enough interval, then Γ is k-convex for all n.

5. FINAL REMARKS

In this paper we have shown that for a wide class of cost functions, the notions of regularity and optimality are related. It seems likely that the methods of section 4 together with some computing could strengthen these results; in particular, they could be used to add weight to the conjecture of Göbel and Hoede that optimal trees are nearly regular.

In Goldschlager and Eades [3], and Eades and Staples [1] similar results are obtained for slightly different cost functions for trees: in particular, the cost of a tree can be taken as the *sum of the costs of all paths from the root to a node* [3], or as the *cost of the most expensive path* [1]. A suitable calculus of trees could possibly produce a general result to cover these different cost functions; however, the author knows of no such calculus.

References

[1] Peter Eades and John Staples, "On Optimal Trees", Technical Report No. 19, Dept. of Computer Science, University of Queensland (1980). (To appear).

[2] F. Göbel and C. Hoede, "On an optimality property of ternary trees", *Inf. and Control* 42, 10-26 (1979).

[3] Leslie M. Goldschlager and Peter Eades, "Cheapsort", (to appear).

[4] D. Knuth, The Art of Computer Programming, Vol. 3, "Sorting and Searching", Addison-Wesley, Mass, 1973.

[5] M. Schumberger and J. Vuillemin, *Acta Informatica* 3 (1973), 25-36.

Department of Computer Science
University of Queensland
St. Lucia
Queensland 4067

SIMPLE AND MULTIGRAPHIC REALIZATIONS OF DEGREE SEQUENCES

05 C 99

R.B. EGGLETON AND D.A. HOLTON

A simple realization of a suitable sequence $\underset{\sim}{d}$ of nonnegative integers is a simple graph having $\underset{\sim}{d}$ as its degree sequence. The graph RS($\underset{\sim}{d}$) of simple realizations of $\underset{\sim}{d}$ has the simple realizations of $\underset{\sim}{d}$ as its vertices, and two vertices are adjacent in RS($\underset{\sim}{d}$) if one can be obtained from the other by a degree-preserving operation called switching. We similarly define RM($\underset{\sim}{d}$), the graph of multigraphic realizations of $\underset{\sim}{d}$, the vertices being the multigraphs with degree sequence $\underset{\sim}{d}$.

We characterize those infinite sequences $\underset{\sim}{d}$ for which RS($\underset{\sim}{d}$) is connected, and those for which RM($\underset{\sim}{d}$) is connected. We also characterize those infinite $\underset{\sim}{d}$ for which there are isolated vertices in RS($\underset{\sim}{d}$) or RM($\underset{\sim}{d}$), and determine the possible numbers of such vertices: in each case there can be zero, one, three, a countable infinity or an uncountable infinity of isolated vertices, and there are no other possibilities.

Those finite sequences $\underset{\sim}{d}$ for which there is a unique realization as a multigraph, that is, for which RM($\underset{\sim}{d}$) is the trivial graph, were completely determined by Hakimi. We obtain a new characterization of these sequences, and the corresponding infinite sequences, in terms of forbidden subgraphs.

1. INTRODUCTION.

In this paper we take up several questions about multigraphs, and simple graphs, which have the same degree sequence. Analogous questions about pseudographs were answered in two previous papers [3, 4]. Much of the notation and terminology of those papers and the earlier paper [2] will be used here. In particular, when referring to cardinality, the word "infinite" used without qualification will mean "countably infinite". Further, for any sequence $\underset{\sim}{d}$ of numbers from $\mathbb{N} = \{0, 1, 2, \ldots\}$, we use $\Sigma\underset{\sim}{d}$ for the sum (possibly infinite) of all the terms of the sequence, and when $\underset{\sim}{d}$ is bounded we use max $\underset{\sim}{d}$ for the largest number occurring as a term of $\underset{\sim}{d}$.

A <u>switching</u> on a multigraph G is a replacement of any two independent edges [a, b] and [c, d] by the edges [a, c] and [b, d]. Switching is a degree-preserving operation, and so yields a multigraph (possibly isomorphic to G) with the same degree sequence as G. If G is a simple graph, the switchings appropriate to obtain other simple graphs with the same degree sequence are those which do not duplicate edges already present in G. We study the relationships between various graphs with the same degree sequence by determining how switching connects one with another.

This amounts to studying the structure of graphs RS($\underset{\sim}{d}$) and RM($\underset{\sim}{d}$), which we now define.

The graph RS($\underset{\sim}{d}$) of simple realizations of $\underset{\sim}{d}$ has as its vertices the simple graphs with degree sequence $\underset{\sim}{d}$, two vertices being adjacent in RS($\underset{\sim}{d}$) if one can be obtained from the other by a switching. We similarly define the graph RM($\underset{\sim}{d}$) of multigraphic realizations of $\underset{\sim}{d}$, which has as its vertices the multigraphs with degree sequence $\underset{\sim}{d}$. These notations are a little more convenient than the corresponding R($\underset{\sim}{d}$; (0, 0, 1)) and R($\underset{\sim}{d}$; (0, 0, ∞)) introduced in [2].

2. SEQUENCES WITH SIMPLE OR MULTIGRAPHIC REALIZATIONS.

We begin by characterizing those sequences $\underset{\sim}{d}$ which arise as degree sequences of simple graphs or of multigraphs. The characterizations for finite $\underset{\sim}{d}$ are due to Erdős and Gallai [5] in the case of simple graphs, and to Senior [10] in the case of multigraphs. We merely extend the characterizations to include infinite $\underset{\sim}{d}$.

Recall the definition from [4] that a sequence $\underset{\sim}{d}$ of nonnegative integers is infinitary if it has infinitely many terms greater than 1, and otherwise is finitary. It was proved in [4] that any infinitary sequence $\underset{\sim}{d}$ has a realization as an acyclic simple graph. This readily extends to any sequence $\underset{\sim}{d}$ with infinite sum. For suppose $\Sigma\underset{\sim}{d}$ is infinite but $\underset{\sim}{d}$ is not infinitary: then $\underset{\sim}{d}$ has infinitely many terms equal to 1 and at most finitely many terms greater than 1. For each term $a > 1$ in $\underset{\sim}{d}$, we can form the star $K_{1,a}$. The union of all such stars with infinitely many paths P_2 gives an acyclic simple realization of the positive subsequence of $\underset{\sim}{d}$. Adjoining a set of isolated vertices of the same cardinality as the set of zero terms of $\underset{\sim}{d}$ gives an acyclic simple realization of $\underset{\sim}{d}$. Thus we have the following result.

Lemma 0. If $\Sigma\underset{\sim}{d}$ is infinite, then $\underset{\sim}{d}$ has a realization as an acyclic simple graph.

Incorporating this lemma with a result of Senior [10], for the case when $\Sigma\underset{\sim}{d}$ is finite, gives the following characterization of multigraphic sequences.

Theorem 1. The sequence $\underset{\sim}{d}$ is realizable as a multigraph if and only if

(i) $\Sigma\underset{\sim}{d}$ is infinite, or

(ii) $\Sigma\underset{\sim}{d}$ is finite and even, and $\Sigma\underset{\sim}{d} \geq 2 \max \underset{\sim}{d}$.

Again, incorporating the lemma with a result of Erdős and Gallai [5], in the case when $\Sigma\underset{\sim}{d}$ is finite, gives the following characterization of simple graphic sequences.

Theorem 2. The sequence $\underset{\sim}{d}$ is realizable as a simple graph if and only if

(i) $\Sigma\underset{\sim}{d}$ is infinite, or

(ii) $\Sigma\underset{\sim}{d}$ is finite and even, and if the terms of $\underset{\sim}{d}$ are arranged in nonincreasing order, then

$$\sum_{i\leq k} d_i \leq k(k-1) + \sum_{i>k} \min\{k, d_i\}$$

holds for every $k \geq 1$.

A reduction of the set of inequalities specified in case (ii) was described in [1]. There are alternative characterizations for case (ii) of Theorem 2, such as the recursive characterization due independently to Havel [8] and Hakimi [7].

3. CONNECTED GRAPHS OF REALIZATIONS.

We now consider those graphic sequences wich give rise to graphs of realizations which are connected. The results obtained are analogous to those obtained for pseudographs as Theorems 2 and 3 of [3].

Two realizations of a degree sequence $\underset{\sim}{d}$ are associates if there is a degree-preserving bijection between their vertex sets which identifies all but at most finitely many of their edges. (This formulation removes some unintended ambiguities in earlier less formal statements which we have given for this notion, such as describing two realizations as associates "if they differ at only a finite number of vertices" [3], or "if they differ only in their realization of some finite subsequence" [4]. The ambiguities present in the former were clearly brought to attention by an example due to Billington, as indicated in [4]. This example is discussed by Taylor in [11].)

The following theorem includes a proof of the result that if $\underset{\sim}{d}$ is any finite sequence realizable as a simple graph, then RS($\underset{\sim}{d}$) is connected. This result was announced in [1]. A different proof is supplied by Taylor [11].

Theorem 3. The classes of associates of simple graphs with degree sequence $\underset{\sim}{d}$ comprise the components of the graph RS($\underset{\sim}{d}$).

Proof. Let G_1 and G_2 be two simple graphs with degree sequence $\underset{\sim}{d}$. If G_1 and G_2 are in the same component of RS($\underset{\sim}{d}$), there is a finite sequence of switchings which transforms G_1 into G_2. Each switching preserves the degrees of all vertices, and replaces two edges by two others, so is a degree-preserving bijection between vertex sets which identifies all but finitely many of the edges in the two graphs in question. Since G_1 transforms into G_2 by the composite of these bijections, it follows that G_1 and G_2 are associates.

Now suppose G_1 and G_2 are associates. We shall show that there is a finite sequence of switchings which transforms G_1 into G_2, whence G_1 and G_2 are in the same component of $RS(\underset{\sim}{d})$. Since there is a degree-preserving bijection between the vertex sets of G_1 and G_2 which identifies all but at most finitely many of their edges, for convenience we shall regard them as having the same vertex set, and edge sets with symmetric difference Δ which is finite. Let V be the set of vertices incident with edges in Δ. For $i = 1, 2$ let G_i' be the induced subgraph of G_i with vertex set V. Note that G_1' and G_2' are finite simple graphs with the same degree sequence $\underset{\sim}{d}'$.

Let $\prec$ be a linear ordering of V such that for any $u, v \in V$ we have $u \prec v$ exactly when $\deg u > \deg v$ in each G_i'. This induces a linear ordering of the edges in the complete graph on V, as follows: if $[u, v]$, $[x, y]$ are any edges in the complete graph on V, with $u \prec v$ and $x \prec y$, then

$$[u, v] \prec [x, y] \Leftrightarrow u \prec x, \text{ or } u = x \text{ and } v \prec y.$$

With $a \prec b$, let $[a, b]$ be the earliest edge in Δ which is in G_2'. Then $[a, b]$ is not an edge of G_1', and deg a is the same in G_1' and G_2', so there is an earliest edge $[a, c]$ in Δ which is in G_1'. We can suppose the graphs have been named so that $b \prec c$. Hence $\deg b > \deg c$. For any vertex $v \in V$, let $N(v)$ denote the set of vertices in V which are adjacent to v in G_1'. Since $a \in N(c) \setminus N(b)$, and $\deg b = |N(b)| > \deg c = |N(c)|$, it follows that $|N(b)| = |N(b) \setminus \{a\}| > |N(c) \setminus \{a\}|$. Moreover $c \in N(b)$ if and only if $b \in N(c)$, so $|N(b) \setminus \{a, c\}| > |N(c) \setminus \{a, b\}|$. Hence there is some vertex d which is in $N(b) \setminus \{a, c\}$ but not in $N(c) \setminus \{a, b\}$, that is, $d \in N(b) \setminus (N(c) \cup \{c\})$. Thus $[a,c]$ and $[b, d]$ are edges which are in G_1', while $[a, b]$ and $[c, d]$ are not edges in G_1'. The switching

$$[a, c], [b, d] \rightarrow [a, b], [c, d]$$

transforms G_1' into a simple graph H which, like G_2', has an edge $[a, b]$ and does not have an edge $[a, c]$. Thus if Δ' is the symmetric difference of the edge sets of H and G_2, and V' is the vertex set of Δ', then $V' \subseteq V$, $|\Delta'| < |\Delta|$ and the earliest edge in Δ' which is in G_2' is later than $[a, b]$. Since Δ is finite, iteration of this construction yields a finite sequence of switchings which transforms G_1 into G_2 as required.

With only minor modifications, this argument also applies to multigraphs, giving the following result.

Theorem 4. The classes of associates of multigraphs with degree sequence $\underset{\sim}{d}$ comprise the components of the graph $RM(\underset{\sim}{d})$.

We omit the proof.

We can now characterize the sequences $\underset{\sim}{d}$ for which $RM(\underset{\sim}{d})$ is connected, that is, the sequences $\underset{\sim}{d}$ for which any multigraphic realization can be obtained from any other by a finite sequence of switchings.

Theorem 5. The graph $RM(\underset{\sim}{d})$ is connected if and only if $\underset{\sim}{d}$ is finitary. If $\underset{\sim}{d}$ is infinitary then $RM(\underset{\sim}{d})$ has uncountably many components.

Proof. If $\Sigma\underset{\sim}{d}$ is finite, then $RM(\underset{\sim}{d})$ is connected, by Lemma 1 of Hakimi [6]. If $\Sigma\underset{\sim}{d}$ is infinite but $\underset{\sim}{d}$ is finitary, then $\underset{\sim}{d}$ has infinitely many terms equal to 1 and at most finitely many terms greater than 1. Let $\underset{\sim}{d}'$ be the subsequence of $\underset{\sim}{d}$ comprising all terms greater than 1, together with s terms equal to 1, where s is the sum of all terms greater than 1. Let $\underset{\sim}{d}''$ be the subsequence of $\underset{\sim}{d}$ complementary to $\underset{\sim}{d}'$. Clearly $\underset{\sim}{d}''$ has a unique multigraphic realization, comprising infinitely many paths P_2 and an appropriate number of isolated vertices. Also $\underset{\sim}{d}'$ is realizable as a (possibly empty) multigraph, by Theorem 1, and [7] ensures that $RM(\underset{\sim}{d}')$ is connected. Moreover, every multigraphic realization G of $\underset{\sim}{d}$ is the disjoint union of a realization of $\underset{\sim}{d}'$ and the unique realization of $\underset{\sim}{d}''$. This holds since G_1, the induced subgraph of G on all vertices of degree greater than 1 and all vertices adjacent to them, is a realization of a subsequence of $\underset{\sim}{d}'$; by Theorem 1, the complementary subsequence of $\underset{\sim}{d}'$ comprises an even number of terms, all equal to 1, and this complementary subsequence has a unique multigraphic realization G_2 comprising a suitable number of paths P_2. Thus $G_1 \cup G_2$ is an induced subgraph of G which realizes $\underset{\sim}{d}'$, and $G \setminus (G_1 \cup G_2)$ is a disjoint subgraph which realizes $\underset{\sim}{d}''$. Thus, the connectedness of $RM(\underset{\sim}{d}')$ ensures the connectedness of $RM(\underset{\sim}{d})$, since the two graphs are clearly isomorphic. This completes the discussion for finitary sequences.

Now suppose $\underset{\sim}{d}$ is infinitary. We shall construct a realization of $\underset{\sim}{d}$ which identifiably encodes into its structure any given set S of natural numbers greater than 2. By showing that the realizations corresponding to different choices of S cannot be associates, so must be in different components of $RM(\underset{\sim}{d})$ by Theorem 4, we shall have completed the proof.

Let $\underset{\sim}{e}$ be any infinite subsequence of $\underset{\sim}{d}$, with no term less than 2. For any given natural number $a > 2$ we can form the cycle C_a and add further edges until we have a multigraph E_1 with degree sequence $(e_1, \ldots, e_{i-1}, e_i', e_{i+1}, \ldots, e_a)$ in which $e_i' < e_i$. (If any two vertices had lesser degree than the corresponding terms in $\underset{\sim}{e}$, a further edge could be added between them: E_1 is a multigraph obtained from C_a when no such additional edges can be inserted.) Let E_2 be an acyclic simple realization of the sequence $(e_i - e_i', e_{a+1}, e_{a+2}, \ldots)$, guaranteed to exist by Lemma 0. Let E result from E_1 and E_2 by identifying the vertex of E_1 corresponding to e_i', with the vertex of E_2 corresponding to $e_i - e_i'$, so E is a realization of $\underset{\sim}{e}$ with largest cycle C_a.

Now let S be any set of natural numbers greater than 2. Let $\underset{\sim}{f}$ be any infinite subsequence of $\underset{\sim}{d}$ with no term less than 2, and such that the complementary subsequence of $\underset{\sim}{d}$ has infinite sum. Partition $\underset{\sim}{f}$ into infinitely many infinite subsequences $\underset{\sim}{e}(a, r)$, where a runs through the elements of S and independently r runs through the natural numbers. Realize each $\underset{\sim}{e}(a, r)$ by a multigraph with largest cycle C_a, as already described. The desired realization G of $\underset{\sim}{d}$ is now obtained by taking the disjoint union of all these realizations, together with an acyclic simple realization of the subsequence of $\underset{\sim}{d}$ complementary to $\underset{\sim}{f}$. For each $a \in S$, there are infinitely many components of G with largest cycle C_a. Moreover, any finite sequence of switchings applied to G can only change the largest cycle size in finitely many components, so any associate of G has infinitely many components with largest cycle C_a if and only if $a \in S$. Thus, realizations of $\underset{\sim}{d}$ corresponding in this way to different choices of S cannot be associates. Since there are uncountably many choices for S, Theorem 4 ensures that $RM(\underset{\sim}{d})$ has uncountably many components.

Theorem 6. The graph $RS(\underset{\sim}{d})$ is connected if and only if $\underset{\sim}{d}$ is finitary. If $\underset{\sim}{d}$ is infinitary then $RS(\underset{\sim}{d})$ has uncountably many components.

Proof. The proof is essentially that of Theorem 5, but for infinitary $\underset{\sim}{d}$ it is necessary to modify the construction for encoding any set S, of natural numbers greater than 2, into a simple realization of $\underset{\sim}{d}$. For any natural number $a > 2$ and any infinite subsequence $\underset{\sim}{e}$ of $\underset{\sim}{d}$, with no term less than 2, we can form the cycle C_a and add further edges (as chords) until we have a simple graph E_1 with degree sequence $(e_1', e_2', \ldots, e_a')$ in which $e_i' \leq e_i$ for $i = 1, 2, \ldots, a$. There may be several vertices with degree less than the corresponding term in $\underset{\sim}{e}$; we complete the construction by adjoining at each such vertex of E_1 an appropriate acyclic simple graph, so that the resultant graph E is a connected simple realization of $\underset{\sim}{e}$ with largest cycle C_a. The rest of the details follow as in the proof of Theorem 5.

4. UNIQUE REALIZATIONS.

In this section we take up the question of degree sequences $\underset{\sim}{d}$ which have a unique multigraphic or simple graphic realization, that is, those $\underset{\sim}{d}$ for which $RM(\underset{\sim}{d})$ or $RS(\underset{\sim}{d})$ is the trivial graph, with just one vertex. Hakimi [7] determined all finite sequences for which $RM(\underset{\sim}{d})$ is the trivial graph, and Koren [9] determined the finite sequences for which $RS(\underset{\sim}{d})$ is the trivial graph. Here we obtain the corresponding results for infinite sequences, and give a result which includes a new characterization of the finite sequences with unique multigraphic realization, in terms of forbidden subgraphs.

Theorem 7. Let $\Sigma\underset{\sim}{d}$ be infinite. Then the graph $RM(\underset{\sim}{d})$ is trivial if and only if $\underset{\sim}{d}$ has at most one term greater than 1.

Proof. If $RM(\underset{\sim}{d})$ is trivial, then $\underset{\sim}{d}$ is finitary by Theorem 5, so $\underset{\sim}{d}$ has at most finitely many terms greater than 1. Suppose $\underset{\sim}{d}$ has at least two terms greater than 1. Without loss of generality we may suppose the first two terms of $\underset{\sim}{d}$ exceed 1, so $\underset{\sim}{d}$ can be partitioned into complementary subsequences $\underset{\sim}{e}$, $\underset{\sim}{e}'$ with infinite sum and $e_1 \geq 2$, $e_1' \geq 2$. By Lemma 0, the sequence $\underset{\sim}{e} - (2) = (e_1 - 2, e_2, e_3, \ldots)$ has a realization E_1 as an acyclic simple graph; similarly $\underset{\sim}{e}' - (2)$ has such a realization E_2. We can identify the vertex of E_1 corresponding to $e_1 - 2$ with one vertex of C_2, the unique multigraphic realization of (2, 2), and the vertex of E_2 corresponding to $e_1' - 2$ with the other vertex of C_2. Then the resultant graph E is a multigraphic realization of $\underset{\sim}{d}$ containing the cycle C_2. Moreover, Lemma 0 ensures that $\underset{\sim}{d}$ has an acyclic simple realization G, so $RM(\underset{\sim}{d})$ has at least two vertices, contrary to our hypothesis. It follows that $\underset{\sim}{d}$ has at most one term greater than 1.

Conversely, if $\Sigma\underset{\sim}{d}$ is infinite but $\underset{\sim}{d}$ is finitary, let $\underset{\sim}{d}'$ be the subsequence of $\underset{\sim}{d}$ comprising all terms greater than 1, together with s terms equal to 1, where s is the sum of all terms greater than 1. As shown in the proof of Theorem 5, every multigraphic realization of $\underset{\sim}{d}$ is the disjoint union of a realization of $\underset{\sim}{d}'$ and the unique realization of the complementary subsequence $\underset{\sim}{d}''$. Now if $\underset{\sim}{d}$ has at most one term greater than 1, then $\underset{\sim}{d}'$ is either empty or has just one term d_1' greater than 1 and $s = d_1'$ other terms equal to 1. In either case, $\underset{\sim}{d}'$ has a unique multigraphic realization, and hence so does $\underset{\sim}{d}$. It follows that $RM(\underset{\sim}{d})$ is the trivial graph.

Theorem 8. Let $\Sigma\underset{\sim}{d}$ be infinite. Then the graph $RS(\underset{\sim}{d})$ is trivial if and only if $\underset{\sim}{d}$ has at most one term greater than 1.

Proof. The proof is essentially the same as for Theorem 7, except the details must be modified to show that if $\underset{\sim}{d}$ is finitary but has at least two terms greater than 1, then it has at least two nonisomorphic simple realizations. First suppose $\underset{\sim}{d}$ is finitary but has its first $a \geq 3$ terms greater than 1. Partition $\underset{\sim}{d}$ into subsequences $\underset{\sim}{e}(r)$, $1 \leq r \leq a$, each with infinite sum and first term $e_1(r) \geq 2$. By Lemma 0, the sequence $\underset{\sim}{e}(r) - (2)$ has a realization E_r as an acyclic simple graph. Form the cycle C_a, and with its rth vertex identify the vertex of E_r corresponding to $e_1(r) - 2$, for every r. The resultant graph E is a simple realization of $\underset{\sim}{d}$ containing the cycle C_a. Lemma 0 also ensures that $\underset{\sim}{d}$ has a simple acyclic realization G, so RS($\underset{\sim}{d}$) has at least two vertices.

Now it suffices to deal with the case in which $\underset{\sim}{d}$ has precisely two terms greater than 1, these being the first two terms. As before, partition $\underset{\sim}{d}$ into subsequences $\underset{\sim}{e}$ and $\underset{\sim}{e}'$, each with infinite sum and $e_1 \geq 2$, $e_1' \geq 2$. Lemma 0 ensures each has an acyclic simple realization, so the disjoint union of the two is a simple realization of $\underset{\sim}{d}$ in which the two vertices of degree greater than 1 are in different components. A single switching between two edges of this graph, each incident with one of the vertices of degree greater than 1, yields a simple realization of $\underset{\sim}{d}$ in which the two vertices of degree greater than 1 are in the same component. (In fact, the proof given in [4] for Lemma 0 yields such a realization.) Thus, we again conclude RS($\underset{\sim}{d}$) has at least two vertices. It follows that if RS($\underset{\sim}{d}$) is trivial, $\underset{\sim}{d}$ has at most one term greater than 1. The rest of the proof is as for Theorem 7.

Before passing to our characterization of finite-sum sequences with unique multigraphic realization, we wish to point out that various results in this paper, including the two theorems just established, have a rather natural formulation in terms of dominance between certain sequences. This idea was used to some advantage in [3], and in particular, in the characterization given there for sequences with unique pseudographic realization. Here we adopt a rather wider notion of dominance: given two sequences $\underset{\sim}{d}$, $\underset{\sim}{e}$ of natural numbers, we shall say that $\underset{\sim}{d}$ dominates $\underset{\sim}{e}$ if $\underset{\sim}{d}$ has a subsequence which can be permuted into a sequence $\underset{\sim}{d}'$ so that every corresponding pair of terms in $\underset{\sim}{d}'$ and $\underset{\sim}{e}$ satisfies $d_i' \geq e_i$. (For the narrower notion of dominance used in [3], we had to say "$\underset{\sim}{d}$ has a subsequence equivalent to one which dominates $\underset{\sim}{e}$", where we can now say "$\underset{\sim}{d}$ dominates $\underset{\sim}{e}$".) With this viewpoint, we have the following alternative formulations, using the sequences $\underset{\sim}{1} = (1^\infty) = (1, 1, 1, \ldots)$ and $\underset{\sim}{2} = (2^\infty) = (2, 2, 2, \ldots)$.

Theorem 5a. The graph RM($\underset{\sim}{d}$) is connected if and only if $\underset{\sim}{d}$ does not dominate $\underset{\sim}{2}$. If $\underset{\sim}{d}$ does dominate $\underset{\sim}{2}$, then RM($\underset{\sim}{d}$) has uncountably many components.

Theorem 7a. If $\underset{\sim}{d}$ dominates $\underset{\sim}{1}$, then the graph RM($\underset{\sim}{d}$) is trivial if and only if $\underset{\sim}{d}$ does not dominate (2, 2).

Evidently Theorems 6 and 8 admit corresponding formulations, as Theorems 6a and 8a.

Loosely speaking, if the sequence $\underset{\sim}{d}$ does not dominate a given sequence $\underset{\sim}{e}$, then the various realizations of $\underset{\sim}{d}$ do not contain certain subgraphs; in other words, we have a corresponding result in terms of forbidden subgraphs. A number of the results in [2] illustrate this relationship. Theorems 7 and 8 can be given alternative formulations from this viewpoint. In particular, the proof of Theorem 7 shows that if $\underset{\sim}{d}$ dominates (2, 2), it has a realization containing the cycle C_2 as a subgraph; the converse clearly holds so we have

Theorem 7b. Let $\Sigma\underset{\sim}{d}$ be infinite. Then the graph RM($\underset{\sim}{d}$) is trival if and only if no multigraphic realization of $\underset{\sim}{d}$ contains the subgraph C_2.

Likewise, the proof of Theorem 8 shows that if $\underset{\sim}{d}$ dominates (2, 2), either it has a realization which contains a cycle C_a with $a \geq 3$, or it has a realization with two vertices of degree at least 2 in one component. In the latter case, or in the former case if $a \geq 4$, there is a subgraph P_4, the path with 4 vertices. In the former case with $a = 3$, switching an edge of the C_3 with an edge not incident with the cycle again yields a realization with P_4 as a subgraph. Conversely, if $\underset{\sim}{d}$ has a realization with P_4 as a subgraph, then $\underset{\sim}{d}$ dominates (2, 2), so we have

Theorem 8b. Let $\Sigma\underset{\sim}{d}$ be infinite. Then the graph RS($\underset{\sim}{d}$) is trivial if and only if no simple graphic realization of $\underset{\sim}{d}$ contains the subgraph P_4.

Similarly, switching an edge of C_2 with a disjoint edge yields a P_4, so we also have

Theorem 7c. Let $\Sigma\underset{\sim}{d}$ be infinite. Then the graph RM($\underset{\sim}{d}$) is trivial if and only if no multigraphic realization of $\underset{\sim}{d}$ contains the subgraph P_4.

We now pass to a characterization of the sequences with finite sum which have unique multigraphic realization. As remarked earlier, the argument provides an alternative way of obtaining these sequences, which were first considered by Senior [10] and determined by Hakimi [7]. Our characterization concerns certain <u>edge-induced subgraphs</u>, that is, subgraphs whose edges are taken with the full multiplicity of the corresponding edges in the relevant multigraph.

Let G be any multigraph having a degree sequence with finite sum, and let m be the maximum multiplicity of G, that is, G contains some edge with multiplicity m, but none with multiplicity greater than m. We are interested in the possible presence, in G, of subgraphs P and Q, which we now define. The subgraph P has four vertices u, v, w, x and edges $[u, v]^a$, $[v, w]^m$, $[w, x]^b$, where a, b are any positive integers not exceeding m. The subgraph Q has five vertices u, v, w, x, y and edges $[u, v]^1$, $[u, w]^{m-1}$, $[v, w]^{m-1}$, $[w, x]^m$, $[w, y]^c$, where c is any positive integer strictly less than m. These graphs are shown in Figure 1.

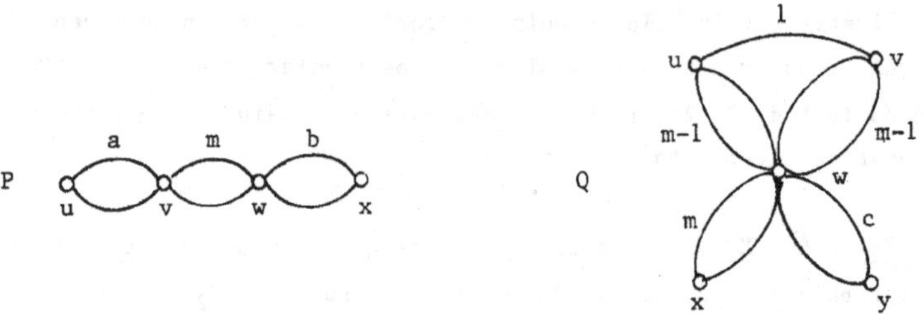

FIGURE 1.

The presence of P or Q in G (with the specified multiplicities coinciding with those in G) is enough to guarantee that G is not the unique multigraphic realization of its degree sequence. We shall prove this in the following theorem.

Theorem 9. Let $\Sigma \underset{\sim}{d}$ be finite. Then the graph RM($\underset{\sim}{d}$) is trivial if and only if no realization of $\underset{\sim}{d}$ contains P or Q as an edge-induced subgraph.

Proof. First, suppose $\underset{\sim}{d}$ has a multigraphic realization G which contains P as an edge-induced subgraph. The switching [u, v], [w, x] $\rightarrow$ [u, x], [v, w] results in [v, w] having multiplicity m + 1, greater than the maximum multiplicity of G, so the resultant realization is not isomorphic to G, and RM($\underset{\sim}{d}$) is nontrivial.

Next suppose that $\underset{\sim}{d}$ has a multigraphic realization G which contains Q as an edge induced subgraph. Note that m $\geq$ 2 in this case, since Q contains an edge of multiplicity c which is required to be a positive integer strictly less than m. The switching [u, v], [w, y] $\rightarrow$ [u, w], [v, y] results in [u, w] having multiplicity m, while the multiplicities of [u, v] and [w, y], which are the only multiplicities decreased by the switching, are necessarily less than m in G. Hence the switching increases the total number of edges with multiplicity m and the resultant realization is not isomorphic to G. Hence RM($\underset{\sim}{d}$) is non-trivial.

Henceforth we can suppose no realization of $\underset{\sim}{d}$ contains P or Q as an edge-induced subgraph. In particular, this implies that all realizations of $\underset{\sim}{d}$ have the same maximum multiplicity. For if there were a sequence of switchings which transforms some realization with maximum multiplicity m into one with maximum multiplicity m + 1, within the sequence there would be one switching which transforms a realization G with maximum multiplicity m into H, with maximum multiplicity m + 1. Then G would necessarily contain P as an edge-induced subgraph, contrary to our supposition about $\underset{\sim}{d}$. Since RM($\underset{\sim}{d}$) is connected when $\Sigma\underset{\sim}{d}$ is finite, by Theorem 5, all realizations of $\underset{\sim}{d}$ must have the same maximum multiplicity, say m. If m = 0, clearly the only realization of $\underset{\sim}{d}$ is a set of isolated vertices. Henceforth, we suppose m ≥ 1.

Suppose $\underset{\sim}{d}$ has a realization G in which no edge is independent of a particular edge [v, w] of multiplicity m. Then the absence, from G, of any edge-induced subgraph P ensures that if there is a vertex u which is adjacent to both v and w, then G can have no other vertices of positive degree. If there is no such vertex u, then the absence of any edge-induced subgraph P forces all other vertices of positive degree to be adjacent to just one of the vertices v, w. Thus the <u>skeleton</u> of G (the simple subgraph comprising just the vertices of positive degree, with an edge between any two vertices just if there is at least one such edge in G) is either K_3 or else is a star. In all such cases, there is no edge-induced subgraph Q, and RM($\underset{\sim}{d}$) is trivial.

So now we can suppose that every realization of $\underset{\sim}{d}$ has an edge independent of some edge of multiplicity m. Let G be a realization of $\underset{\sim}{d}$ with least number of edges of multiplicity m. In G let [w, x] be an edge of multiplicity m and let [u, v] be an independent edge. Neither of the switchings between [u, v] and [w, x] can reduce the total number of edges of multiplicity m, though each reduces the multiplicity of [w, x]. Thus, without loss of generality, we can suppose [u, w] has multiplicity m - 1 in G, and either [u, x] or [v, w] also has multiplicity m - 1 in G. Denote the corresponding subgraphs of G by A and B respectively, as shown in Figure 2, with the multiplicity of [u, v] equal to the positive number r.

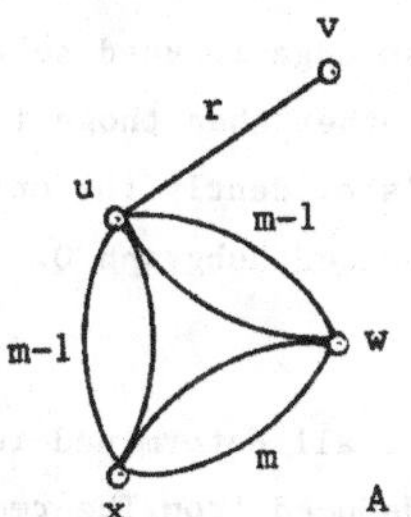

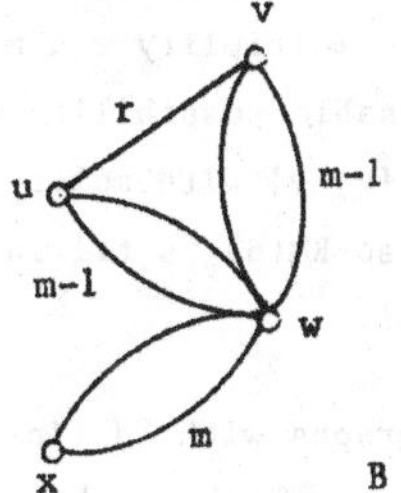

FIGURE 2.

If m = 1, the subgraphs A and B are identical. The graph G cannot contain $2P_3$ (the graph comprising two independent paths P_3) as a subgraph, since switching between two edges in such a subgraph yields P_4, which is the forbidden edge-induced subgraph P when m = 1. Hence, if m = 1 the skeleton of G is necessarily a subgraph of the disjoint union of a star and a finite number of copies of P_2. In all such cases RM($\underset{\sim}{d}$) is trivial. Recall also that when m = 1 there cannot be an edge-induced subgraph Q.

We now take m > 1. If G has an edge-induced subgraph A, then no other edge can be incident with w or x, since G has no edge-induced subgraph P. Also the multiplicity r of [u, v] in G must be 1, for otherwise the switching [u, v], [w, x] → [u, x], [v, w] would raise the multiplicity of [u, x] to m and produce an edge-induced subgraph P on the vertices v, u, x, w. There cannot be any other edge in G which is not incident with u, for otherwise switching such an edge with [x, w] would reduce the total number of edges of multiplicity m present, contrary to the choice of G. If G does contain an edge [u, y], where y is a vertex not in A, the switching [u, y], [w, x] → [u, x], [w, y] increases the multiplicity of [u, x] to m, and produces an edge-induced subgraph P on the vertices v, u, x, w, contrary to hypothesis. Hence G = A, with r = 1. Evidently RM($\underset{\sim}{d}$) is trivial in this case, and G contains no edge-induced subgraph Q.

Now, with m > 1, it remains to consider the case where G contains an edge-induced subgraph B. No other edge can be incident with x, since there is no edge-induced subgraph P. No other edge can be incident with u, since the switching, [u, y], [w, x] → [u, w], [x, y] would produce an edge-induced subgraph P on the vertices v, u, w, x, contrary to hypothesis. Similarly, no other edge can be incident with v and the multiplicity r of [u, v], must be 1. If there is any edge [y, z] present, different from [u, v] and independent of [w, x], then it must also have multiplicity r = 1 and [w, y], [w, z] must be present with multiplicity m - 1.

Thus contrary to hypothesis G contains an edge-induced subgraph Q, with [w, y] having multiplicity c = m - 1. Moreover, if [y, z] were absent from G, but [w, y] present with multiplity c < m, we would still have an edge-induced subgraph Q. The only admissable possibility is that all edges of G other than those in B are edges of the form [w, y] with multiplicity m. But then G is evidently the unique realization of $\underset{\sim}{d}$, so RM($\underset{\sim}{d}$) is trivial, and G has no edge-induced subgraph Q. This completes the proof.

The multigraphs with $\Sigma\underset{\sim}{d}$ finite and RM($\underset{\sim}{d}$) trivial are all determined in the course of this proof. Those with $\Sigma\underset{\sim}{d}$ infinite are readily deduced from Theorem 7. So we have the following result.

Corollary. The multigraphs which are the unique realizations of their degree sequences, up to isomorphism, are (i) all those whose skeleton is a star or K_3, (ii) all simple graphs which are subgraphs of the disjoint union of a star and infinitely many copies of P_2, (iii) the graph A with $m > 1$ and $r = 1$, together with any number of isolated vertices, and (iv) the graph B with $m > 1$ and $r = 1$, together with any finite set of edges, of multiplicity m, incident with the vertex w, and any number of isolated vertices.

It may be noted that if $\underset{\sim}{d}$ has more than one multigraphic realization, then necessarily $\underset{\sim}{d}$ dominates (2, 2, 1, 1). This follows from Theorems 7a and 9. It is straightfoward to combine this observation with the above Corollary to deduce a variant of Hakimi's classification [7], where infinite sequences are also included. Here and subsequently, for any given sequence $\underset{\sim}{d}$ we shall refer to the subsequence comprising every positive term of $\underset{\sim}{d}$ as the positive subsequence of $\underset{\sim}{d}$.

Theorem 10. The graph $RM(\underset{\sim}{d})$ is non-trivial if and only if $\underset{\sim}{d}$ dominates (2, 2, 1, 1) and either (i) $\underset{\sim}{d}$ dominates $\underset{\sim}{1}$ or (ii) $\underset{\sim}{d}$ does not dominate $\underset{\sim}{1}$ but $2 \max \underset{\sim}{d} < \Sigma \underset{\sim}{d}$ and the positive subsequence of $\underset{\sim}{d}$ is not equivalent to $(2m - 1, 2m - 1, 2m - 1, 1)$ or to $(mn - 2, m^n)$, where m, n are positive integers.

Note that if $\underset{\sim}{d}$ dominates (2, 2, 1, 1), its positive subsequence could only be equal to one of the sequences explicitly specified in (ii) if $m \geq 2$ and $n \geq 3$, so it is unnecessary to include these conditions in the statement of the theorem.

5. ISOLATED VERTICES.

In this concluding section we are concerned with finding those graphs of realizations which contain components which are isolated vertices. In the previous section we determined all sequences $\underset{\sim}{d}$ with infinite sum having $RM(\underset{\sim}{d})$ or $RS(\underset{\sim}{d})$ trivial, thereby complementing finite results by Hakimi [7] and Koren [9]. So here we take up the question of isolated vertices when the graph of realizations has more than one component. In view of Theorems 5 and 6, this amounts to treating the case in which $\underset{\sim}{d}$ is infinitary.

Recall from [4] that an infinitary sequence $\underset{\sim}{d}$ is a focal sequence if each of its positive terms is equal to infinitely many other terms. The multigraphic focus $FM(\underset{\sim}{d})$ of a focal sequence $\underset{\sim}{d}$ is the multigraphic realization of $\underset{\sim}{d}$ comprising infinitely many disjoint copies of all nonisomorphic multigraphs whose degree sequence is a finite subsequence of positive terms of $\underset{\sim}{d}$, together with an isolated vertex for each zero term of $\underset{\sim}{d}$. The simple focus $FS(\underset{\sim}{d})$ of a focal sequence is made up similarly from isolated vertices and all simple graphs whose degree sequence is a finite subsequence of positive terms of $\underset{\sim}{d}$. Analogous to Theorem 6 for pseudographs in [3], we have the following two results. Here we say that G is a component subgraph of H if G is isomorphic to some subset of the components of H.

Theorem 11. If $\underset{\sim}{d}$ is a focal sequence, the multigraphic focus $FM(\underset{\sim}{d})$ is a multigraphic realization of $\underset{\sim}{d}$ which is invariant under switching. Every multigraphic realization of $\underset{\sim}{d}$ which is invariant under switching contains $FM(\underset{\sim}{d})$ as a component subgraph.

Proof. It is straightforward to see that $FM(\underset{\sim}{d})$ is switching invariant. Now let G be any switching invariant multigraphic realization of $\underset{\sim}{d}$. If a is any positive term of $\underset{\sim}{d}$, G contains vertices of degree a, and by switching pairs of edges incident with two such vertices, we could obtain a component multigraph realizing the sequence (a, a). Since G is switching invariant, it must already contain infinitely many such components, for each a. Let H be any multigraph whose degree sequence is a finite subsequence of the positive terms of $\underset{\sim}{d}$. Then 2H, comprising two disjoint copies of H, has degree sequence $\underset{\sim}{d}'$ which is the union of a finite number of disjoint subsequences of the form (a, a). We have just seen that G has one component subgraph G' which is a multigraphic realization of $\underset{\sim}{d}'$. Moreover, $RM(\underset{\sim}{d}')$ is connected, by Theorem 5, so a finite sequence of switchings applied to G' yields 2H; but G is switching invariant, so G already contains 2H as a component subgraph. Hence it contains H as a component subgraph. Switching invariance of G in fact guarantees that it contains infinitely many copies of H as component subgraphs, for otherwise we could alter the number of such component subgraphs in G by a suitable switching. It follows that G contains $FM(\underset{\sim}{d})$ as a component subgraph.

Theorem 12. If $\underset{\sim}{d}$ is a focal sequence, the simple focus $FS(\underset{\sim}{d})$ is a simple realization of $\underset{\sim}{d}$ which is invariant under switching. Every simple realization of $\underset{\sim}{d}$ which is invariant under switching contains $FS(\underset{\sim}{d})$ as a component subgraph.

Proof. The proof is essentially that for Theorem 11, except that we need a different starting point to ensure that if H is any simple graph whose degree sequence is a finite subsequence of the positive terms of $\underset{\sim}{d}$, and G is any switching invariant realization of $\underset{\sim}{d}$, then G contains H as a component subgraph. Let m be the least common multiple of the numbers a + 1, where a runs through the terms of the degree sequence of H. For any such a, we can carry out switchings on G to obtain the complete graph K_{a+1} as a component, and switching invariance ensures that G already has such a component, and so infinitely many such components. Then G has a component subgraph which realizes the degree sequence $\underset{\sim}{d}'$ of mH, the graph comprising m disjoint copies of H. Theorem 6 ensures connectedness of RS($\underset{\sim}{d}$'), so G contains mH as a component subgraph, and therefore H itself as a component subgraph. It follows that G contains FS($\underset{\sim}{d}$) as a component subgraph.

We can now deal with infinitary sequences with isolated vertices in their graphs of realizations.

Theorem 13. If $\underset{\sim}{d}$ is an infinitary degree sequence, the graph RM($\underset{\sim}{d}$) has an isolated vertex if and only if (i) $\underset{\sim}{d}$ is a focal sequence, or (ii) the positive subsequence of $\underset{\sim}{d}$ is $(1, 2^{\infty})$.

Proof. If $\underset{\sim}{d}$ is a focal sequence, the multigraphic focus FM($\underset{\sim}{d}$) is an isolated vertex of RM($\underset{\sim}{d}$). If $\underset{\sim}{d}$ has positive subsequence $(1, 2^{\infty})$, that is $(1) \cup \underset{\sim}{2}$, then the graph $P'_{\infty} \cup FM(\underset{\sim}{2})$ is an isolated vertex of RM($\underset{\sim}{d}$), where P'_{∞} is the one-way infinite path, discussed in [3].

Now suppose that $\underset{\sim}{d}$ is infinitary but not focal, and RM($\underset{\sim}{d}$) has an isolated vertex. Since $\underset{\sim}{d}$ is not focal, it has some positive term a occurring with only finite multiplicity. If a has multiplicity at least 2, we can switch any realization of $\underset{\sim}{d}$ to obtain an associate in which no two vertices of degree a are adjacent, and another associate in which at least two vertices of degree a are adjacent, so by Theorem 4, every component of RM($\underset{\sim}{d}$) has at least two vertices, contrary to the assumption concerning RM($\underset{\sim}{d}$). It follows that each positive term of $\underset{\sim}{d}$ with finite multiplicity has multiplicity 1. If $\underset{\sim}{d}$ has two different positive terms with finite multiplicity, switching any realization will give an associate in which two corresponding vertices are adjacent, and another associate in which they are not adjacent. Once again, this is a contradiction, so $\underset{\sim}{d}$ has just one positive term a with finite multiplicity. If $\underset{\sim}{d}$ has two different positive terms b, c with infinite multiplicity, we can switch any realization to obtain an associate in which the vertex of degree a is only adjacent to vertices of degree b, and another in which the vertex of degree a is only adjacent to vertices of degree c. Again this is a contradiction, so $\underset{\sim}{d}$ has only one positive term with infinite multiplicity, and the positive subsequence of $\underset{\sim}{d}$ is (a, b^{∞}). As $\underset{\sim}{d}$ is infinitary, we have $b \geq 2$. If $a \geq 2$, we can find an associate of any realization of $\underset{\sim}{d}$ which has a multiple edge between the vertex of degree a and some vertex of degree b, and another associate in which no such multiple edge occurs. This contradiction shows we must actually have $a = 1$. Finally, if $b \geq 3$, we can find an associate of any realization of $\underset{\sim}{d}$ which has the vertex of degree 1 adjacent to a vertex of degree b which belongs to a cycle C_3, and another associate in which this is not so. This final contradiction shows we must actually have $b = 2$, so $\underset{\sim}{d} = (1, 2^{\infty})$.

Theorem 14. If $\underset{\sim}{d}$ is an infinitary sequence, the graph RS($\underset{\sim}{d}$) has an isolated vertex if and only if (i) $\underset{\sim}{d}$ is a focal sequence, or (ii) the positive subsequence of $\underset{\sim}{d}$ is $(1, 2^{\infty})$.

The proof follows that of Theorem 13 in all but a few minor details, so we omit the discussion.

We finally decide the cardinalities of the sets of isolated vertices in RM($\underset{\sim}{d}$) and RS($\underset{\sim}{d}$).

Theorem 15. Let $\underset{\sim}{d}$ be an infinitary sequence.

(i) If $\underset{\sim}{d}$ is a focal sequence with some term $a > 2$, then RM($\underset{\sim}{d}$) has uncountably many isolated vertices.

(ii) If $\underset{\sim}{d}$ is a focal sequence with positive subsequence $(1^{\infty}, 2^{\infty})$, then RM($\underset{\sim}{d}$) has precisely three isolated vertices.

(iii) If $\underset{\sim}{d}$ has positive subsequence (2^{∞}) or $(1, 2^{\infty})$ then the isolated vertices of RM($\underset{\sim}{d}$) comprise a countably infinite set.

The proof follows that of Theorem 8 for pseudographs in [3], in all but minor details, so we omit the discussion.

Theorem 16. Let $\underset{\sim}{d}$ be an infinitary sequence.

(i) If $\underset{\sim}{d}$ is a focal sequence with some term $a > 2$, then RS($\underset{\sim}{d}$) has uncountably many isolated vertices.

(ii) If $\underset{\sim}{d}$ is a focal sequence with positive subsequence $(1^{\infty}, 2^{\infty})$, then RS($\underset{\sim}{d}$) has precisely three isolated vertices.

(iii) If $\underset{\sim}{d}$ has positive subsequence (2^{∞}) or $(1, 2^{\infty})$, then the isolated vertices of RS($\underset{\sim}{d}$) comprise a countably infinite set.

Proof. Again the details are similar to those of Theorem 8 for pseudographs in [3], with the exception that to deal with case (i) of the present theorem we need simple graphs to play the role of the multigraphs $H_n(a)$ in [3]. For any positive integers a, n with $a > 2$, let $J_n'(a)$ be the simple graph with vertex set $\mathbf{Z}$ and edge set E, where E comprises $[i, i + 1]$ for each $i \in \mathbf{Z}$, together with $[(2n + 1)j - k, (2n + 1)j + k]$ for each $j, k \in \mathbf{Z}$ with $1 \leq k \leq n$. If $a > 3$, we form $J_n(a)$ from $J_n'(a)$ by identifying each vertex of degree 2 in $J_n'(a)$ with the vertex of degree $a - 2$ in a corresponding copy of a given simple acyclic realization $T_2(a)$ of $(a - 2, a^{\infty})$, and identifying each vertex of degree 3 in $J_n'(a)$ with the vertex of degree $a - 3$ in a corresponding copy of a given simple acyclic realization $T_3(a)$ of $(a - 3, a^{\infty})$. The resultant graph $J_n(a)$ is a regular simple graph of degree a which encodes the positive integer n into its structure. Similarly, if $a = 3$ we form

$J_n(3)$ for $J_n'(3)$ by identifying each vertex of degree 2 in $J_n'(3)$ with the vertex of degree 1 in a corresponding copy of a given acyclic realization $T_2(3)$ of $(1, 3^\infty)$. (We illustrate $J_2(3)$ in Figure 3.) The rest of the proof follows.

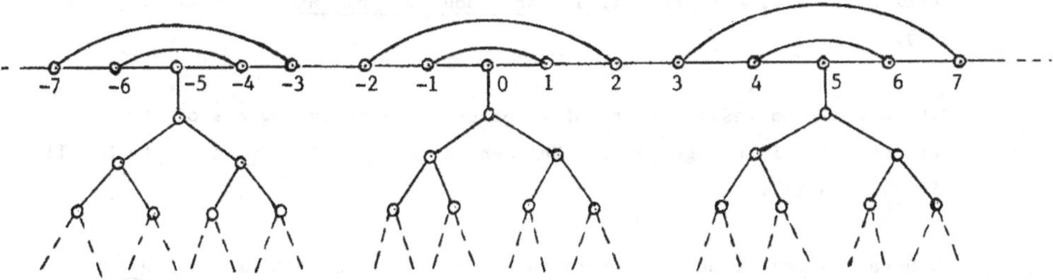

FIGURE 3. - The graph $J_2(3)$.

In [3, 4] we were able to answer further questions about pseudographs for which the multigraph and simple graph analogues remain open. In particular, for which sequences $\underset{\sim}{d}$ does $RM(\underset{\sim}{d})$ or $RS(\underset{\sim}{d})$ have a component which is a nontrivial path? For which $\underset{\sim}{d}$ does $RM(\underset{\sim}{d})$ or $RS(\underset{\sim}{d})$ have a vertex of degree 1?

REFERENCES

[1] R.B. Eggleton, Graphic sequences and graphic polynomials: a report, in Infinite and Finite Sets, Vol. 1, ed. A. Hajnal et al, Colloqu. Math. Soc. J. Bolyai 10 (North Holland, Amsterdam, 1975), 385-392.

[2] R.B. Eggleton and D.A. Holton, Graphic sequences, Combinatorial Math. VI, Proc. 6th Australian Conf. on Combinatorial Math., Armidale, 1978 (Springer-Verlag, L.N.M. 748, 1979), 1-10.

[3] R.B. Eggleton and D.A. Holton, The graph of type $(0, \infty, \infty)$ realizations of a graphic sequence, op. cit., 41-54.

[4] R.B. Eggleton and D.A. Holton, Pseudographic realizations of an infinitary degree sequence, Combinatorial Math. VII, Proc. 7th. Australian Conf. on Combinatorics Math., Newcastle, 1979 (Springer-Verlag, L.N.M. 829, 1980), 94-109.

[5] P. Erdős and T. Gallai, Graphs with prescribed degrees of vertices (in Hungarian), _Mat. Lapok,_ 11 (1960), 264-274.

[6] S.L. Hakimi, On realizability of a set of integers as degrees of the vertices of a linear graph I, _J. Soc. Indust. Appl. Math._, 10 (1962), 496-506).

[7] S.L. Hakimi, On realizability of a set of integers as degrees of the vertices of a linear graph II: Uniqueness, _J. Soc. Indust. Appl. Math._, 11 (1963), 135-147.

[8] V. Havel, A remark on the existence of finite graphs (in Czech), _Casopis Pest. Mat._, 80 (1955), 477-480.

[9] M. Koren, Sequences with a unique realization by simple graphs, _J. Combinatorial Theory,_ 21B (1976), 235-244.

[10] J.K. Senior, Partitions and their representative graphs, _Amer. J. Math._, 73 (1951), 663-689.

[11] R. Taylor, Constained switchings in graphs, this volume.

Department of Mathematics,
University of Newcastle,
New South Wales, 2308,
Australia.

Department of Mathematics,
University of Melbourne,
Parkville, Vic., 3052,
Australia.

CRITICAL LINK IDENTIFICATION IN A NETWORK

L.R. FOULDS

There exist systems which can be usefully described by a network containing links through which a commodity flows. This paper is concerned with finding a solution procedure for a particular multi-commodity flow network design problem. The problem is to identify a set of arcs in the network such that if travel is prohibited in them all, flow travels by feasible paths and its total cost is minimal. The total flow in both directions of each link may not exceed the capacity of the link, which is a known constant. Each arc and each vertex of the network has a non-negative unit traversal cost related nonlinearly to the amount of flow passing through it. Between each pair of distinct vertices there is a given non-negative rate of flow from the first vertex to the second which travels by the shortest available path. The optimality criterion is the total traversal cost of all flow, which is to be minimized.

The principle contribution of this paper is the presentation of a solution procedure for the above problem based on branch and bound enumeration.

1. INTRODUCTION

The graph-theoretic notation and terminology used in this paper conforms with that of Busacker and Saaty [1], except for the following changes. The branches of an undirected graph are here called <u>links</u>. The branches of a network, which is a directed graph are still called <u>arcs.</u> The <u>associated undirected graph</u> of a network is defined to be the graph which results from removing the orientation from the arcs of the network and replacing multiple links by a single link connecting the same vertices.

An arc from vertex i to vertex j is denoted by the ordered pair (i,j). A link between vertices i and j is denoted by the unordered pair {i,j}.

The problem that this paper deals with is concerned with a network with the following characteristics:

(i) Each arc and each vertex has a non-negative unit traversal cost which is dependent upon the amount of flow passing through it.

(ii) The total flow in the two oppositely directed arcs of each link may not exceed a constant known capacity of flow for the link.

(iii) Between each ordered pair of distinct vertices there is a given non-negative rate of flow from the first vertex to the second vertex which travels by the shortest available path.

(iv) Flow may be controlled by prohibiting travel in any of the arcs.

(v) The optimality criterion of the problem is the total traversal cost of all flow which is to be minimized.

(vi) Each vertex has an upper limit of flow it can accommodate.

2. THE FLOW ASSIGNMENT PROCEDURE

The first question that must be answered by a solution procedure for the given problem is the following: given a network, its capacities, initial traversal costs, and origin-destination demands; how does the flow distribute itself over the network? The distribution is found by assigning each unit of flow the shortest path between its origin and destination. These shortest paths are based upon initial traversal cost estimates which must be supplied. The amount of flow in each link and vertex is totalled up and this total is used to update the traversal costs. The units of flow are again assigned shortest paths and the process is repeated. This is continued until either there is no change in shortest path assignments from one iteration to the next (equilibrium has been reached) or some termination criterion is satisfied.

Typical relationships between arc flow and traversal cost are [4]:

$$a_{ij} = \bar{a}_{ij} \exp(g_{ij}/e_{ij} - 1)$$

$$a_{ij} = \bar{a}_{ij} [2^{(g_{ij}/e_{ij} - 1)}]$$

$$a_{ij} = \bar{a}_{ij} [1 + 0.15(g_{ij}/e_{ij})^4]$$

where

g_{ij} = the flow assigned to arc (i,j)

e_{ij} = the capacity of link {i,j}

$\bar{a}_{ij}$ = the original traversal cost estimate for arc (i,j)

a_{ij} = the adjusted traversal cost estimate for arc (i,j)

Similar relationships exist for vertex flow and traversal cost. Once a flow distribution has been achieved by using the flow assignment process with initial traversal cost estimates the amount of total flow in each arc and vertex is calculated. These values are substituted into the relationship and new traversal costs are found. Flow is once more distributed and the process is repeated until equilibrium is obtained.

The flow assignment procedure requires that a shortest path be found between every pair of vertices that have a positive flow demand. If relatively few of the flow demands are actually positive the shortest paths can be found efficiently one at a time by the method of Dijkstra [2]. If most of the flow demands are positive the shortest paths can be found efficiently all at once by Floyd's method [3].

3. THE BRANCH AND BOUND METHOD

In this section the Branch and Bound method for solving combinatorial problems is described in general. Next the specialized Branch and Bound algorithm which makes up part of the solution procedure for the problem is outlined. The components of the algorithm are: the solution, elimination and partitioning routines. (The partitioning routine has two subroutines - the detection routine and the branching routine).

The Branch and Bound Method

The Branch and Bound method is a sequential technique for solving integer programming problems which produces a directed tree. The first iteration produces the node at which the tree is rooted. Any subsequent iteration produces a number of nodes which are connected to the existing tree by directed arcs which all emanate from one existing node. A set of decisions representing a solution to the problem is associated with each node along with a bound on the value of any solution which contains this set. The algorithm starts by generating the root of the directed tree. The solution routine calculates a bound on all solutions which are associated with this first node. If the initial solution is not feasible, the partitioning routine replaces it by a set of related solutions. This is represented in the tree by a number of nodes which stem from the first node. The solution routine then sequentially calculates one bound for each new node of the tree. The elimination routine discards a solution if it can be shown that this solution can make no contribution to the solution of the original problem. The method continues generating new nodes at each iteration. Termination occurs when finally the optimal solution to the original problem, or evidence that no such solution exists, has been obtained. This evidence is provided by the elimination routine when all possible partial solutions have been eliminated.

The Branch and Bound Algorithm

The algorithm which applies the Branch and Bound method to the specific problem at hand is described next. During the Branch and Bound algorithm the final traversal cost estimates as supplied by the assignment process are <u>assumed fixed</u>.

The algorithm begins by solving the problem with no arcs prohibited. This creates the initial node at which the tree is rooted. This problem is solved by assigning all origin-destination flow demands their shortest path. This is achieved using the shortest path algorithm described earlier using the final traversal cost estimates supplied by the assignment process. Next the flow levels in each arc and each vertex are calculated. Then the flow in both directions of each link is compared with the capacity of the link. Similarly the total flow through each vertex is compared with the capacity of the vertex. If there are no capacity violations the initial solution with no arcs prohibited is optimal and the algorithm is terminated. If not, the largest capacity violation is identified. If it involves a link {i,j} then the partitioning routine produces three new nodes to the tree representing three new solutions. The three solutions are: prohibit arc (i,j), prohibit arc (j,i), and protect link {i,j}. When an arc is prohibited, no flow can travel along it. When a link is protected, neither of its arcs can be prohibited. If the largest capacity violation involves a vertex i then the set of all links incident with vertex i is identified. The link in this set whose flow most violates its capacity constraint or most nearly violates its capacity constraint is identified, {i,j} say. Then three new solutions are created as before. If it has been found during the solution routine for the initial problem that no path exists for certain origin-destination flow demands, then the problem is not well formulated and no solution exists. In this case the algorithm and indeed the whole procedure is terminated.

The solution and elimination routines then provide a value for each of the solutions described above. (An algorithm for the calculation of the value of a general solution by the solution and elimination routines is described later.) Next the values of the three solutions are compared and the solution with the lowest value (bound) is selected for partitioning.

The process continues, building up a directed tree, until either:

1. A feasible solution is identified which has a value no larger

than all other solutions.

2. No feasible solution exists.

The algorithm is then terminated and the optimal solution (if it exists) and its value, the list of prohibited arcs, and the levels of flow for the optimal solution are recorded.

The solution and elimination routines. The input to these routines is a set of prohibited arcs (which specify a solution) and a set of protected links, together with the traversal cost estimates, the link capacities, and the origin-destination flow demands. The output of the routines is the value of the solution to which the routines are applied.

For each prohibited arc in the solution the traversal cost of the arc is set equal to a relatively large number. Then all flow is assigned a shortest path as described earlier. The elimination routine involves the following two strategies. If no path exists for one or more flow demands then the value of the solution is set equal to a relatively large number. Suppose the set of protected links includes a set of arcs comprising a shortest path and the flow demand for that path is greater than the capacity of one of the links. Then the value of the solution is set equal to the relatively large number. This is because a solution containing such a set of prohibited arcs could never be feasible - the capacity constraint for the link in question would always be violated. Otherwise the solution routine totals up the traversal costs of all flow demand to yield a value of the solution.

The partitioning routine. The partitioning routine has two parts - the detection subroutine and the branching subroutine. If a given solution is infeasible the detection routine determines which link is critical. A link is deemed critical if it contains a flow level violating its capacity constraint by the largest amount (relative to all links and vertices), or if it is incident with a vertex which has a flow level violating its capacity by the largest amount and the link has the most flow relative to its capacity constraint. Of course, a protected link cannot be declared critical.

The branching routine replaces an infeasible solution by a number of new solutions concerning the critical link as determined by the detection routine. The new solutions are of one of the following forms:

{original solution} ∪ {prohibit one arc of the critical link}

{original solution} ∪ {protect the critical link}.

The number is usually three as described in the main algorithm outlined earlier. However it may happen that the critical link for an infeasible solution will contain an arc that is already prohibited in that solution. For example suppose a solution contains a prohibited arc (i,j) and the critical link for the solution is link {i,j}. That is, the link is a candidate for prohibition of flow in both directions. In this case the branching routine replaces the solution by just two solutions:

{original solution} ∪ {prohibit arc (j,i)}

{original solution} ∪ {protect arc (j,i)}.

This is carried out because if the usual third solution was specified as:

{original solution} ∪ {prohibit arc (i,j)}

this new solution would be identical to the original solution as arc (i,j) has already been prohibited.

It may occur that a particular solution is infeasible and has been selected for partitioning, but the links which are candidates for branching are all protected. In this case the term <u>critical</u> must be extended to include a link which contains a flow which most nearly violates its capacity.

4. THE COMPLETE PROCEDURE

The procedure begins by removing all arcs, (i,j) whose corresponding link capacity is less than the minimum positive flow demand. Next an estimated level of arc flow, G1 is produced using the flow assignment process. The estimate is assumed fixed and used in the Branch and Bound algorithm. This generates a solution, P; its value, z; and a new flow level estimate G2. G1 and G2 are compared to find if the procedure should be terminated. If so, the solution P is declared the final solution. If not, the traversal costs are revised. Then the above procedure is repeated until some termination criterion is met, whereupon the list P is the final solution.

5. THE CONVERGENCE OF THE PROCEDURE

The procedure that has been described is of the iterative type. First the initial traversal cost estimates are used to produce flow levels. These flow levels are then used to update the cost estimates. This cycle is continued until some termination criterion is reached. We now attempt to find conditions under which the procedure

converges. Towards this end, some necessary concepts are now introduced. Next a theorem is proven stating conditions under which the procedure will converge. The Branch and Bound algorithm produces an optimal solution for given traversal costs. This fact is proven at the end of this section.

Consider the flow in one arc of the network of a given problem. For brevity of notation, assume that current flow in the arc is defined as g units. The traversal cost of the arc is updated by the equation:

$$c = a1 + a2(g/e)^4 \qquad g \geq 0 \qquad (1)$$

where c = the current unit traversal cost of the arc.

a1 = the unit traversal cost for zero flow.

a2 = the increase in the unit traversal cost when the flow level is increased from zero to capacitated flow.

e = the capacity of the corresponding link.

Let

$$f(g) = a1 + a2(g/e)^4 \qquad g \geq 0$$

Thus $$c = f(g) \qquad g \geq 0.$$

It is assumed here that flow is continuous rather than discrete. Because f is differentiable for all non-negative g:

$$f'(g) = 4(a2/e)(g/e)^3 \qquad g \geq 0.$$

Thus $$f'(g) > 0 \qquad g > 0$$

since a2 and e are assumed to be positive.

Consider now the relationship that defines flow as a function of unit cost. Let h_1 and h_2 be the algorithmic maps for the traffic assignment process and the branch and bound algorithm respectively. Let $h_1(c)$ represent the flow in a specific arc which has cost c. Similarly for $h_2(c)$.

For a relatively high unit cost (large c), the flow level in the arc will be relatively small (small g). For a relatively small cost, the flow level will be relatively large. This is true for both h_1 and h_2, i.e. h_1 and h_2 are decreasing functions. It is assumed that both h_1 and h_2 are differentiable.

Consider the effects of the solution procedure on the traversal cost and flow level of the one arc being studied. The procedure begins by using the initial cost estimate, c_1 say, to generate a final flow level, g_1, with the traffic assignment process. Next g_1 is used to produce an updated cost estimate, c_2, with equation (1). Then c_2 is used to generate a new flow level, g_2, with the branch and bound algorithm. Finally g_2 is used to produce an updated traversal cost with equation (1). This new cost is used in place of c_1 and the cycle is repeated until a given number of iterations have been performed or the convergence criterion has been satisfied.

When will the procedure converge? To answer this question, the following discussion of exactly what is meant by convergence is presented.

Suppose that at the beginning of the cycle, c_i is the input unit traversal cost for a particular arc k. Let the flow level generated by the traffic assignment process be designated by g_i.

Thus $$g_i = h_1(c_i) \tag{2}$$

Let the unit traversal cost produced by equation (1) using g_i be designated by c_{i+1}.

Thus $$c_{i+1} = f(g_i) \tag{3}$$

Let the flow level generated by the branch and bound algorithm using c_{i+1} be designated by g_{i+1}.

Thus $$g_{i+1} = h_2(c_{i+1}) \tag{4}$$

<u>Definition</u>. If for arc k

(a) $g_i = g_{i+1}$

and (b) $|g_j - g^*| < |g_{j-2} - g^*|$ $\quad j = 3,4,5,\ldots,\leq i,$

then the procedure is said to have <u>converged to g* with respect to arc k</u> where g^* denotes the common value of g_i and g_{i+1}.

<u>Definition</u> If the procedure has converged to g^* with respect to each arc in the network, then the procedure is said to have converged.

<u>Theorem 1.</u>

If in the network of any problem, a traversal cost, c^*, can be found for each arc such that

$$c^* = f(h_1(c^*)) \tag{5}$$

$$1 > h'_1(c^*)f'(h_1(c^*)) > -1 \tag{6}$$

$$1 > h'_2(c^*)f'(h_2(c^*)) > -1 \tag{7}$$

$$h_1(c^*) = h_2(c^*) \tag{8}$$

then the procedure will converge.

<u>Proof</u>.

The notation of equations (2), (3), and (4) is adopted for the remainder of this section. The three arguments (a), (b), and (c) that follow are each concerned with just one arbitrary arc of the network.

(a) Suppose at the end of some cycle of the procedure a flow value g_{n-1} is generated. The use of equation (1) and the traffic assignment process generates a new flow value g_n. If equations (5) and (6) hold, g_n will be closer in magnitude to g^* than g_{n-1}.

Let $\quad dg = g_{n-1} - g^*$

$$\text{Thus} \quad g_{n-1} = g^* + dg \tag{9}$$

Then the updated traversal cost, c_n is found:

$c_n = f(g_{n-1})$ by assumption

$= f(g^* + dg)$ by (9)

$= f(g^*) + f'(g^*)dg$ as f is differentiable and assuming second order terms negligible.

$= f(h_1(c^*)) + f'(g^*)dg$ by (2)

$= c^* + f'(g^*)dg$ by (5)

Now g_n, the new flow value, is found:

$g_n = h_1(c_n)$ by assumption

$= h_1(c^* + f'(g^*)dg)$ from above

$= h_1(c^*) + h'_1(c^*)f'(g^*)dg$ as h is differentiable and assuming second order terms negligible.

$= g^* + h'_1(c^*)f'(g^*)dg$ by (2)

Thus $g^* - g_{n-1} = dg$ by (9)

$> h'_1(c^*)f'(g^*)dg$ by (6) if magnitudes only are considered

$= g^* - g_n$ from above.

(b) Suppose during some cycle of the procedure a flow value g_{n-1} is generated by the traffic assignment process. The use of equation (1) and the branch and bound algorithm generates a new flow value, g_n. If equations (5) and (7) hold, g_n will be closer in magnitude to g* than g_{n-1}. The proof of this fact can be constructed by replacing h_1 by h_2 in part (a).

(c) It has been shown in (a) and (b) that successive values of g, whether generated by the traffic assignment process or by the branch and bound algorithm, are successively closer in magnitude to g*. Because of equation (8), the procedure is declared to have converged when it reaches g* with respect to the one arc under study. The above proof can be repeated for all arcs in the network.

This completes the proof.

Theorem 2.

The branch and bound algorithm produces an optimal solution for given traversal costs.

Proof.

(a) To show that the partitioning routine is exhaustive. Suppose an infeasible solution, P, is specified during the procedure whose value is the smallest of all existing solutions. Let the critical link of P be designated by (i,j). There are three cases:

1. P does not specify the prohibition of either arc (i,j) or arc (j,i).
2. P does specify the prohibition of arc (i,j).
3. P does specify the prohibition of arc (j,i).

(P cannot specify the prohibition of both arcs (i,j) and (j,i) because if this was so there would be no flow in link (i,j) and thus link (i,j) would not be critical).

Then the partitioning routine will replace P by a set of new solutions of the form:

1. P ∪ {prohibit arc (i,j)}
   P ∪ {prohibit arc (j,i)}
   P ∪ {protect link (i,j)}

2. P ∪ {prohibit arc (j,i)}

   P ∪ {protect arc (i,j)}

3. P ∪ {prohibit arc (i,j)}

   P ∪ {protect arc (j,i)}

The purpose of this argument is to show that for any case a feasible solution (to the original problem) which includes P, includes at least one of the set of new solutions above.

Now any feasible solution which includes P must contain one of:

1. P ∪ {prohibit arc (i,j)}

   P ∪ {prohibit arc (j,i)}

   P ∪ {prohibit neither arc (i,j) nor arc (j,i)}

2. P ∪ {prohibit arc (i,j)}

   P ∪ {do not prohibit arc (i,j)}

3. P ∪ {prohibit arc (j,i)}

   P ∪ {do not prohibit arc (j,i)}

There are no other possibilities.

However these are exactly the new solutions outlined. Hence the partitioning routine is exhaustive.

(b) The partitioning routine produces mutually exclusive solutions. Suppose an infeasible solution, P is specified during the procedure whose value is the smallest of all existing solutions. Let the critical link of P be designated by {i,j}. Then the partitioning routine will replace P by a solution with one of the forms given in the previous section. For any of the three cases given previously all the solutions are mutually different. Hence the partitioning routine is mutually exclusive.

(c) The solution routine produces a genuine lower bound. The solution routine produces the actual value of the solution at each node. It would not be possible to reduce this value by introducing additional prohibitions (constraints). Hence the actual value is a genuine lower bound.

(d) The algorithm must terminate in a finite number of steps. Because the network contains a finite number of nodes it must contain a finite number of arcs. Each iteration of the solution procedure reduces the problem of one arc. Hence the algorithm must terminate in a finite number of steps.

This completes the proof.

6. AN APPLICATION

The procedure finds application in traffic engineering. In that situation the vertices and links represents intersection and streets respectively with traversal costs in terms of time or distance and capacities in terms of numbers of vehicles per time unit. The flow demand represents origins and destinations of drivers. Controlling flow by prohibiting travel in an arc is equivalent to preventing travel in one direction of a street; i.e. making a street "one-way". The optimality criterion of total traversal cost is a commonly-used index when measuring congestion which traffic engineers often attempt to minimize.

REFERENCES

[1] R.G. Busacker and T.L. Saaty, Finite Graphs and Networks (McGraw-Hill, New York, 1965).

[2] E. Dijkstra, A Note on Two Problems in Connection with Graphs, Numberische Mathematik 1 (1959) 269-271.

[3] R.W. Floyd, Algorithm 97 - Shortest Path, Comm. ACM 5 (1962) 345.

[4] National Cooperative Highway Research Program Report 58 (1968).

Operations Research
University of Canterbury
Christchurch
New Zealand.

ENUMERATION OF BINARY PHYLOGENETIC TREES

L.R. FOULDS AND R.W. ROBINSON*

Evolutionary trees of biology are represented by a special class of labelled trees, termed phylogenetic *trees. These are characterised by having disjoint subsets of the labelling set assigned to the points of a tree, in such a way that no point of degree less than 3 is assigned an empty set of labels. By a* binary *tree is meant one in which every point has degree 1 or 3. The exact and asymptotic numbers of binary phylogenetic trees are determined under the presence or absence of two additional conditions on the labelling. The optional constraints studied require nonempty label sets to be singletons, and that only endpoints be labelled.*

1. INTRODUCTION

Biologists often generate phylogenetic (evolutionary) trees from protein sequence data; see [2],[3],[4] and [8]. Determination of a common ancestor, or root, is based on separate criteria. It is common for the directed rooted tree so created to be binary. The trees considered here are binary trees, in which every point has degree 1 or 3. An evolutionary tree is labelled with the names of known species. If two species are not distinguished by the protein sequences under study then they are assigned to the same point in the tree. Conversely, it is often convenient to hypothesize common ancestors for whom no direct evidence is known, and these become points to which no name at all is assigned. Of course, a hypothetical point with no name would not have degree 1, because including such a point would serve no purpose in explaining the biological evidence.

As a mathematical model of an evolutionary tree containing n known species we take a binary tree labelled by an assignment of the label set $\{1,\ldots,n\}$ to points so that every endpoint receives at least one label. Such a tree is called a *binary phylogenetic tree*. The number n of labels is termed the *magnitude* of the phylogenetic tree, and the number of points in it is termed the *order*. A *planted* binary tree is one with a distinguished endpoint called the *root*. This corresponds to the common ancestor in an evolutionary tree. Since the root is already distinguished we never assign it a label.

The exact and asymptotic numbers of phylogenetic trees with given magnitude, along with the average and variance of their orders, were determined in [5]. This

* The second author is grateful for the support of the Australian Research Grants Committee for the project "Numerical Implementation of Unlabelled Graph Counting Algorithms", under which research and computing for this paper were performed.

was also accomplished under the restriction to 1-1 labellings. In the present paper we perform the same analysis for binary phylogenetic trees, but in addition, study the effect of restricting the labelling to endpoints. This gives a total of four cases to study. The calculations are facilitated by the particularly simple form of the most restricted case, and the fact that the other three can then be obtained by algebraic transformations of the exponential generating functions.

2. LABELLING RESTRICTED TO ENDPOINTS AND 1-1

The simplest case is the most restricted, being binary phylogenetic trees with 1-1 labelling and no interior points labelled. It is a standard result that the number of such trees with n endpoints is $\prod_{i=0}^{n-3}(2i+1)$. Direct combinatorial proofs of this fact appear in [2,p.241] and [3,p.28]; a more elaborate proof involving generating functions is given in [8,pp.51-52]. In [9,p.72] it is pointed out that this is a corollary of Prüfer's proof of Cayley's result that there are n^{n-2} labelled trees. Here the labelling is 1-1 onto all the points, and the proof gives a 1-1 correspondence between trees and sequences of length n-2 from the label set. It was noted by Prüfer [10] that each point of degree d is represented in its sequence exactly d-1 times. Thus if 1,...,n are reserved for labelling the endpoints of a binary tree and n+1,...,2n-2 for labelling the interior points, then there are $(2n-4)!/2^{n-2}$ sequences in which the latter occur exactly twice each. We simply divide by (n-2)! for the labellings on the interior points, and find that the number with just endpoints labelled is

$$(2n-4)!/(n-2)!2^{n-2} = 1.3 \cdot \ldots \cdot (2n-5),$$

as claimed.

If we denote by T_n the number of binary trees of magnitude n in this case we have, by the above discussion, $T_1 = 0$, $T_2 = 1$, and

$$T_n = \prod_{i=0}^{n-3}(2i+1) \tag{2.1}$$

for $n \geq 3$. It is clear that in a binary tree the number of endpoints exceeds the number of interior points by exactly 2. Thus each binary tree of magnitude n has order 2n-2. We denote by R_n the sum of the orders p of the binary trees of magnitude n, so that

$$R_n = (2n-2)T_n \tag{2.2}$$

for $n \geq 1$. Also, denote by S_n the sum of p(p-1), which gives

$$S_n = (2n-2)(2n-3)T_n \tag{2.3}$$

for $n \geq 1$. In general, R_n and S_n are needed in order to determine the average μ_n

and variance ν_n of the order for the trees of magnitude n. The relations for $n \geq 1$ are:

$$\mu_n = R_n/T_n ,$$

$$\nu_n = S_n/T_n + \mu_n - \mu_n^2 . \qquad (2.4)$$

Of course in this case $\mu_n = 2n-2$ and $\nu_n = 0$, but R_n and S_n will be useful for the three cases considered later.

Let P_n denote the number of planted binary trees of magnitude n. Since the root is not labelled this means there are exactly n+1 endpoints in such a planted tree. If the root point is labelled with the number n+1 and then unrooted, the result is an ordinary binary tree of magnitude n+1. This process gives a 1-1 correspondence, so that

$$P_n = T_{n+1} \qquad (2.5)$$

for $n \geq 1$.

In order to obtain asymptotic estimates we need the exponential generating functions T(x), R(x), S(x) and P(x) for these four sequences. From (2.1) and the binomial theorem we have

$$T(x) = -\frac{1}{3} + x + \frac{1}{3}(1-2x)^{3/2}. \qquad (2.6)$$

By Stirling's formula it can be seen that the coefficient of x^n in $(1-x)^{-s}$ is

$$\frac{n^{s-1}}{\Gamma(s)}\left(1 + \frac{s(s-1)}{2n} + O\left(\frac{1}{n^2}\right)\right) \qquad (2.7)$$

as long as $s \neq 0,-1,-2,\ldots$ Thus

$$T_n = \frac{n!\,2^n}{4\pi^{1/2}n^{5/2}}\left(1 + \frac{15}{8n} + O\left(\frac{1}{n^2}\right)\right). \qquad (2.8)$$

Similarly one has

$$P(x) = 1 - (1-2x)^{1/2}, \qquad (2.9)$$

and

$$P_n = \frac{n!\,2^n}{2\pi^{1/2}n^{3/2}}\left(1 + \frac{3}{8n} + O\left(\frac{1}{n^2}\right)\right). \qquad (2.10)$$

From (2.2) it follows that

$$R(x) = 2xT'(x) - 2T(x),$$

so from (2.6) we have

$$R(x) = \frac{2}{3} - \frac{2}{3}(1+x)(1-2x)^{1/2}. \qquad (2.11)$$

Likewise, (2.3) gives

$$S(x) = 4x^2T''(x) - 6xT'(x) + 6T(x),$$

and combining with (2.6) yields

$$S(x) = -2 + 2(1-x)(1-2x)^{-1/2}. \qquad (2.12)$$

3. LABELLING RESTRICTED TO ENDPOINTS

This case differs from the previous one in allowing more than one label to be assigned to an endpoint. Interior points are still not allowed labels. Thus whereas the exponential generating function for labelling a single endpoint was x, it is now e^x-1. This is because an endpoint may receive 1,2,3,... labels. For any $k \geq 1$ there is just one way to assign k labels to a point. Interleaving of label sets is accounted for in multiplying exponential generating functions together; see [6,Chapter 1] for an account of the uses of exponential generating functions in labelled enumeration. Thus $T(e^x-1)$ is the exponential generating function by magnitude for binary phylogenetic trees in which interior points are not labelled. Similarly $R(e^x-1)$, $S(e^x-1)$ and $P(e^x-1)$ give exponential generating functions for the sum of the order p, the sum of p(p-1) and the number of planted trees, respectively. We denote these generating functions by $\tilde{T}(x)$, $\tilde{R}(x)$, $\tilde{S}(x)$ and $\tilde{P}(x)$.

Substitition of e^x-1 for x in (2.6), (2.11), (2.12) and (2.9) gives

$$\begin{aligned} \tilde{T}(x) &= e^x - \frac{4}{3} + \frac{1}{3}(3-2e^x)^{3/2}, \\ \tilde{R}(x) &= \frac{2}{3} - \frac{2}{3}e^x(3-2e^x)^{1/2}, \\ \tilde{S}(x) &= -2 + (4-2e^x)(3-2e^x)^{-1/2} \\ \tilde{P}(x) &= 1 - (3-2e^x)^{1/2}. \end{aligned} \qquad (3.1)$$

Recurrence relations for $\tilde{T}_n$, $\tilde{R}_n$, $\tilde{S}_n$ and $\tilde{P}_n$ could be deduced directly from these equations. However for numerical purposes it is easier to start with the simple expression for T_n, R_n, S_n and P_n provided in Section 2. From the fact that $(e^x-1)^k/k!$ is the exponential generating function

$$\sum_{n=0}^{\infty} S(n,k)x^n/n!$$

for Stirling numbers of the second kind, it follows that

$$\tilde{T}_n = \sum_{k=2}^{n} S(n,k)T_k \tag{3.2}$$

for $n \geq 2$. The analogous equations hold for $\tilde{R}_n$, $\tilde{S}_n$ and $\tilde{P}_n$. In the latter case the sum must include $k = 1$, and the result is also valid for $n = 1$. Stirling numbers are readily calculated from the recurrence

$$S(n+1,k) = S(n,k-1) + kS(n,k),$$

which holds for $k \geq 1$, and from the boundary conditions $S(0,0) = 1$, $S(0,k) = 0$ if $k > 0$ and $S(n,0) = 0$ if $n > 0$.

Roughly speaking, the asymptotic behaviour of $\tilde{T}_n$ is determined by the radius of convergence $\tilde{\rho}$ of the exponential generating function $\tilde{T}(x)$. From (3.1) it is evident that $\tilde{\rho} = \ln 3/2$, that $\tilde{\rho}$ is also the radius of convergence of $\tilde{R}(x)$, $\tilde{S}(x)$ and $\tilde{P}(x)$, and that in each case the point $x = \tilde{\rho}$ is the sole singularity on the circle of convergence. It is then classical (see [1,Theorem 4] or [7,p.489]) that an expansion of the generating function in powers of $(1-x/\tilde{\rho})^{\frac{1}{2}}$ can be used in conjunction with (2.7) to determine the precise asymptotic growth rate of the coefficients. The first two odd powers are sufficient to give the nth coefficient with a factor of $(1+O(\frac{1}{n^2}))$.

Because $e^{\tilde{\rho}} = 3/2$ we have

$$3 - 2e^x = 3\tilde{\rho}(1-x/\tilde{\rho})(1 - \tfrac{\tilde{\rho}}{2}(1-x/\tilde{\rho})\pm\ldots),$$

and so from (3.1)

$$\tilde{T}(x) = 3^{1/2}\tilde{\rho}^{3/2}(1-x/\tilde{\rho})^{3/2} - \frac{3^{3/2}\tilde{\rho}^{5/2}}{4}(1-x/\tilde{\rho})^{5/2} \pm\ldots$$

Summing the contributions of these two terms according to (2.7) yields

$$\tilde{T}_n = \frac{3^{3/2}\tilde{\rho}^{3/2}}{4\pi^{1/2}} \cdot \frac{n!}{n^{5/2}\tilde{\rho}^n}\left(1 + \frac{15(1+\tilde{\rho})}{8n} + O(\tfrac{1}{n^2})\right). \tag{3.3}$$

In the same fashion, the other three expressions in (3.1) can be expanded, with the following results:

$$\tilde{R}_n = \frac{3^{1/2}\tilde{\rho}^{1/2}}{2\pi^{1/2}} \cdot \frac{n!}{n^{3/2}\tilde{\rho}^n}\left(1 + \frac{3(1+5\tilde{\rho})}{8n} + O(\tfrac{1}{n^2})\right); \tag{3.4}$$

$$\tilde{S}_n = \frac{3^{-1/2}\tilde{\rho}^{-1/2}}{\pi^{1/2}} \quad \frac{n!}{n^{1/2}\tilde{\rho}^n}\left(1 - \frac{1+13\tilde{\rho}}{8n} + O(\tfrac{1}{n^2})\right); \tag{3.5}$$

$$\tilde{P}_n = \frac{3^{1/2}\tilde{\rho}^{1/2}}{2\pi^{1/2}} \quad \frac{n!}{n^{3/2}\tilde{\rho}^n}\left(1 + \frac{3(1+\tilde{\rho})}{8n} + O(\tfrac{1}{n^2})\right). \tag{3.6}$$

The mean $\tilde{\mu}_n$ and variance $\tilde{\nu}_n$ of the number of points in the binary trees of magnitude n in this case are given by the obvious analogue of (2.4)

$$\tilde{\mu}_n = \tilde{R}_n/\tilde{T}_n,$$
$$\tilde{\nu}_n = (\tilde{S}_n/\tilde{T}_n) + \tilde{\mu}_n - \tilde{\mu}_n^2 . \qquad (3.7)$$

Thus (3.3), (3.4) and (3.5) can be immediately applied, resulting in

$$\tilde{\mu}_n = \frac{2n}{3\tilde{\rho}}\left(1 - \frac{3}{2n} + O\left(\frac{1}{n^2}\right)\right)$$
$$\tilde{\nu}_n = \frac{4(1-2\tilde{\rho})n}{9\tilde{\rho}^2}\left(1 + O\left(\frac{1}{n}\right)\right). \qquad (3.8)$$

4. 1-1 LABELLING

This case differs from the first case in allowing interior points to be labelled. Let $\overline{T}_n$ denote the number of trees of magnitude n under this labelling convention. Likewise, let $\overline{R}_n$ and $\overline{S}_n$ denote the totals of the order p and of p(p-1) respectively, over these $\overline{T}_n$ trees. Finally, let $\overline{P}_n$ be the number of planted trees of magnitude n. As usual, we denote the exponential generating functions of these four sequences by $\overline{T}(x)$, $\overline{R}(x)$, $\overline{S}(x)$ and $\overline{P}(x)$.

Since labelling is optional for interior points, and at most one label can be assigned to each, the exponential generating function of the labelling possibilities for an interior point is $1+x$. For a tree with n endpoints there are $n-2$ interior points. Each endpoint is labelled, so labelling possibilities for an endpoint has x as its exponential generating function. Thus each 1-1 labelled basic tree with magnitude n and only endpoints labelled gives rise to a number of compatible versions in which interior points may be labelled, and these have $x^n(1+x)^{n-2}$ as exponential generating function. Summing over all T_n basic trees and then over all $n \geq 2$, this gives

$$\overline{T}(x) = T(x+x^2)/(1+x)^2. \qquad (4.1)$$

It is now easy to obtain a recurrence for $\overline{T}_n$. Putting the equation in the form

$$\overline{T}(x) = -2x\overline{T}(x) - x^2\overline{T}(x) + T(x+x^2)$$

and comparing coefficients of $x^n/n!$ yields

$$\overline{T}_n = -2n\overline{T}_{n-1} - n(n-1)\overline{T}_{n-2} + \sum_{k=0}^{\lfloor n/2 \rfloor} \binom{n}{k}\binom{n-k}{k} k!\,T_{n-k} \qquad (4.2)$$

for $n \geq 2$. Here $T_0 = 0$ and $T_1 = 0$ are needed as boundary conditions. Exactly the same transformation gives $\overline{R}(x)$ from $R(x)$ and $\overline{S}(x)$ from $S(x)$, so recurrence relations analagous to (4.2) are valid for $\bar{R}_n$ and $\bar{S}_n$.

In a planted tree the root is an endpoint which is not labelled, so with n labelled endpoints there are n - 1 interior points which might be labelled. This gives

$$\bar{P}(x) = P(x+x^2)/(1+x), \tag{4.3}$$

and

$$\bar{P}_n = -n\bar{P}_{n-1} + \sum_{k=0}^{\lfloor n/2 \rfloor} \binom{n}{k}\binom{n-k}{k} k! P_{n-k} \tag{4.4}$$

for $n \geq 2$ with $\bar{P}_1 = 1$.

Explicit expressions for the exponential generating functions can be found at once from (2.6), (2.9), (2.11) and (2.12):

$$\begin{aligned}
\bar{P}(x) &= \frac{1-(1-2x-2x^2)^{1/2}}{1+x}; \\
\bar{T}(x) &= \frac{-1+3x+3x^2+(1-2x-2x^2)^{3/2}}{3(1+x)^2}; \\
\bar{R}(x) &= \frac{2-2(1+x+x^2)(1-2x-2x^2)^{1/2}}{3(1+x)^2}; \\
\bar{S}(x) &= \frac{-2+2(1-x-x^2)(1-2x-2x^2)^{-1/2}}{(1+x)^2}.
\end{aligned} \tag{4.5}$$

In each case the radius of convergence is $\bar{\rho} = (\sqrt{3}-1)/2$, and $x = \bar{\rho}$ is the sole singularity on the circle of convergence. We have $1+\bar{\rho} = 1/2\bar{\rho}$, so that

$$1 + x = \frac{1}{2\bar{\rho}}(1-2\bar{\rho}(\bar{\rho}-x))$$

and

$$1 - 2x - 2x^2 = 2\sqrt{3}(\bar{\rho}-x)(1 - \frac{1}{\sqrt{3}}(\bar{\rho}-x)).$$

Substituting into (4.5), one finds the first two odd powers of $(1-x/\bar{\rho})^{\frac{1}{2}}$. Finally, (2.7) is applied, to give the following asymptotic estimates:

$$\begin{aligned}
\bar{P}_n &= \frac{2^{1/2}3^{1/4}\bar{\rho}^{-3/2}}{\pi^{1/2}} \frac{n!}{n^{3/2}\bar{\rho}^{-n}} \left(1 + \frac{11\sqrt{3}-18}{8n} + O(\tfrac{1}{n^2})\right); \\
\bar{T}_n &= \frac{2^{1/2}3^{3/4}(2-\sqrt{3})\bar{\rho}^{-3/2}}{\pi^{1/2}} \frac{n!}{n^{5/2}\bar{\rho}^{-n}} \left(1 + \frac{5(7\sqrt{3}-10)}{8n} + O(\tfrac{1}{n^2})\right); \\
\bar{R}_n &= \frac{2^{1/2}3^{1/4}(2-\sqrt{3})\bar{\rho}^{-1/2}}{\pi^{1/2}} \frac{n!}{n^{3/2}\bar{\rho}^{-n}} \left(1 + \frac{19\sqrt{3}-30}{8n} + O(\tfrac{1}{n^2})\right); \\
\bar{S}_n &= \frac{2^{1/2}3^{1/4}(2-\sqrt{3})\bar{\rho}^{-1/2}}{\pi^{1/2}} \frac{n!}{n^{1/2}\bar{\rho}^{-n}} \left(1 - \frac{90-37\sqrt{3}}{24n} + O(\tfrac{1}{n^2})\right).
\end{aligned} \tag{4.6}$$

The mean $\bar{\mu}_n$ and the variance $\bar{\nu}_n$ of the order for trees of magnitude n are found from $\bar{T}_n$, $\bar{R}_n$ and $\bar{S}_n$ just as in (3.7) for the previous case. Asymptotic estimates then follow from (4.6);

$$\bar{\mu}_n = \frac{n}{\sqrt{3}\,\bar{\rho}}\left(1 - \frac{4\sqrt{3}-5}{2n} + O\left(\frac{1}{n^2}\right)\right),$$

and

$$\bar{\nu}_n = \frac{(2\sqrt{3}-3)n}{9\bar{\rho}^2}\left(1 + O\left(\frac{1}{n}\right)\right). \tag{4.7}$$

The recurrence relations (4.2) and (4.3) for $\overline{T}_n$ and $\overline{P}_n$ require $O(n^2)$ arithmetic operations to compute the values up to n, even given that T_k is available already for $k \leq n$. Improved recurrences can be obtained directly from (4.5) by differentiating the explicit expressions for the generating functions, simplifying and then comparing coefficients of $x^n/n!$. In this way one finds:

$$\overline{P}_n = (n-3)\overline{P}_{n-1} + (n-1)(4n-9)\overline{P}_{n-2} + 2(n-1)(n-2)(n-3)\overline{P}_{n-3} \tag{4.8}$$

for $n \geq 3$, with $\overline{P}_0 = 0$, $\overline{P}_1 = 1$ and $\overline{P}_2 = 1$;

$$\overline{T}_n = \overline{P}_{n-1} + 2(n-1)\,\overline{P}_{n-2} - (n+1)\,\overline{T}_{n-1} \tag{4.9}$$

for $n \geq 3$, with $\overline{T}_2 = 1$;

$$\overline{R}_n = -n\overline{R}_{n-1} + \frac{2}{3}(n(n-1)\overline{P}_{n-2} + n\overline{P}_{n-1} + \overline{P}_n) \tag{4.10}$$

for $n \geq 3$, with $R_2 = 2$;

$$\overline{S}_n = n\overline{S}_{n-1} + 4n(n-1)\overline{S}_{n-2} + 2n(n-1)(n-2)\overline{S}_{n-3} - 2(\overline{P}_n - n\overline{P}_{n-1} - n(n-1)\overline{P}_{n-2}) \tag{4.11}$$

for $n \geq 3$, with $\overline{S}_2 = 2$. These relations only require $O(n)$ arithmetic operations in order to calculate values of $\overline{P}_k$, $\overline{T}_k$, $\overline{R}_k$ or $\overline{S}_k$ for $k \leq n$.

5. UNRESTRICTED LABELLING

The final case allows all binary labelled trees, including the possibility of multiple labels and labels for interior points. As for any phylogenetic trees, it is still the case that each endpoint must be assigned at least one label. This differs from the previous case only in allowing multiple labels, so the relation of this section to the previous section is exactly the same as the relation of Section 3 to Section 2. We denote the number of trees of magnitude n by $\hat{T}_n$, and the number of planted trees by $\hat{P}_n$. Similarly, the sum of the order p and the sum of p(p-1) for magnitude n trees are denoted $\hat{R}_n$ and $\hat{S}_n$. The exponential generating functions are $\hat{T}(x)$, $\hat{P}(x)$, $\hat{R}(x)$ and $\hat{S}(x)$ respectively. These are obtained from $\overline{T}(x)$, $\overline{P}(x)$, $\overline{R}(x)$ and $\overline{S}(x)$ by replacing x with $e^x - 1$. Thus the exact numbers are related by

$$\hat{T}_n = \sum_{k=2}^{n} S(n,k)\overline{T}_k \tag{5.1}$$

for $n \geq 2$, which is similar to (3.2). Of course $\hat{R}_n$, $\hat{S}_n$ and $\hat{P}_n$ are calculated analogously from the corresponding numbers determined in the previous section. In the case of $\hat{P}_n$ the sum starts at $k = 1$ and the result is valid for $n = 1$.

To obtain the exponential generating functions explicitly one need only substitute $e^x - 1$ for x in (4.5). The results are:

$$\begin{aligned}
\hat{P}(x) &= e^{-x} - e^{-x}(1+2e^x-2e^{2x})^{1/2};\\
\hat{T}(x) &= 1 - e^{-x} - \tfrac{1}{3}e^{-2x} + \tfrac{1}{3}e^{-2x}(1+2e^x-2e^{2x})^{3/2};\\
\hat{R}(x) &= \tfrac{2}{3}e^{-2x} - \tfrac{2}{3}e^{-2x}(1-e^x+e^{2x})(1+2e^x-2e^{2x})^{1/2};\\
\hat{S}(x) &= -2e^{-2x} + 2e^{-2x}(1+e^x-e^{2x})(1+2e^x-2e^{2x})^{-1/2}.
\end{aligned} \tag{5.2}$$

In each of these generating functions the radius of convergence is $\hat{\rho} = \ln((\sqrt{3}+1)/2)$, and $x = \hat{\rho}$ is the only singularity on the circle of convergence. As in the previous three sections we expand the generating functions in terms of $(1-x/\hat{\rho})^{\frac{1}{2}}$, and apply (2.7) to the first two odd powers. The asymptotic estimates so obtained are:

$$\begin{aligned}
\hat{P}_n &= \frac{(\sqrt{3}-1)(3+\sqrt{3})^{1/2}\hat{\rho}^{1/2}}{2\pi^{1/2}} \cdot \frac{n!}{n^{3/2}\hat{\rho}^n}\left(1 + \frac{3-\hat{\rho}(6-\sqrt{3})}{8n} + O(\tfrac{1}{n^2})\right);\\
\hat{T}_n &= \frac{3^{1/2}(3-\sqrt{3})^{1/2}\hat{\rho}^{3/2}}{2^{1/2}\pi^{1/2}} \cdot \frac{n!}{n^{5/2}\hat{\rho}^n}\left(1 + \frac{5(3-\hat{\rho}(2-\sqrt{3}))}{8n} + O(\tfrac{1}{n^2})\right);\\
\hat{R}_n &= \frac{(2-\sqrt{3})(3+\sqrt{3})^{1/2}\hat{\rho}^{1/2}}{\pi^{1/2}} \cdot \frac{n!}{n^{3/2}\hat{\rho}^n}\left(1 + \frac{3+\hat{\rho}(5\sqrt{3}-6)}{8n} + O(\tfrac{1}{n^2})\right);\\
\hat{S}_n &= \frac{2(2-\sqrt{3})(3+\sqrt{3})^{-1/2}\hat{\rho}^{-1/2}}{\pi^{1/2}} \cdot \frac{n!}{n^{1/2}\hat{\rho}^n}\left(1 - \frac{3+\hat{\rho}(66+13\sqrt{3})}{24n} + O(\tfrac{1}{n^2})\right).
\end{aligned} \tag{5.3}$$

Now the mean $\hat{\mu}_n$ and the variance $\hat{\sigma}_n$ of the order for trees of magnitude n depend on $\hat{T}_n$, $\hat{R}_n$ and $\hat{S}_n$ as in (3.7). From the asymptotic estimates above one then calculates

$$\hat{\mu}_n = \frac{2(2-\sqrt{3})n}{\hat{\rho}(3-\sqrt{3})}\left(1 - \frac{3-\hat{\rho}}{2n} + O(\tfrac{1}{n^2})\right),$$

and

$$\hat{\sigma}_n = \frac{2n}{9\hat{\rho}^2}(6-(3+\hat{\rho})\sqrt{3})(1+O(\tfrac{1}{n})). \tag{5.4}$$

6. NUMERICAL RESULTS

The values of P_n, $\tilde{P}_n$, $\bar{P}_n$ and $\hat{P}_n$ for $1 \leq n \leq 10$ and $n = 15,20,25,30,35$ and 40 are presented in Table 1. The corresponding values of T_n, $\tilde{T}_n$, $\bar{T}_n$ and $\hat{T}_n$ are presented in Table 2, those of μ_n, $\tilde{\mu}_n$, $\bar{\mu}_n$ and $\hat{\mu}_n$ in Table 3, and those of $\tilde{\nu}_n$, $\bar{\nu}_n$ and $\hat{\nu}_n$ in Table 4. The full range of values for $1 \leq n \leq 40$ in all cases is available from the second author. The calculations are based on the equations giving recurrences for P_n, T_n, R_n, S_n, $\tilde{P}_n$, $\tilde{T}_n$, $\tilde{R}_n$, $\tilde{S}_n$, $\bar{P}_n$, $\bar{T}_n$, $\bar{R}_n$, $\bar{S}_n$, $\hat{P}_n$, $\hat{T}_n$, $\hat{R}_n$ and $\hat{S}_n$ in the previous four sections. Then μ_n and ν_n are computed from T_n, R_n and S_n by (2.4), and $\tilde{\mu}_n$, $\tilde{\nu}_n$, $\bar{\mu}_n$, $\bar{\nu}_n$, $\hat{\mu}_n$ and $\hat{\nu}_n$ are found in the same way. Since $\nu_n = 0$ for all n, those values are omitted.

Asymptotic estimates for all of these quantities are derived in the preceeding four sections. In Table 5 the relative errors of the estimates are presented. If E is an estimate for the quantity Q, we define the relative error to be (E-Q)/Q. The estimates for P_n, T_n, μ_n, $\tilde{P}_n$, $\tilde{T}_n$, $\tilde{\mu}_n$, $\bar{P}_n$, $\bar{T}_n$, $\bar{\mu}_n$, $\hat{P}_n$, $\hat{T}_n$ and $\hat{\mu}_n$ are to second order in 1/n, so that the relative errors are $O(1/n^2)$. The estimates for $\hat{\nu}_n$, $\bar{\nu}_n$ and $\tilde{\nu}_n$ are only to first order in 1/n, since the leading terms added out exactly, and so the relative errors are O(1/n). Since the estimate for μ_n is exact the relative error is always zero, and those values have been omitted.

The computations were programmed on a PDP11/45 by A. Nymeyer while employed under an A.R.G.C. grant. Multiple precision integer arithmetic was employed for the exact results, so no errors should have been introduced by arithmetic operations in the course of the computations.

n	P_n / $\tilde{P}_n$ / $\bar{P}_n$ / $\hat{P}_n$
2	1 2 1 2
3	3 7 6 10
4	15 41 39 83

Table 1. Exact Numbers of Planted Trees

Value	n
105	5
346	
390	
946	
945	6
3797	
4815	
13772	
10395	7
51157	
73080	
2 44315	
1 35135	8
8 16356	
13 04415	
51 13208	
20 27025	9
150 50581	
268 47450	
1233 42166	
344 59425	10
3147 26117	
6255 28575	
33695 68817	
21345 80466 76875	15
5 52134 63465 43307	
18 60930 96068 88000	
222 49860 76835 28550	
82 00794 53263 78915 59375	20
6030 06160 20371 20901 60876	
34162 88904 24238 71173 19375	
9 07848 89714 77640 85142 98308	
1 19256 81927 74434 12353 99076 40625	25
249 60937 77924 41570 77429 36703 12521	
2369 38660 81672 49563 88051 41829 06250	
1 40020 31727 81284 38254 77005 19138 11921	
4951 79769 00801 98183 90136 61171 60891 40625	30
29 52055 00537 50105 69454 33962 29379 21916 46717	
468 78078 22176 09468 90311 78233 86499 38771 09375	
61622 23725 36992 56577 22418 48868 47524 62475 40292	
488 96013 03686 63401 54392 27834 73071 78464 62136 71875	35
8 30555 12190 54585 07530 50112 37947 61629 76250 03010 43432	
220 45043 83215 09719 97185 12351 30927 75473 42274 65921 87500	
64469 94776 42821 80627 78241 11507 01198 15512 97655 72710 58200	
100 98473 64737 86927 09053 02433 22159 25040 62302 66320 27246 09375	40
4 88849 59488 48468 12925 66057 13518 13962 35628 62289 55846 36023 59701	
216 76354 20265 56596 78464 48363 20829 15417 44296 91092 26936 17246 09375	
1 41043 29642 04081 01696 47860 11232 21536 87652 57986 44515 59972 16529 95683	

Table 1 concluded

n	T_n	$\tilde{T}_n$	$\bar{T}_n$	$\hat{T}_n$
2	1	1	1	1
3	1	4	1	4
4	3	16	7	20
5	15	85	45	155
6	105	646	465	1716
7	945	6664	5775	24654
8	10395	86731	88515	4 34155
9	1 35135	13 54630	15 88545	90 43990
10	20 27025	246 07816	328 52925	2174 57456
15	790 58535 80625	26178 52548 56584	62467 65138 36375	8 98476 50088 10504

Table 2. Exact Numbers of Free Trees

2 21643 09547 66997 71875	20
205 49784 36233 45183 15851	
842 45483 44899 82007 04375	
26682 69789 26535 31333 74235	
2537 37913 35626 25794 76576 09375	25
6 64178 57351 59954 30396 13789 93560	
46 17076 71760 51374 73244 85886 40625	
3235 11311 52182 42235 34626 34054 77330	
86 87364 36856 17511 99826 95810 02822 65625	30
64436 62064 76134 50310 45988 54233 29473 09166	
7 55042 77813 17552 31463 00849 57120 96622 65625	
1172 90718 32815 25159 49765 43242 31672 39084 02156	
7 29791 23935 62140 32155 10863 20493 60872 60628 90625	35
15368 80705 80797 59840 80975 60737 16744 65780 96218 14259	
3 02580 48379 86006 84198 80394 40559 05889 76656 22199 21875	
1043 30147 98740 01154 14781 09291 78957 78854 81627 17151 95629	
1 31149 00840 75154 89727 96135 49638 43182 34575 35926 23730 46875	40
7850 74372 68277 04558 58143 14560 59876 66156 89466 40712 27350 48676	
2 59200 36605 50234 64023 20968 12167 26967 31842 95195 74464 69511 71875	
1985 12629 86757 57145 96126 26837 39795 87614 49844 23808 78544 49936 30160	

Table 2 concluded

μ_n	$\tilde{\mu}_n$	$\bar{\mu}_n$	$\hat{\mu}_n$	n
2	2.00000 00000 00000	2.00000 00000 00000	2.00000 00000 00000	2
4	2.50000 00000 00000	4.00000 00000 00000	2.50000 00000 00000	3
6	3.50000 00000 00000	4.85714 28571 42857	3.60000 00000 00000	4
8	5.05882 35294 11765	6.66666 66666 66667	4.96774 19354 83871	5
10	6.87616 09907 12074	8.06451 61290 32258	6.34149 18414 91841	6
12	8.67647 05882 35294	9.67272 72727 27273	7.70017 03577 51278	7
14	10.4124 93802 67724	11.2099 64412 81139	9.05386 78582 53389	8
16	12.1104 72970 47902	12.7804 87804 87805	10.4063 90321 08616	9
18	13.7896 80969 65615	14.3440 24162 23213	11.7588 57383 11957	10
28	22.0911 28106 38671	22.1987 85618 08196	18.5248 31560 70087	15
38	30.3436 73372 50291	30.0699 42364 27419	25.2950 68606 39104	20
48	38.5816 66700 12733	37.9475 81912 66192	32.0672 91540 57386	25
58	46.8133 11991 00284	45.8283 62637 25108	38.8405 35372 58541	30
68	55.0416 12908 97430	53.7109 00716 17330	45.6143 66223 76070	35
78	63.2679 34337 78013	61.5945 20432 18619	52.3885 64087 49734	40

Table 3. Mean Order of Trees with Fixed Magnitude

$\tilde{\nu}_n$	$\bar{\nu}_n$	$\hat{\nu}_n$	n
0.00000 00000 00000	0.00000 00000 00000	0.00000 00000 00000	2
0.75000 00000 00000	0.00000 00000 00000	0.75000 00000 00000	3
2.25000 00000 00000	0.97959 18367 34694	1.84000 00000 00000	4
3.82006 92041 52248	0.88888 88888 88888	2.54734 65140 47868	5
4.55122 73672 70844	1.67325 70239 33404	3.12114 55599 56749	6
4.82810 18289 66879	1.85652 89256 19834	3.71248 71941 46887	7
5.13028 35400 88531	2.35093 27389 47078	4.31499 16329 39918	8
5.52204 49273 73383	2.69839 38132 06426	4.92129 87937 60629	9
5.96404 81520 58989	3.10673 00739 16410	5.52838 11503 75196	10
8.38429 32248 87521	5.04591 99744 39442	8.55839 33406 31767	15
10.8925 61215 87343	6.98141 89245 42325	11.5800 79720 01440	20
13.4236 40224 09535	8.91280 60113 55968	14.5975 78794 19007	25
15.9642 03499 83919	10.8420 07039 11216	17.6128 72291 44108	30
18.5096 36929 00962	12.7699 13435 93193	20.6268 81592 54253	35
21.0579 06897 65325	14.6969 92847 02091	23.6400 81773 08085	40

Table 4. Variance of the Order of Trees with Fixed Magnitude

Quantity	10	20	30	40
P_n	-.00198255	-.000491952	-.000218101	-.000122529
$\tilde{P}_n$	-.00566099	-.00134128	-.000586665	-.000327466
$\bar{P}_n$	.00130770	.000311656	.000136469	.0000761654
$\hat{P}_n$	.000919071	.000202635	.0000861797	.0000474024
T_n	-.0290372	-.00739040	-.00330385	-.00186380
$\tilde{T}_n$	-.0716409	-.0173360	-.00765566	-.00429481
$\bar{T}_n$	-.0107192	-.00277488	-.00125092	-.000708669
$\hat{T}_n$	-.0296345	-.00734383	-.00326257	-.00183542
$\tilde{\mu}_n$	.0134911	.00244101	.000992320	.000535098
$\bar{\mu}_n$	-.00636135	-.00145205	-.000623642	-.000344765
$\hat{\mu}_n$	-.00251239	-.000600032	-.000259226	-.000143563
$\tilde{\nu}_n$	-.142979	-.0615036	-.0394795	-.0290935
$\bar{\nu}_n$	.238924	.102642	.0650247	.0475617
$\hat{\nu}_n$	.0892020	.0399797	.025646	.0188668

Table 5. Relative Error of Asymptotic Estimates

7. RELATED RESULTS

In [5] the numbers of phylogenetic trees were studied with no restriction on the degrees of the points. There it was noted that the mean and the variance of the order were both $O(n)$. Therefore as $n \to \infty$ the distribution of orders in trees of magnitude n becomes gradually sharper as a percentage of mean value. This is also true of all four cases considered in Sections 2-5.

The methods of the present paper have been applied to other classes of trees which are relevant to the formation of phylogenetic diagrams in biology. These classes are determined by applying certain combinations of the following conditions: no points of degree 2 are allowed; no interior points are labelled; the labelling is 1-1. It is planned to present those results elsewhere.

REFERENCES

[1] E.A. Bender, Asymptotic methods in enumeration, *SIAM Rev.* 16 (1974), 485-515.

[2] L.L. Cavalli-Sforza and A.W.F. Edwards, Phylogenetic analysis. Models and estimation procedures, *Amer. J. Human Genet.* 19 (1967), 233-257 and *Evolution* 21 (1967), 550-570.

[3] J. Felsenstein, The numbers of evolutionary trees, *Syst. Zool.* 27 (1978), 27-33.

[4] L.R. Foulds, David Penny and M.D. Hendy, A Graph Theoretic Approach to the Development of Minimal Phylogenetic Trees, *J. Mol. Evol.* 13 (1979), 127-150.

[5] L.R. Foulds and R.W. Robinson, Determining the asymptotic numbers of phylogenetic trees. *Combinatorial Mathematics VII*, Lecture Notes in Mathematics 829 (Springer, Berlin, 1980), 110-126.

[6] F. Harary and E.M. Palmer, *Graphical Enumeration* (Academic Press, New York, 1973).

[7] F. Harary, R.W. Robinson and A.J. Schwenk, Twenty step algorithm for determining the asymptotic number of trees of various species, *J. Austral. Math. Soc. Ser. A* 20 (1975), 483-503.

[8] E.F. Harding, The probabilities of rooted tree-shapes generated by random bifurcation, *Adv. Appl. Prob.* 3 (1971), 44-77.

[9] J.W. Moon, Various proofs of Cayley's formula for counting trees. *A Seminar on Graph Theory* (F. Harary, ed, Holt, Rinehart and Winston, New York, 1967), 70-78.

[10] H. Prüfer, Neuer Beweis eines Satzes über Permutationen, *Arch. Math. Phys.* 27 (1918), 742-744.

Operations Research
University of Canterbury
Christchurch
New Zealand

Department of Mathematics
University of Newcastle
New South Wales 2308

MINIMISATION OF MULTIPLE ENTRY FINITE AUTOMATA

W. HAEBICH AND J-L. LASSEZ

A multiple entry finite automaton (mefa) can be viewed as a set of finite automata acting in parallel but in a compacted form. Mefas are defined in a similar manner to finite automata except that any state can be initial. Unlike finite automata, they cannot be minimised in a unique way. We show that the usual minimisation process applied to mefas is unnecessarily weak. We propose a more natural alternative. This solves a current problem and provides a unique (in a restricted sense) minimal structure.

1. Introduction

In this note we generalise the minimisation process for finite automata (fa's) to (connected) multiple entry finite automata (mefa's) in a new way. Much effort is being directed towards investigating the properties of Moore reduced mefas. (See Gill and Veloso [4] and Valk [2]). We show that the concept of Moore reduced is not the most useful or natural technique for mefa minimisation.

Mefa's differ from fa's in that they do not have a *single* initial state. A word is recognised by a mefa if it takes *any state* to a final state (see Definition 2.1). These structures were first defined and investigated by A. Gill and L-T. Kou in 1974 [1] with subsequent work by P.A.S. Veloso [4] and others.

Of the fa's which recognise a given language there exists one, unique up to isomorphism, which has the least possible number of states and is also a homomorphic image of all the others. It is naturally called the minimal fa with respect to the language.

The situation is far more complicated for mefa's. To make sense we must distinguish between two potentially different types of minimality, which happen to coincide for fa's. A machine is combinatorially minimal (c-minimal) if, of all the machines recognising a language, it has the minimum possible number of states. It is algebraically minimal (a-minimal) if it has no proper homomorphic image which recognises the same language. Note that this notion depends on the particular definition of homomorphism used.

According to the definition which has been employed so far, 'Moore reduced' is 'a-minimal' (Veloso [3] lemma 2.2(c)).

Gill and Kou (Theorem 2) give examples of

1. a pair of non-isomorphic c-minimal mefa's, and
2. Moore reduced mefa which is not c-minimal.

Veloso [3], shows that the situation is even worse by presenting, for each finite alphabet, an infinite family of non-isomorphic, Moore reduced mefa's, each recognising the universal language on the alphabet. It is obvious that the mefa consisting of a single state (which is final) is c-minimal in this case.

We demonstrate that this odd effect arises because the definition of mefa homomorphism [1], [3], which is adopted from the classical fa homomorphism is too restrictive. We propose more general definitions of homomorphism under which this problematic infinite family collapses to the trivial mefa (see homomorphism, L-homomorphism; Definitions 2.2, 3.3). *From now on the term a-minimal will relate to L-homomorphism.* Section 5 gives an example to show that an a-minimal mefa is not necessarily c-minimal and a further example that Moore reduced does not imply a-minimality.

Relationships between congruences and homomorphisms are established analogously to those for fa's. Direct products are then introduced to provide a framework for a *unique* minimisation process similar to but more restricted in scope than the fa case.

2. Basic Definitions

2.1 Definition (a) *A mefa is a quadruple* $M = (A, S, \Delta, F)$, *where* A *is a finite non-empty set, the alphabet* S *is a finite non-empty set, the set of states,* F *is a subset of* S, *the final states,* Δ *is a mapping from* A *to the set of maps from* S *to* S.

(b) $A(M)$, *the language recognised by* M, *is* $\{u \in A^* | s(u\Delta) \in F$ *for some* $s \in S\}$.

The set of maps from S to S (S^S) forms a monoid under map composition. A^*, the set of words on A is also a monoid under concatenation. Δ can be extended to a monoid homomorphism from A^* to S^S. That is, $(uv)\Delta = (u\Delta) \circ (v\Delta)$ where $u,v \in A^*$, and 'o' denotes composition. When the meaning is unambiguous we will abbreviate $s(u\Delta)$ to su, $s \in S$, $u \in A^*$.

Our definition of mefa homomorphism is crucial to what follows. The Universal Algebra concept of a homomorphism is that it is a mapping which preserves algebraic structure. Given two mefa's $M = (A, S, \Delta, F)$ and $M' = (A, S', \Delta', F')$, for a map $\lambda: S \to S'$ to be a homomorphism it should satisfy $(s(u\Delta))\lambda = (s\lambda)(u\Delta')$ or, more loosely, $(su)\lambda = (s\lambda)u$. However, as well as the action of A or S, the embedding of F in S is part of the structure of M. To complete the definition of a homomorphism, it would seem reasonable to restrict $F\lambda$ to lie in F'. The adequacy of these two conditions is supported by the fact that $(A, S\lambda, \Delta', F\lambda)$ can be then made into a mefa in a natural way.

2.2 Definition *If* $M = (A, S, \Delta, F)$ *and* $M' = (A, S', \Delta', F')$ *are mefa's then a map* $\lambda: S \to S'$ *is a mefa homomorphism, iff* $(s\lambda)(u\Delta) = (s(u\Delta'))\lambda \ \forall s \in S, u \in A^*$ *and* $F\lambda \subseteq F'$.

The classical definition referred to in the introduction is equivalent to insisting, in addition to the above, that $s\lambda \in F'$ only if $s \in F$. This is a special case of our

L-homomorphism below.

3. The Construction of a-Minimal Mefa's

Firstly we need to establish a relationship between congruences and homomorphisms. From now on, M will be a mefa, (A, S, Δ, F).

3.1 Definition *A congruence, θ, on* M *is an equivalence relation on* S *such that,* $s_1 \theta s_2$ *implies* $(s_1u)\ \theta\ (s_2u)$ *for all* $s_1, s_2 \in S$, $u \in A^*$.

3.2 Proposition *There is a 1-1 correspondence between the congruences on* M *and the homomorphisms of* M.

Proof Let θ be a congruence on M. Consider its equivalence classes, [s], s ε S. By definition if $[s_1] = [s_2]$ then $[s_1u] = [s_2u]$. The relation $a\Delta_\theta$: $[s] \to [su]$, $a \in A$ is therefore a mapping on $S_\theta = \{[s] \mid s \in S\}$. If $F_\theta = \{[f] \mid f \in F\}$ then these maps make $M_\theta = (A, S_\theta, \Delta_\theta, F_\theta)$ into a mefa. It is straightforward to check that

$\Lambda(\theta)$: $s \to [s]$, $s \in S$ is a homomorphism.

Conversely if $\lambda: M \to M'$ is a homomorphism the equivalence relation $\Theta(\lambda)$ given by $s_1\ \Theta(\lambda)\ s_2$ iff $s_1\lambda = s_2\lambda$ is a congruence. Moreover, $\Lambda(\Theta(\lambda)) = \lambda$.

For any homomorphism $\lambda: M \to M$, $A(M) \subseteq A(M\lambda)$. Which homomorphic images of a mefa recognise exactly the same language? In other words when does $A(M) = A(M\lambda)$? This can be expressed as a condition on the congruence for λ. Let $\tilde{A}(M)$ denote the set complement of A(M) in A* and $S\tilde{A}(M)$ have the obvious meaning as $\{su \mid s \in S,\ u \in \tilde{A}(M)\}$. Furthermore if θ is the congruence of λ and [s] the equivalence class of s under θ, let <F> denote $\bigcup_{f \in F} [f]$.

3.3 Proposition $A(M) = A(M\lambda)$ *if and only if* $S\tilde{A}(M) \cap \langle F \rangle = \emptyset$.

Proof Firstly note that $u \in A(M\lambda)$ *if and only if* $Su \cap \langle F \rangle \neq \emptyset$. That is $u \notin A(M)$ *if and only if* $Su \cap \langle F \rangle = \emptyset$.

If $A(M) = A(M\lambda)$ *then* $u \notin A(M)$ implies $u \notin A(M\lambda)$ and hence $su \cap \langle F \rangle = \emptyset$. Conversely if $Su \cap \langle F \rangle = \emptyset$ for all $u \notin A(M)$ then $u \notin A(M)$ implies $u \notin A(M\lambda)$ or, $A(M) = A(M\lambda)$.

All this condition means is that factoring out equivalence classes should not introduce new maps from states to final states by words whose original maps avoided final states.

3.4 Definition *A momomorphism* $\lambda: M \to M'$ *is an* L*-homomorphism (for language preserving) if and only* $A(M) = A(M\lambda)$. *The corresponding congruence is an* L*-congruence.*

It is the L-homomorphisms which can be used to construct a-minimal mefa's.

3.5 Definition *Given two congruence* θ_1 *and* θ_2 *of* M, $\theta_1 > \theta_2$ (θ_1 *contains* θ_2) *if the equivalence classes of* θ_2 *are subsets of those of* θ_1.

The 'L-lattice' of a mefa maps into the 'L-lattice' of its homomorphic images as follows.

3.6 Proposition *For an* L-*homomorphism* $\lambda: M \to M'$ *with corresponding* L-*congruence* $\Theta(\lambda)$, *there is a* 1-1 *correspondence between the* L-*congruences of* $M\lambda$ *and those of* M *containing* $\Theta(\lambda)$.

Proof (a) Let θ be an L-congruence containing $\Theta(\lambda)$. Denote the θ-equivalence classes in S by $[s]$, $s \in S$. If the images $[s_1]\lambda$ and $[s_2]\lambda$ intersect non-trivially then there exists $t_i \in [s_i]$ such that $t_1\lambda = t_2\lambda$. Now t_1 and t_2 must be in the same equivalence class of $\Theta(\lambda)$. Since $\Theta(\lambda) \leq \theta$, this forces $[s_1] = [s_2]$ and hence $[s_1]\lambda = [s_2]\lambda$. The $[s]\lambda$, $s \in S$ must therefore partition $M\lambda$ into disjoint subsets which defines an equivalence relation θ' say on $M\lambda$. In fact we have proved that $s_1 \theta s_2$ *if and only if* $(s_1\lambda)\ \theta'\ (s_2\lambda)$ for all $s_1, s_2 \in S$. This will be sufficient to make θ' a congruence since

$$
\begin{aligned}
(s_1\lambda)\ \theta'\ (s_2\lambda) &\Rightarrow s_1\ \theta\ s_2 \\
&\Rightarrow s_1u\ \theta\ s_2u \\
&\Rightarrow ((s_1u)\lambda)\ \theta'\ ((s_2u)\lambda) \\
&\Rightarrow ((s_1\lambda)u)\ \theta'\ ((s_2\lambda)u)
\end{aligned}
$$

By a similar chain of reasoning, θ' is an L-congruence.

(b) Conversely, given an L-congruence θ' of $M\lambda$, let $[t]'$ be the equivalence class of $t \in S\lambda$ under θ'. Put $[t]'\lambda^{-1} = \{s \in S \mid s\lambda \in [t]'\}$. The proof is dual to part (a). If $s \in ([t_1]'\lambda^{-1}) \cap ([t_2]'\lambda^{-1})$ then $(s\lambda)\ \theta'\ t_1$ and $(s\lambda)\ \theta'\ t_2$. This implies that $t_1\ \theta'\ t_2$ and hence $[t_1]' = [t_2]'$. Again $s_1\ \theta\ s_2$ *if and only if* $(s_1\lambda)\ \theta'\ (s_2\lambda)$. θ must therefore be an L-congruence.

We will refer to a congruence or homomorphism as proper if it has at least one-trivial equivalence class.

Since a mefa is finite it must have a proper maximal L-congruence or no proper congruences at all. In the first instance if one of these, θ, is 'factored out' (that is its corresponding L-homomorphism is applied to M) the resulting mefa, N, cannot have any proper L-homomorphic images. For, it if did, there would be, by proposition 3.6, a proper L-congruence containing θ in M. N is therefore 'minimal' subject to the constraint of having its structure 'condensed' from M, yet still recognising A(M).

This can be formalised by defining a-minimal mefa's and stating a theorem.

3.7 Definition *A mefa* M *is* a-*minimal (with respect to* A(M)*) if it has no proper* L-*homomorphisms.*

3.8 Theorem *Every mefa is either* a-*minimal or it has a proper* L-*homomorphism from it onto an* a-*minimal mefa.*

To justify that a classical homomorphism is an L-homomorphism observe that $\lambda: M \to M'$ is classical if $(su)\lambda = (s\lambda)u$ and $s\lambda \in F\lambda$ implies $s \in F$.

Under these conditions, $u \in A(M\lambda) \Rightarrow s\lambda u \in F\lambda$

$$\Rightarrow (su)\lambda \in F\lambda$$

$$\Rightarrow su \in F$$

$$\Rightarrow u \in A(M).$$

Thus $A(M) = A(M\lambda)$ and λ is an L-homomorphism. A counter example to the converse is provided by Veloso's construction of infinitely many Moore reduced connected mefa's [3], which all accept A^*.

Each member of this infinite family maps L-homomorphically onto the trivial mefa. The family has two components. Firstly, let the mefa M_n have $S = \{s_0, s_1, \ldots, s_{n-1}\}$, $F = \{s_{n-1}\}$, $s_i a = s_{i+1}$ for $0 \leq i \leq n-2$ and $s_{n-1}a = s_{n-1}$, for all $a \in A$. The trivial mefa, T, has one state s which is a final state and $sa = s$, $a \in A$. The map $s_i \to s$, $0 \leq i \leq n$ is a homomorphism since

$$(s_i a)\lambda = s_{i+1}\lambda = s = sa = (s_i)\lambda a, \quad 0 \leq i \leq n$$

and

$$(s_{n-1}a)\lambda = s_{n-1}\lambda = s = (s_{n-1})\lambda a.$$

It is an L-homomorphism because $A(M_n) = A^* = A(T)$.

A member of the second component $M_n(D)$ is formed for each n and each proper subset D of A by taking S as above and F as $\{s_0\}$. In this case $sa = s$ for each $s \in S$, $a \in A-D$ and $s_i d = s_{i+1}$, $i < n-1$, $s_{n-1}d = s_0$ for each $d \in D$. Similarly the whole of S can be taken as the single equivalence class of an L-congruence. $M_n(D)$ maps L-homomorphically onto T.

These example indicate clearly that L-homomorphism is a definition better adapted to mefa's than the definition used in the previous papers.

4. Direct Product of a-minimal Mefa's

A mefa $\bar{M}$ may have several distinct maximal L-congruences. These would give rise, via Theorem 3.7, to a-minimal mefa's of perhaps different order and with structure differently related to M. These relationships can be expressed in terms of direct products.

4.1 Definition *A mefa* $\bar{M} = (A, \bar{S}, \bar{\Delta}, \bar{F})$ *is a sub-mefa of* $M = (A, S, \Delta, F)$ *where* $\bar{S} \subseteq S$, $\bar{F} \subseteq F$ *and each* $a\bar{\Delta}$ *is the restriction of* $a\Delta$ *to* $\bar{S}$, $a \in A$. $\bar{M}$ *is an L-sub mefa of* M *if it is a sub-mefa such that* $A(M) = A(\bar{M})$.

Note that for $a\bar{\Delta}$ to be $a\Delta$ restricted to $\bar{S}$ and $\bar{M}$ to be a mefa it is necessary and sufficient that $\bar{S}(a\Delta) \subseteq \bar{S}$.

4.2 Definition *The direct product,* $X_{i=1}^{n}M_i$ *of mefa's* $M_i = (A, S_i, \Delta_i, F_i)$, $i=1,\ldots,n$ *is the mefa* $(A, X_{i=1}^{n}S_i, \Delta^X, X_{i=1}^{n}F_i)$ *where,*

(a) $X_{i=1}^{n}S_i$, *is the cartesian product of the sets,* S_i.

(b) $X_{i=1}^{n}F_i$ *is the cartesian product of the sets,* F_i.

(c) For each $a \in A$, $a\Delta^X$ *maps* $X_{i=1}^{n} S_i$ *to itself according to the rule,*

$$(s_1, s_2, \ldots, s_n)(a\Delta^X) = (s_1(a\Delta_1), s_2(a\Delta_2), \ldots, s_n(a\Delta_n)).$$

$X_{i=1}^{n}M_i$ *will be abbreviated to* XM_i.

4.3 Proposition $A(XM_i) = \bigcap_{i=1}^{n} A(M_i)$.

4.4 Corollary *If* $A(M_1) = A(M_i)$, $i = 2, \ldots n$, *then* $A(XM_i) = A(M_i)$, $i = 1, 2, \ldots n$.

Proof $u \in A(M_1)$ *if and only if* $(s_1, s_2, \ldots s_n)u \in XF_i$ for all $s_i \in S_i$, $i = 1, 2, \ldots n$. By definition, this is equivalent to $s_iu \in F_i$ for $i = 1, 2, \ldots n$.

4.5 Lemma *Let* M *have* L*-congruences,* $\theta_1, \theta_2, \ldots, \theta_n$ *yielding mefa's* $M_1, M_2, \ldots, M_n$. *Then* M *can be mapped* L*-homomorphically onto an* L*-sub mefa of* XM_i. *The* L*-congruence corresponding to this* L*-homomorphism has equivalence classes formed from the intersection of the equivalence classes of the* θ_i.

Proof Denote by $[s]_i$, the equivalance class of $s \in S$ under θ_i. From proposition 3.2 the states of M_i can be identified with the classes $[s]_i$, $s \in S$. Consider the map $\psi : M \to XM_i$ given by $s \to ([s]_1, [s]_2, \ldots [s]_n)$. Then

$$(su)\psi = ([su]_1, [su]_2, \ldots, [su]_n)$$
$$= ([s]_1u, [s]_2u, \ldots, [s]_nu)$$
$$= (s\psi)u.$$

The final states F_i of M_i are the $[f]_i$, $f \in F$, so that $F\psi \subseteq XF_i$. ψ is therefore a homomorphism.

Now $s\psi = s'\psi \iff [s]_i = [s']_i \quad i = 1, \ldots, n$

$$\iff s' \in [s]_i \quad i = 1, \ldots, n$$

$$\iff s' \in \bigcap_{i=1}^{n}[s]_i$$

Hence $\{s' \in S \mid s'\psi = s\psi\} = \bigcap_{i=1}^{n}[s]_i$. If we write $[s] = \bigcap_{i=1}^{n}[s]_i$ then the $[s]$ partition S into disjoint subsets which define the congruence corresponding to ψ.

In addition if $u \notin A(M)$ then

$$Su \cap \bigcup_{f \in F}[f] = Su \cap \bigcup_{f \in F}\left(\bigcap_{i=1}^{n}[f]_i\right)$$

$$\subseteq Su \cap \bigcup_{f \in F}[f]_i \text{ for each } i$$

$= \emptyset$ Since θ_i is an L-congruence.

Thus θ is an L-congruence and ψ an L-homomorphism. $M\psi$ is clearly an L-sub mefa of XM_i.

A restriction of this lemma gives a canonical method of relating M to its a-minimal mefa's. The M_i will be a -minimal if and only if the θ_i are L-maximal. Clearly in that case $M\psi$ is a L-homomorphic image of any L-homomorphic image of M having the same a-minimal mefa's.

4.6 Theorem *Suppose* M *has* a*-minimal mefa's* $M_1, M_2, \ldots, M_n$. *Then* M *maps homomorphically onto an* L*-sub mefa* $M\psi$ *of* XM_i. $M\psi$ *is the smallest* L*-homomorphic image of* M *which has the same* a*-minimal mefa's* $M_1, \ldots, M_n$.

If XM_i is connected in the sense that every state is the image of some fixed state $s_0 \in s$, then XM_i doesn't have any proper sub-mefa's containing s_0. Any M which maps into XM_i, and whose image contains s_0, must map *onto* XM_i. Similarly if XM_i is strongly connected then every M with minimal mefa's $M_1, \ldots, M_n$ must map onto XM_i.

5. Examples

Each of the three mefa's below recognises the language $U = b^* + A^*bb$ where $A = \{a,b\}$. In each case, s_1 represents the final state.

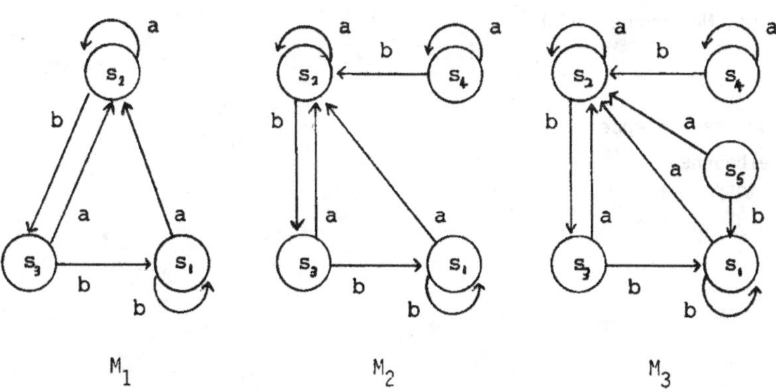

M_1 M_2 M_3

A simple enumeration of possibilities demonstrates that U is not recognised by any two state mefa. M_1 *is therefore* c*-minimal.*

M_2 cannot have an L-homomorphism to a two state mefa because no two state machine recognises U. It has no three state homomorphic image (let alone an L-homomorphic image) because this would require the merging of one pair of states. M_2 *is therefore* a*-minimal* *but not* c-*minimal.*

M_3 can be checked to be Moore reduced by examining each of its states. The map from M_3 to M_2 given by $s_i \to s_i$, $i \neq 5$; $s_5 \to s_1$ is a homomorphism. Since $A(M_2) = U = A(M_3)$ it is also an L-homomorphism. M_3 is therefore Moore reduced but not a-minimal.

References

[1] A. Gill and L-T. Kou Multiple entry finite automata, Journal of Computer and System Sciences, 9, (1974), 1-19.

[2] R.Valk Minimal machines with several initial states are not unique, Information and Control 31, (1976), 193-196.

[3] P.A.S. Veloso Networks of finite state machines. Doctoral dissertation, University of California, Berkeley, May 1975.

[4] P.A.S. Veloso and A.Gill On mimimal finite automata with several initial states, Information and Control, submitted for publication.

Operations Research Department
National Mutual Life Association
447 Collins Street, Melbourne 3000

Department of Computer Science
University of Melbourne
Parkville Vic. 3052

A SINGULAR DIRECT PRODUCT FOR QUADRUPLE SYSTEMS*

A. HARTMAN

A Steiner quadruple system is an ordered pair (X,Q) *where* X *is a finite set and* Q *is a set of 4-subsets of* X *such that every 3-subset of* X *is contained in a unique member of* Q.

This paper gives a structure for studying all the known recursive constructions for quadruple systems. The structure is then applied to existence problems for quadruple systems with subsystems.

A *Steiner quadruple system of order* v, denoted $QS(v)$, is an ordered pair (X,Q) where X is a set of cardinality v, whose elements are called *points*, and Q is a set of 4-subsets of X called *blocks*, with the property that every set of three distinct points is contained in a unique block. We admit as trivial systems all cases with $v \le 2$ and Q empty.

The existence of quadruple systems was originally postulated by Woolhouse [9] and by Steiner [8]. Both authors asked, among other things, for the determination of conditions on v which are necessary and sufficient for the existence of a $QS(v)$.

Simple counting arguments show it is necessary that $4|\binom{v}{3}, 3|\binom{v-1}{2}$ and $2|\binom{v-2}{1}$, whence $v \equiv 2$ or $4 \pmod 6$ is necessary for the existence of a $QS(v)$. In fact this condition is also sufficient.

For over a century, sporadic constructions of various $QS(v)$ expanded the list of known systems without fully settling the existence problem. (See Lindner and Rosa [5] for a bibliography.) Finally, a complete solution to the existence problem was given by Hanani [2], when he gave constructions for quadruple systems of all orders $v \equiv 2$ or $4 \pmod 6$.

Hanani's paper contains six recursive constructions which may be described as follows:

(A) $QS(v) \to QS(2v)$

(B) $QS(v) \to QS(3v - 2)$

(C) $QS(v) \to QS(3v - 8)$, for $v \equiv 8 \pmod{12}$

(D) $QS(v) \to QS(3v - 4)$, for $v \equiv 10 \pmod{12}$

(E) $QS(v) \to QS(4v - 6)$

(F) $QS(v) \to QS(12v - 10)$

---

*This work forms part of the author's research towards a Ph.D. at the University of Newcastle.

Generalizations of some of these constructions have been made by Rokowska [7]. These generalizations are:

(G1) $QS(u)$, $QS(v) \to QS((u-1)(v-1)+1)$

(G2) $QS(u)$, $QS(v) \to QS((u-1)(v-4)+4)$ for $v \equiv 8 \pmod{12}$

(G3) $QS(u)$, $QS(v) \to QS((u-1)(v-2)+2)$ for $v \equiv 10 \pmod{12}$

Aliev [1] and Phelps [6] also independently obtained the first of these generalizations.

A quadruple system (X,Q) of order V has a *subsystem* of order v if there exists sets $x \subseteq X$ and $q \subseteq Q$ such that (x,q) is a quadruple system of order v. This relationship is written $(x,q) \le (X,Q)$. Clearly every non-trivial $QS(v)$ contains subsystems of orders zero, one, two, and four.

Our aim in this paper is to give a generalized framework for the description of all the above constructions. We also use this framework to derive new results on the existence of quadruple systems with subsystems.

The following theorem gives a necessary condition for the existence of a $QS(V)$ with a subsystem of order v.

Theorem 1. *Let (X,Q) be a $QS(V)$. If $(x,q) \le (X,Q)$ and (x,q) has order $v < V$, then $2v \le V$.*

Proof. Since $(x,q) \le (X,Q)$, no block in Q contains precisely three points of x; therefore the number of blocks containing precisely two points of x is $\frac{1}{2}(V-v)\binom{v}{2}$. Hence the number of blocks containing precisely one point of x is

$$\frac{1}{3}\left[v\binom{V-v}{2} - (V-v)\binom{v}{2}\right].$$

Thus
$$v\binom{V-v}{2} - (V-v)\binom{v}{2} \ge 0,$$

so
$$V \ge 2v \quad \text{when} \quad V > v.$$
□

Note that $2v = V$ in this theorem exactly when (X,Q) contains two disjoint subsystems of order v.

Various authors (see for example [4]) have adapted construction (A) to obtain the following result.

Theorem 2. *If there exists a $QS(V)$ with a subsystem of order v then there exists a $QS(2V)$ with two disjoint subsystems of order V, and a subsystem of order $2v$ with two disjoint subsystems of order v.*

In [3] the author proves the following generalization of Constructions (C) and (D).

Theorem 3. *If there exists a $QS(V)$ with a subsystem of order v then there also exists a $QS(3V-2v)$ with subsystems of orders V and v, provided $V \equiv 2v \pmod 6$ and $v \ge 2$.*

The construction given in [3] also ensures the presence of three subsystems of order V in the $QS(3V-2v)$. These three subsystems intersect one another precisely in the subsystem of order v, as indicated in Diagram 1.

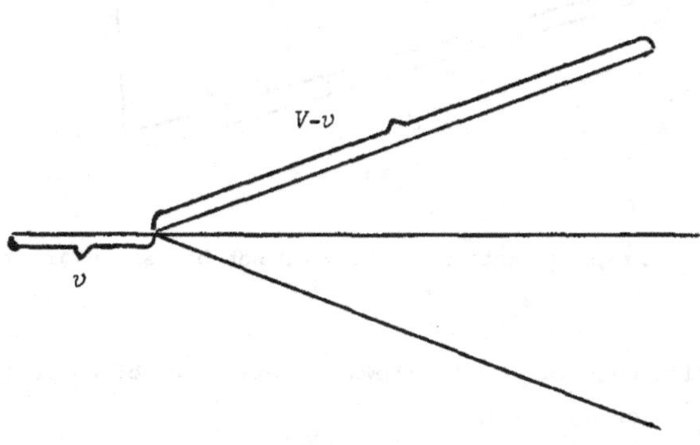

Diagram 1. Schematic representation of the point set of the quadruple system constructed in Theorem 3.

This relationship between subsystems occurs so frequently in the constructions described above that we were motivated to consider the following definition.

For non-negative integers n,g,v, we define a *singular direct product quadruple system*, denoted by SDP(n,g,v), to be a quadruple system (X,Q) of order $ng+v$, with a subsystem (x_o,q_o) of order v and n subsystems $(x_i \cup x_o, q_i \cup q_o)$ of order $g+v$ satisfying $X = x_0|x_1|x_2|\ldots|x_n$ and $(x_o,q_o) \le (x_i \cup x_o, q_i \cup q_o) \le (X,Q)$ for $i = 1,2,\ldots,n$.

Blocks in Q not contained in some $q_i \cup q_o$ for $i = 1,\ldots,n$ are called *cross blocks*, i.e. a cross block contains points from at least two distinct sets x_i with $i > 0$. A design SDP(n,g,v) may be thought of as a quadruple system induced by taking n copies of a quadruple system of order $g+v$ and amalgamating them on a subsystem of order v (see Diagram 2). Naturally many cross blocks have to be added to complete the system, and this is the major problem in constructing an SPD(n,g,v). Some trivial examples of singular direct products of order V include systems SDP$(n,0,V)$ and SDP$(0,g,V)$.

The following are some less trivial examples.

Example 1. A quadruple system of order V with a subsystem of order v is an SDP$(1,V-v,v)$.

Example 2. The quadruple system described in Theorem 2 is an SDP$(2,V,0)$.

Example 3. The quadruple system described in Theorem 3 is an SDP$(3,V-v,v)$.

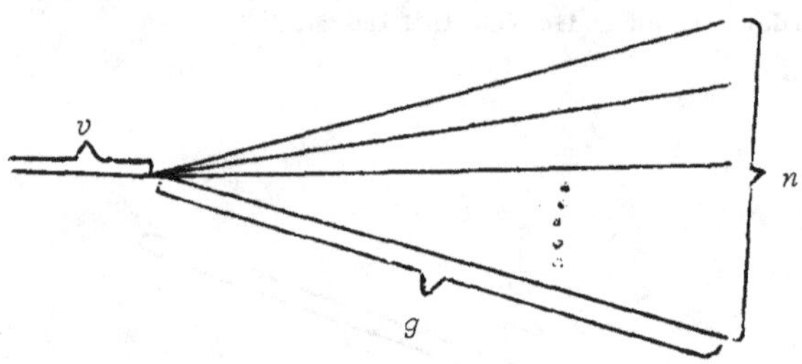

Diagram 2. Pictorial representation of the point set of a singular direct product SDP(n,g,v).

Example 4. Hanani's construction (E) shows the existence of an SDP$(4,V-2,2)$ for every $V \equiv 2$ or 4 (mod 6).

Example 5. Any non-trivial QS(V) is an SDP$(\frac{V-2}{2},2,2)$, taking x_0 as any set of two points and $\{x_i \cup x_0 : 1 \le i \le \frac{V-2}{2}\}$ as the set of all blocks containing x_0.

Example 6. A quadruple system of order V containing a set of $\frac{V-1}{3}$ blocks whose intersection is a single point is an SDP$(\frac{V-1}{3},3,1)$.

Example 7. As part of his construction (F), Hanani has given a direct construction of an SDP(3,12,2).

Example 8. The version of Hanani's construction (B) due to Phelps [6] shows the existence of an SDP$(3,V-1,1)$ for every $V \equiv 2$ or 4 (mod 6).

Examples 2, 3 and 4 demonstrate that Hanani's constructions (A), (C), (D) and (E) are just direct constructions of singular direct product quadruple systems. After a short technical lemma we give a construction which unites the methods used in all other constructions referred to above.

Lemma 4. *If there exists an* SDP(n,g,v) *then there exists an* SDP(n,g,v) *with* n *isomorphic subsystems of order* $g+v$.

Proof. Let (X,Q) be an SDP(n,g,v) with $X=x_0|x_1|\dots|x_n$ and $(x_0,q_0)\le(x_i\cup x_0,q_i\cup q_0)\le(X,Q)$ for $i = 1,2,\dots,n$. Let Q' be a set of blocks obtained from Q by replacing each subset q_i by the image of q_1 induced by any bijection from x_1 to x_i for $i = 1,2,\dots,n$. The blocks of q_0 and all cross blocks are left unaltered. Then (X,Q') is the required homogeneous SDP(n,g,v). □

Construction (SDP). Let $(Z_n\cup\{\infty\},B)$ be a QS$(n+1)$. We shall regard this system as an SDP$(1,n,1)$ amalgamated on the trivial subsystem $\{\infty\}$.

Let (X,Q) be a homogeneous SPD$(3,g,v)$ with $X = x_0|x_1|x_2|x_3$ and

$(x_o,q_o) \leq (x_i \cup x_o,\ q_i \cup q_o) \leq (X,Q)$ for $i = 1,2,3$.

Let $X' = (Z_n \times Z_g) \cup Z_v$. We shall now define a set Q' of blocks on the points of X' so that (X',Q') is an SDP(n,g,v). For each block $[\infty,a,b,c] \in B$ containing the point ∞, form the block set Q of our homogeneous SDP$(3,g,v)$ on the points $X = (\{a,b,c\} \times Z_g) \cup Z_v$, with

$$x_o = Z_v$$
$$x_1 = \{a\} \times Z_g$$
$$x_2 = \{b\} \times Z_g$$
$$x_3 = \{c\} \times Z_g.$$

By the homogeneity of (X,Q) we have so far constructed precisely

$$|q_o| + n|q_1| + \frac{1}{3}\binom{n}{2}(|Q| - 3|q_1| - |q_o|)$$

distinct blocks.

We now complete Q' with the following set of blocks. For each block $[a,b,c,d] \in B$ which does not contain ∞ construct the g^3 blocks $[(a,k),(b,l),(c,m),(d,n)]$, where $k + l + m + n \equiv 0 \pmod g$. The number of quadruples constructed in this manner is $[\frac{1}{4}\binom{n+1}{3} - \frac{1}{3}\binom{n}{2}]g^3$. (Note that the zero in the congruence may be replaced by any constant.) We now check that every 3-subset of X' is contained in some block of Q'.

Every triple contained in some set $(\{a,b\} \times Z_g) \cup Z_v$ with $a,b \in Z_n$, is contained in a block of the first type. Every triple of the form $\{(a,i),(b,j)(c,k)\}$, with a,b and c all distinct, is contained in a block of the first type if $[\infty,a,b,c] \in B$, and otherwise in a block of the second type. The total number of blocks is given by:

$$|Q'| = \frac{1}{4}\binom{v}{3} + n[\frac{1}{4}\binom{g+v}{3} - \frac{1}{4}\binom{v}{3}] + \frac{1}{3}\binom{n}{2}[\frac{1}{4}\binom{3g+v}{3} - \frac{3}{4}\binom{g+v}{3} + \frac{1}{2}\binom{v}{3}] + [\frac{1}{4}\binom{n+1}{3} - \frac{1}{3}\binom{n}{2}]g^3$$
$$= \frac{1}{4}\binom{ng+v}{3}, \text{ as required.}$$

Thus (X',Q') is an SDP(n,g,v). □

Theorem 5. *If there exist an* SDP$(3,g,v)$ *and an* SDP$(1,n,1)$ *then there exists an* SDP(n,g,v).

Proof. If $n = 0,1$ then an SDP(n,g,v) is contained in the structure of an SDP$(3,g,v)$. The remaining cases are covered by Lemma 4 and Construction (SDP). □

Corollary 6 [Construction (B)]. *If there exists a* QS(V) *then there exists a* QS$(3V-2)$.

Proof. Hanani's construction (B) implicitly contains the construction of a QS(10), which is also an SDP$(3,3,1)$. If we consider the QS(V) as an SDP$(1,V-1,1)$, and use Theorem 5 we deduce the existence of an SDP$(V-1,3,1)$. □

Corollary 7 [Construction (F)]. *If a* QS(V) *exists then a* QS$(12V-10)$ *exists*.

Proof. As noted in Example 7 Hanani has constructed an SDP(3,12,2). Theorem 5 then guarantees the existence of an SDP(V-1,12,2). □

Corollary 8 [Construction (G1)]. *If there exist quadruple systems of orders u and v, then there exists a* QS($(u-1)(v-1)+1$).

Proof. As noted in Example 8, Phelps has constructed an SDP($3,v-1,1$) for all $v \equiv 2$ or 4 (mod 6). Hence Theorem 5 ensures the existence of an SDP($u-1,v-1,1$). □

Corollary 9. *If there exist quadruple systems of orders u and v, then if $v \equiv 2$ (mod 6) there exists a* QS($(u-1)(v-4)+4$) *and if $v \equiv 4$ (mod 6) there exists a* QS($(u-1)(v-2)+2$).

Proof. By Example 3 and Theorem 3 there exists an SDP($3,v-4,4$) if $v \equiv 2$ (mod 6) and an SDP($3,v-2,2$) if $v \equiv 4$ (mod 6). The result then follows from Theorem 5. □

Corollary 9 is a stronger version of Rokowska's constructions (G2) and (G3).

We now show that Construction (SDP) contains additional information about subsystems.

Theorem 10. *If there exists an* SDP($3,g,v$) *and an* SDP($1,f,u$) *with $f \neq 0$, then there exists a* QS($(f+u-1)g+v$) *with at least one subsystem of order u, one subsystem of order v, $f+u-1$ subsystems of order $g+v$ and $\frac{1}{3}(\frac{f+u-1}{2})$ subsystems of order $3g+v$.*

Proof. Suppose (X,Q) is an SDP($1,f,u$) with partition $X = x_0|x_1$. Then $x_1 \neq \emptyset$ since $f \neq 0$ and we may take the point ∞ of construction (SDP) to be an element of x_1. The subsystems of order $v, g+v$ and $3g+v$ are clearly shown in that construction. The subsystem of order u is on the point set $x_0 \times \{0\}$, and other subsystems of order u will be found on the point set $x_0 \times \{i\}$ for every i such that $4i \equiv 0$ (mod g). □

We now use construction (SDP) to generate two further recursive constructions of singular direct product quadruple systems.

Theorem 11. *If there exists an* SDP($3,g,v$) *and an* SDP($n,3,1$), *then there exists an* SDP($n,3g,v$).

Proof. Suppose the system $(Z_{3n} \cup \{\infty\}, B)$ is an SDP($n,3,1$) with partition $Z_{3n} \cup \{\infty\} = \{\infty\}|x_1|x_2|\ldots|x_n$, then it may also be considered as an SDP($1,3n,1$) with partition $\{\infty\}|Z_{3n}$. Application of construction (SDP) yields an SDP($3n,g,v$) on the point set $(Z_{3n} \times Z_g) \cup Z_v$, which may also be considered as an SDP($n,3g,v$) because the construction builds subsystems of order $3g+v$ on each set $(x_i \times Z_g) \cup Z_v$ for $i = 1,2,\ldots,n$, and $(Z_{3n} \times Z_g) \cup Z_v = Z_v|x_1 \times Z_g|x_2 \times Z_g|\ldots|x_n \times Z_g$. □

Corollary 12. *If there exists an* SDP($3,g,v$) *then there exists an* SDP($3,3^k g,v$) *for all $k \geq 0$.*

Proof. As noted in the proof of Corollary 6, the QS(10) is an SDP(3,3,1). The result then follows by induction on k, using Theorem 11. □

In the light of Theorem 5 and its corollaries it is clear that every instance of a design SDP$(3,g,v)$ gives rise to a recursive construction for quadruple systems of the form

$$QS(n+1) \Rightarrow QS(ng+v).$$

It is therefore natural to ask for which integers g,v does an SDP$(3,g,v)$ exist, and more generally, for which integers n,g,v does an SDP(n,g,v) exist. The following Theorem gives the necessary conditions on n, g and v.

Theorem 13. *Let* $A = \{0,1\} \cup \{a: a \equiv 2 \text{ or } 4 \pmod 6\}$. *If an* SDP$(n,g,v)$ *exists then*

(i) $\{v, g+v, ng+v\} \subseteq A$

(ii) $ng = 0$ or $g \geq v$

(iii) $ng = 0$ or $n = 1$ or $(n-2)g \geq v$.

Proof. Condition (i) is true since A is the set of admissible orders for quadruple systems, and an SDP(n,g,v) contains subsystems of orders $v, g+v$ and $ng+v$. Conditions (ii) and (iii) are direct consequences of Theorem 1. □

Corollary 14. *If an* SDP$(3,g,v)$ *exists then one of the following conditions holds:*

(i) $v = 1$, $g \equiv 1$ or $3 \pmod 6$;

(ii) $v \equiv 2$ or $4 \pmod 6$, $g \equiv 0$ or $v \pmod 6$ and $g = 0$ or $g \geq v$.

It is not known if these conditions are sufficient for the existence of an SDP$(3,g,v)$. We summarize the known results on existence of an SDP$(3,g,v)$ below.

(a) An SDP$(3,g,1)$ exists whenever $g \equiv 1$ or $3 \pmod 6$ (Phelps [6]).

(b) An SDP$(3,12,2)$ exists (Hanani [2]).

(c) If $g \equiv v \pmod 6$ and there exists an SDP$(1,g,v)$ (i.e. a QS$(g+v)$ with a subsystem of order v) then an SDP$(3,g,v)$ exists (Hartman [3]).

(d) If an SDP$(3,g,v)$ exists then an SDP$(3,3^k g,v)$ exists for all $k \geq 0$) (Corollary 12).

We now apply these results to obtain an effective (i.e. finite) necessary condition for the existence of quadruple systems with subsystems of order eight. The method used is quite general and may be applied to many existence problems for quadruple systems with specified properties.

Theorem 15. *If there exist quadruple systems of orders 34 and 38 with subsystems of order 8, then there exist quadruple systems of order* V *with subsystems of order 8 for all* $V \geq 16$, $V \equiv 2$ *or* $4 \pmod 6$.

Proof. The proof is by induction in seven cases:

(i) $V \equiv 4$ or $8 \pmod{12}$,

(ii) $V \equiv 4$ or $10 \pmod{18}$,

(iii) $V \equiv 16 \pmod{18}$,

(iv) $V \equiv 2$ or $10 \pmod{24}$,

(v) $V \equiv 8 \pmod{18}$,

(vi) $V \equiv 14 \pmod{18}$,

(vii) $V \equiv 38 \pmod{72}$.

(i) If $V \equiv 4$ or $8 \pmod{12}$ then $V = 2(v-1)+2$ for some $v \equiv 2$ or $4 \pmod 6$ and $v \geq 8$. The QS(8) is an SDP(3,2,2) so by construction (SDP) and the existence of SDP$(1,v-1,1)$ for all $v \equiv 2$ or $4 \pmod 6$, there exists an SDP$(v-1, 2,2)$ with $\frac{1}{3}\binom{v-1}{2}$ subsystems of order 8.

(ii) If $V \equiv 4$ or $10 \pmod{18}$ then $V = 3g+1$ for some $g \equiv 1$ or $3 \pmod 6$ with $g \geq 7$. By Phelps' construction (Example 8) there exists, for all such g, an SDP$(3,g,1)$ with a subsystem of order $g+1$. This subsystem may be deleted and replaced by an SDP$(1,g-7,8)$ with a subsystem of order 8. Systems SDP$(1, g-7,8)$ exist for every $g \geq 15$ by the induction hypothesis. If $g = 7$ note that the SDP(3,7,1) contains 3 subsystems of order 8; and for $g = 9$ or 13 note that $V = 28$ or 40 and these cases are covered in (i).

(iii) If $V \equiv 16 \pmod{18}$ then $V = 3g+4$ for some $g \equiv 4 \pmod 6$. By Theorem 3 there exists an SDP$(3,g,4)$ with a subsystem of order $g+4$. As in case (ii) this system may be deleted and replaced by an SDP$(1,g-4,8)$. Such systems exist for every $g \geq 16$ by the induction hypothesis. If $g = 4$ then $V = 16$ and case (i) covers this eventuality, if $g = 10$ then $V = 34$ and this case is covered by hypothesis.

(iv) If $V \equiv 2$ or $10 \pmod{24}$ then $V = 4g+2$ for some $g \equiv 0$ or $2 \pmod 6$ with $g \geq 6$. By Hanani's construction (Example 4) there exists an SDP$(4,g,2)$ with a subsystem of order $g+2$ for all such g. Replacing the subsystem by an SDP$(1,g-6,8)$ yields the result for all $g \geq 14$ by induction. If $g = 6$ then an SDP(4,6,2) contains four subsystems of order 8. If $g = 8$ then $V = 34$, which is covered by hypothesis. If $g = 12$ then $V = 50$, and an SDP(7,7,1) containing 7 subsystems of order 8 exists by Corollary 8.

(v) If $V \equiv 8 \pmod{18}$ then $V = 3g+2$ for some $g \equiv 2 \pmod 6$ with $g \geq 8$. As in case (iii) we use Theorem 3 to recursively construct an SDP$(3,g,2)$ for any $g \geq 14$ replacing the subsystem of order $g+2$. For $g = 8$ we have $V = 26$, which is covered by (iv).

(vi) If $V \equiv 14 \pmod{18}$ then $V = 3g+8$ for some $g \equiv 2 \pmod 6$ with $g \geq 8$. By the induction hypothesis there exists an SDP$(1,g,8)$ for all such g, so by Theorem 3 there exists an SDP$(3,g,8)$ containing a subsystem of order 8.

(vii) If $V \equiv 38 \pmod{72}$ then $V = 12n+2$ for some $n \equiv 3 \pmod 6$. If $n = 3$ then $V = 38$ and there exists an SDP(1,30,8) by hypothesis. For $n > 3$ there exists an SDP$(n,12,2)$ with $\frac{1}{3}\binom{n}{2}$ subsystems of order 38 (Theorem 10). Replacing one of these subsystems with a system SDP(1,30,8) yields the result. □

We remark that the proof of Theorem 15 is very similar in form to Hanani's initial proof of the existence of quadruple systems. The main difference lies in the discussion of case (vi).

We now give two further applications of the results obtained to date. The first application is an analysis of the existence problem for small subsystems. We then use this application to obtain a new result on the existence of large subsystems.

To discuss the existence of subsystems of orders 8, 10 and 14, we need the sets

$$E_8 = \{34, 38, 86, 110, 146\},$$
$$E_{10} = \{32, 38, 86, 110, 146\},$$
and
$$E_{14} = \{32, 44, 46, 52, 58, 62, 70, 122, 124, 178, 206\}.$$

We shall also need the functions $s(V)$ and $g(V)$ defined, for any $V \equiv 2$ or $4 \pmod 6$, by

$$s(V) = \begin{cases} 14, & \text{if } V \equiv 2 \pmod{18}, \\ 1, & \text{if } V \equiv 4 \text{ or } 10 \pmod{18}, \\ 2, & \text{if } V \equiv 8 \pmod{18}, \\ 8, & \text{if } V \equiv 14 \pmod{18}, \\ 4, & \text{if } V \equiv 16 \pmod{18}, \end{cases}$$

and
$$g(V) = \frac{V-s(V)}{3}.$$

Theorem 16. *For $v = 8,10$ or 14 and for any admissible $V \geq 2v$, with the possible exceptions of $V \in E_v$, there exists a QS(V) with a subsystem of order v.*

Proof. For all admissible $V < 620$ the direct constructions of Theorems 2,3 and 10 and Examples 4, 7 and 8 produce systems of order V with subsystems of order v. Details of these constructions are given in Appendix A of [4]. Now let us suppose that $V \geq 620$. Observe that

$$V > g(V) + s(V) \geq 208 > max(E_8 \cup E_{10} \cup E_{14}),$$

so for induction purposes we may assume that there exists a QS$(g(V)+s(V))$ with a subsystem of order v for each $v \in \{1,2,4,8,10,14\}$. Note that if $V \equiv 2,8,14$, or $16 \pmod{18}$ we have $g(V) \equiv s(v) \pmod 6$, and otherwise $g(V) \equiv 1$ or $3 \pmod 6$. Hence either Theorem 3 or Example 8 guarantees the existence of an SDP$(3,g(V),s(V))$ of order V with a subsystem of order $g(V) + s(V)$. Replacing this subsystem with one of the same order containing a subsystem of order 8,10 or 14 yields the result. □

Notice that the induction argument in the proof of Theorem 16 depends only on two recursive constructions, thus simplifying the recursion in Hanani's original existence proof for quadruple systems, at the expense of increasing the number of initial cases. Further simplification of this kind may be achieved by constructing a suitable set E_{16}, a sufficiently large set of initial cases, and omitting the use of Example 8 by defining the functions s' and g' to agree with s and g on all $V \equiv 2,8,14$ or 16 (mod 18) and setting

$$s'(V) = \begin{cases} 10 & \text{if } V \equiv 4 \pmod{18}, \\ 16 & \text{if } V \equiv 10 \pmod{18}, \end{cases}$$

and $$g'(v) = \frac{V-s'(V)}{3}.$$

We now turn our attention to the existence of quadruple systems with large subsystems.

Theorem 17. *For all admissible V except $V = 14$ and possibly $V = 146$ there exists a $QS(V)$ with a subsystem of order greater than $\frac{1}{3}V$.*

Proof. By Theorem 1, no QS(14) can have a subsystem of order greater than 4. For all admissible $V < 620$, except $V = 14$ and 146, Appendix A of [4] lists constructions of QS(V) with subsystems of order greater than $\frac{1}{3}V$. For $V \geq 620$ the induction argument in the proof of Theorem 16 ensures that there exists a quadruple system of order $V = 3g(V) + s(V)$ with a subsystem of order $g(V) + s(V)$. □

From Theorem 1 the best possible result in this direction would be the construction of a QS(V) with a subsystem of order essentially $\frac{1}{2}V$ for all admissible orders V. Theorem 2 guarantees this best result for all $V \equiv 4$ or 8 (mod 12); however no result signigicantly better than $\frac{1}{3}V$ has been obtained for $V \equiv 2$ or 10 (mod 12).

REFERENCES

[1] S.O. Aliev, Symmetric algebras and Steiner systems, *Soviet Math. Dokl.* 8 (1967), 651-653.

[2] H. Hanani, On quadruple systems, *Can. J. Math.* 12 (1960), 145-157.

[3] A. Hartman, Tripling quadruple systems, *Ars Combinatoria* (to appear).

[4] A. Hartman, *Construction and resolution of quadruple systems*, Ph.D. Thesis, University of Newcastle, Australia, 1980.

[5] C.C. Lindner and A. Rosa, Steiner quadruple systems - a survey, *Discrete Math.* 22 (1978), 147-181.

[6] K.T. Phelps, Rotational quadruple systems, *Ars Combinatoria* 4 (1977), 177-185.

[7] B. Rokowska, Some new constructions of 4-tuple systems, *Colloq. Math.* 17 (1967), 111-121.

[8] J. Steiner, Combinatorische Ausgabe, *J. Reine Angew. Math.* 45 (1853), 181-182.

[9] W.S.B. Woolhouse, Prize question 1733, *Lady's and Gentleman's diary* (1844).

Department of Combinatorics and Optimization
University of Waterloo
Waterloo, Ontario N2L 3G1, Canada

THE MAXIMUM NUMBER OF INTERCALATES IN A LATIN SQUARE

KATHERINE HEINRICH AND W.D. WALLIS

An intercalate in a Latin square is a subsquare of order 2; I(n) denotes the maximum number of intercalates in any Latin square of order n.

Upper bounds for I(n) are found, and it is shown that they are attained if and only if $n = 2^\alpha$ or $2^\alpha - 1$. A number of lower bounds are found for I(n).

1. INTRODUCTION

We assume that the reader is familiar with the basic ideas of Latin squares, as contained for example in [2].

Suppose A is a Latin square. We denote by $M_k(A)$ the number of $k \times k$ subsquares of A, and by $M_k(n)$ the maximum number of $k \times k$ subsquares in any Latin square of order n: if $\mathcal{L}(n)$ is the set of all $n \times n$ Latin squares,

$$M_k(n) = \max_{A \in \mathcal{L}(n)} M_k(A).$$

We are particularly interested in the number of 2×2 subsquares, or *intercalates* as they were called by Norton [3]; we write I(A) and I(n) for $M_2(A)$ and $M_2(n)$ respectively.

2. UPPER BOUNDS ON INTERCALATES

Theorem 1. *If* n *is even, then*

$$I(n) \leq \frac{n^2(n-1)}{4}. \tag{1}$$

If n *is odd, then*

$$I(n) \leq \frac{n(n-1)(n-3)}{4}. \tag{2}$$

Proof. Suppose n is even. There are $\frac{1}{2}n(n-1)$ pairs of rows in a square of order n. It is conceivable that two rows might contribute $\frac{1}{2}n$ intercalates (as in

1 2 3 4 5 6 ... n-1 n

2 1 4 3 6 5 ... n n-1),

but no more, so

$$I(n) \leq \frac{n(n-1)}{2} \times \frac{n}{2} = \frac{n^2(n-1)}{4} .$$

Now consider n odd. If two rows contributed $\frac{1}{2}(n-1)$ intercalates, they would look like

1 2 3 4 5 6 ... n-2 n-1 n

2 1 4 3 6 5 ... n-1 n-2 n ,

which is impossible as it leads to a repeated element in a column. So the maximum is $\frac{1}{2}(n-3)$, and

$$I(n) \leq \frac{n(n-1)}{2} \times \frac{(n-3)}{2} = \frac{n(n-1)(n-3)}{4} . \qquad \square$$

Although this theorem was proven using very crude combinatorial arguments, it is best-possible. In fact, we shall show that equality is attained in (1) if and only if $n = 2^{\alpha}$, for some α; and in (2) if and only if $n = 2^{\alpha} - 1$.

We begin with a lemma on the lower bound for I(n).

<u>Lemma 1</u>. *If* $I(n) = k$ *and* $I(m) = \ell$ *then*

$$I(mn) \geq m^2k + n^2\ell + 4k\ell.$$

<u>Proof</u>. Let A and B be Latin squares of order n and m respectively, with $I(A) = k$ and $I(B) = \ell$; let $C = A \times B$, the usual direct product. C contains m^2 copies of A and n^2 copies of B, which contribute $m^2k + n^2\ell$ intercalates. There are also other intercalates introduced. Each intercalate in A gives rise to a "block intercalate" of copies of B. For convenience, suppose the elements of A concerned were 1 and 2; label the relevant copies of B as B_1 and B_2. Consider any intercalate in B_1, with elements a and b say, and suppose the corresponding elements in B_2 are c and d. Then the intercalates give rise to the indicated 4×4 subsquare in C.

B_1		B_2	
$\underline{a}$	b	c	$\underline{d}$
b	a	d	c
B_2		B_1	
c	d	a	b
$\underline{d}$	c	b	$\underline{a}$

The 4×4 square contains 12 intercalates; 8 of these have been counted, but there are 4 more (the underlined entries form a typical one). Each of the $k\ell$ choices of a pair of intercalates, one from each of A and B, yields four new intercalates of C in this way, so we have $4k\ell$. Thus

$$I(mn) \geq I(C) \geq m^2k + n^2\ell + 4k\ell. \qquad \square$$

Theorem 2. $I(n) = n^2(n-1)/4$ if and only if $n = 2^\alpha$, $\alpha \geq 1$.

Proof. First suppose $n = 2^\alpha$. We prove $I(n) = n^2(n-1)/4$ by induction. Clearly $I(2) = 1$. Now suppose $I(2^{\alpha-1}) = 2^{2\alpha-2}(2^{\alpha-1}-1)/4 = 2^{2\alpha-4}(2^{\alpha-1}-1)$. Let A be a Latin square of order $2^{\alpha-1}$ which attains this bound, and let B be a Latin square of order 2. By the Lemma,

$$\begin{aligned} I(A\times B) &\geq 2^{2\alpha-2}\times 1 + 4\times 2^{2\alpha-4}(2^{\alpha-1}-1) + 4\times 2^{2\alpha-4}(2^{\alpha-1}-1)\times 1 \\ &= 2^{2\alpha-2} + 2^{2\alpha-2}(2^{\alpha-1}-1) + 2^{2\alpha-2}(2^{\alpha-1}-1) \\ &= 2^{2\alpha-2}(2\times 2^{\alpha-1}-1) \\ &= 2^{2\alpha}(2^\alpha-1)/4 \end{aligned}$$

and equality must hold, by (1).

Conversely, suppose $A = (a_{ij})$ is an $n\times n$ Latin square which achieves $I(A) = n^2(n-1)/4$. Without loss of generality we may assume that A has first row and column $(1,2,\ldots,n)$. Suppose $a_{ij} = a_{k\ell} = x$, and $a_{i\ell} = y$. Then necessarily $a_{kj} = y$, or else rows i and k would not contain $\frac{1}{2}n$ intercalates. In particular since $a_{1j} = a_{j1}$ for all j, the diagonal must be $(1,1,\ldots,1)$; further, from consideration of the cases with $x = 1$, we see that A must be symmetric.

Let us interpret A as the multiplication table of a quasigroup with operation $\circ$. What we have said amounts to the facts that $(A,\circ)$ is commutative, that 1 is an identity element, that every element has order 2, and that

$$\text{if } i\circ j = k\circ\ell \text{ then } i\circ\ell = k\circ j\ . \qquad (3)$$

Since 1 is an identity

$$1\circ(x\circ y) = x\circ y$$

for all x and y; substituting this in (3),

$$y = 1\circ y = x\circ(x\circ y)\ . \qquad (4)$$

Put $y = b\circ a$, $x = c$. Then

$$b\circ a = c\circ(c\circ(b\circ a))$$

and by commutativity

$$b \circ a = ((a \circ b) \circ c) \circ c.$$

Using (3),

$$b \circ c = ((a \circ b) \circ c) \circ a$$
$$= a \circ ((a \circ b) \circ c).$$

So

$$a \circ (b \circ c) = a \circ (a \circ ((a \circ b) \circ c))$$
$$= (a \circ b) \circ c$$

from (4). So $(A, \circ)$ is associative. Hence it is a group, and therefore an abelian group in which every element has order 2. This means it is an elementary abelian 2-group, and has order 2^α for some α. □

Remark. The main fact of Theorem 2 - that a quasigroup has every 2-element subset as a "translation" of the group table of order 2 if and only if it is an elementary abelian 2-group - is part of the folklore; we know it has been proven by R.A. Bailey, F.P. Hiner and R.B. Kilgrove, S.E. Payne, P.J. Owens, and probably others besides ourselves. However, we do not know that it has appeared in print before.

Theorem 3. $I(n) = n(n-1)(n-3)/4$ *if and only if* $n = 2^\alpha - 1$, $\alpha \geq 1$.

Proof. First, suppose $n = 2^\alpha - 1$. Let A be the Latin square of order 2^α constructed from the elementary abelian 2-group. A is symmetric and has constant diagonal; relabel the elements so that the diagonal is $(n+1, n+1, \ldots, n+1)$, and the last row and column are $(1, 2, \ldots, n, n+1)$. Delete row and column $n+1$ of A, and replace the main diagonal by $(1, 2, \ldots, n)$; call the resulting square B. (B is a one-contraction of A - see [2, p.40].)

A had $(n+1)^2 n/4$ intercalates. In moving to B we have lost all $n(n+1)/2$ intercalates which involved $n+1$ and also all intercalates which used the last row ($n(n-1)/2$ of them, excluding the ones involving element $n+1$ because they have already been counted) and all which used the last column (another $n(n-1)/2$). No more have been lost. So

$$I(B) \geq (n+1)^2 n/4 - n(n+1)/2 - 2n(n-1)/2$$
$$= n(n-1)(n-3)/4.$$

Equality must hold by (2). (In any event it is clear that no new intercalates have been added to B.)

Now suppose A is a Latin square of odd order n and I(A) = n(n-1)(n-3)/4. Any two elements can belong to at most (n-3)/2 intercalates: if they belonged to (n-1)/2 of them, we could place these (n-1)/2 2×2 blocks down the diagonal; both the elements would have to occupy position (n,n), which is impossible. So, to attain I(A), every pair must lie in exactly (n-3)/2 intercalates. By row, column and entry permutation we can put A into the form

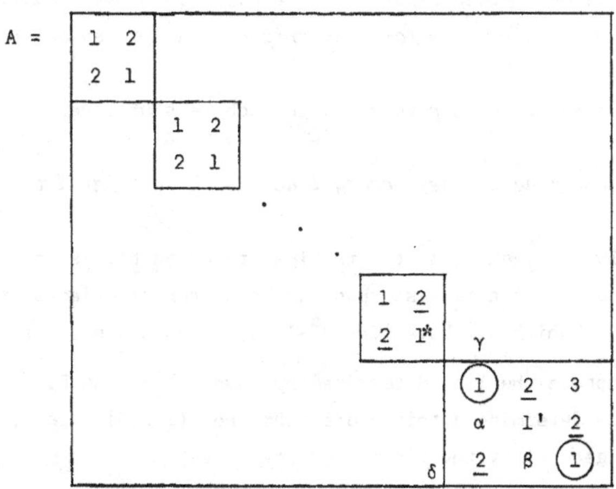

where $\alpha,\beta,\gamma,\delta$ are yet to be determined. Any other two symbols have a similar structure, so somewhere in the square is a 3×3 subarray

$$C = \begin{array}{|ccc|} 1 & 3 & \\ & 1 & 3 \\ 3 & & 1 \end{array}$$

Neither of the circled entries 1 can be in an intercalate with 3, so they must be part of C. To within isomorphism, the third 1 in C can be either 1* or 1'. Suppose it is 1*; then $\gamma = \delta = 3$ and it is clear that none of the underlined entries 2 can form an intercalate with a 3. But there are five underlined entries 2, and all but three 2's must form an intercalate. This is a contradiction. It follows that 1' is in C, that $\alpha = \beta = 3$, and therefore that C is a 3×3 Latin square. The same argument holds for any two elements, not just 1 and 2; so the occurrences of any x and y together form (n-3)/2 2×2 subsquares and one 3×3 subsquare. Thus:

every pair of elements x *and* y *determines exactly one triple* (x,y,z) *and exactly one set of three rows and one set of three columns such that the intersection of those rows and columns form a* 3×3 *subsquare on* (x,y,z). (5)

Given any two rows, they form $(n-3)/2$ intercalates and a 2×3 array on the remaining elements. So they must be two of the three rows corresponding to the triple on the remaining three elements. It follows that we can invert the above statement to read

every pair of rows determines exactly one triple $\{x,y,z\}$ *and exactly one third row such that the three rows contain the triple as a* 3×3 *subsquare.*

We shall refer to the rows as *points* and to such sets of three rows as *lines*. So

any two points determine a line; every line has an associated triple.

Clearly (5) means that the triples, taken as blocks, form a Steiner triple system on n objects. If n is less than 9, the number of objects of such a system can only be 1, 3 or 7, which all have form $2^\alpha-1$, so we assume $n \geq 9$ from here on.

Now suppose the line determined by rows R1, R2 and R3 is {1,2,3}; and suppose R1 and R4 determine a triple disjoint from {1,2,3} - say {4,5,6}, for example. Without loss of generality the first four rows are

```
1 2 3 4 5 6 7 8 9
2 3 1
3 1 2 a b c d e f
7 8 9 6 4 5 1 2 3...
```

(R4 has this form because 17, 28 and 39 are intercalates between R1 and R4). If $d = 8$ then (from rows 1 and 3) $e = 7$. Then R3 and R4 contain as columns 1 and 8

```
3 7
7 2 ;
```

the only way in which this can occur is if {2,3,7} is the triple determined by R3 and R4, but this would give $f = 2$, impossible as 2 already appears in row 3. So $d \neq 8$. The triple for R3 and R4 is {1,8,d} which cannot contain both 2 and 9, so $e = 9$, whence $f = 8$ (rows 1,3); this must mean $d = 3$, and again we have a repetition in row 3. So, if the lines corresponding to two triples contain a common row then the triples contain a common element.

Suppose {R1,R2,R3} determine triple {1,2,3} and {R1,R4,R5} determine {1,4,5}.

We can take A to begin

```
1  2  3  4  5
2  3  1
3  1  2
4        5  1
5        1  4 .
```

Suppose {R2,R4,R6} is a line. We can take its triple to be {1,6,7}, {2,4,6}, {3,4,6} or {3,5,6}.

Case {1,6,7}. The array can be taken as starting

```
1  2  3  4  5     7  8
2  3  1  8  7  6  5  4
3  1  2
4  x  6  5  1  7  y  z
5        1  4
      7     6  1        .
```

Since 24 is an intercalate in R2 and R4, z = 2, so from R1 and R4, x = 8; but from R2 and R4, y = 8 also, which is impossible.

Case {2,4,6}. The elements 2 and 6 must appear in new columns in row 4 - say columns 6 and 7. If 7 is the element which forms an intercalate in R1 and R2 with 4, and we append the column with 7 in row 4, (which cannot be column 2 or 3 since 7 is not in the triple delivered by R2 and R4) we get

```
1  2  3  4  5  z  7  y
2  3  1  7     6  4  x
3  1  2
4        5  1  2  6  7
5        1  4
6              4  2  ...
```

From R2 and R4, x = 5; from R1 and R4, y = 6. So z = 5, which is impossible.

Case {3,4,6} does not arise because 4 is in column 1 of R2 but 3 is not in column 1 of R4 or R6.

So the triple determined by R2 and R4 is {3,5,6}. We assume the third point in the associated line is R6. By the same argument we see that the triple determined by R3 and R5 is {2,4,x} for some x > 5, and the one determined by R3 and R4 is {2,5,y} for some y > 5. So the array A looks like

$$\begin{array}{cccccc}
1 & 2 & 3 & 4 & 5 & \\
2 & 3 & 1 & 6 & & 5 \\
3 & 1 & 2 & y & x & \\
4 & 6 & y & 5 & 1 & 3 \\
5 & & x & 1 & 4 & \\
 & 5 & 4 & 3 & 2 & 6 .
\end{array}$$

Since 1 is not in the triple {2,5,y}, 1 and 6 form an intercalate in rows 3 and 4, so $x = 6$. Therefore the line joining R3 and R5 contains R6.

Since all of the above results hold generally, we have:

given two lines {Ra,Rb,Rc} *and* {Ra,Rd,Re} *which meet in a point* Ra, *the lines joining* Rb *to* Rd *and* Rc *to* Re *also meet in a point.*

This is the usual "closure" axiom for projective spaces [1,p.24], so the points and lines form a projective geometry with 3 points per line. It is well-known that such a geometry has $2^\alpha-1$ points, where $\alpha-1$ is the dimension; so

$$n = 2^\alpha - 1, \quad \text{some } \alpha. \qquad \square$$

Remark. It is not hard to see that one-prolongation of the Latin square derived in this way from a finite geometry will necessarily result in the Latin square from an elementary abelian 2-group, so the square we derived in the first part of Theorem 3 is the only one which attains the bound (up to isomorphism).

3. LOWER BOUNDS ON INTERCALATES

Theorem 4. *For* m *odd,* $I(2m) \geq m^3$.

Proof. Let $A = (a_{ij})$ and $B = (b_{ij})$ be the Latin squares based on $\{1,2,\ldots,m\}$ and $\{m+1,m+2,\ldots,2m\}$ respectively, where

$$a_{ij} \equiv i - j + 1 \pmod m,$$

$$b_{ij} \equiv i + j \pmod m.$$

Let $C = (c_{ij})$ be

$$C = \begin{array}{|c|c|} \hline A & B \\ \hline B & A \\ \hline \end{array}$$

Now consider entry c_{ij}, $1 \leq i,j \leq m$. For any k, $1 \leq k \leq m$,

$$c_{ij} = c_{k+m,j-i+k+m},$$

and also

$$c_{i,j-i+k+m} = b_{i,j-i+k} \equiv j+k,$$

$$c_{k+m,j} = \quad b_{k,j} \quad \equiv k+j$$

so

$$c_{i,j-i+k+m} = c_{k+m,j} .$$

Therefore rows $\{i,k+m\}$ and columns $\{j,j-i+k+m\}$ form an intercalate, so c_{ij} belongs to (at least) m intercalates, for $1 \leq i,j \leq m$. So $I(C) \geq m^3$, and $I(2m) \geq m^3$. □

On checking all Latin squares of order 6 (see, for example, [2, pp.130-137]), we find that $I(6) = 27$, so Theorem 4 is exact for this value.

We have tried variations on the theme of Theorem 4 with various A and B when m is even, but have not done anything better than the following corollary, which is obtained from Lemma 1 using the decomposition $2^\alpha m = 2^{\alpha-1}.2m$:

<u>Corollary 4.1</u>. *For* m *odd*, $\alpha \geq 1$,

$$I(2^\alpha m) \geq (2^\alpha m)^2(2^\alpha m + 2^\alpha - 2)/8. \qquad □$$

This lower bound is about half the upper bound of $(2^\alpha m)^2(2^\alpha m-1)/4$ which comes from Theorem 1, so it is quite a good bound. In general terms, it is about $n^3/8$.

The rest of this section is an account of attempts to find similar "good" bounds - order n^3 - for odd n. We see immediately that $I(3) = 0$ and $I(5) = 4$. We know $I(7) = 42$ from the last section. For order 9, we exhibit a square with 64 intercalates, so $64 \leq I(9) \leq 108$:

9	2	3	4	5	6	8	7	1
2	1	4	3	6	5	9	8	7
3	4	1	2	8	7	5	9	6
4	9	2	1	7	8	6	5	3
5	6	8	7	1	9	3	4	2
6	5	7	9	2	1	4	3	8
8	7	9	6	3	4	1	2	5
7	8	6	5	9	3	2	1	4
1	3	5	8	4	2	7	6	9

Theorem 5. *If* m *is odd and* $\alpha \geq 2$, *then*

$$I(2^{\alpha}m+1) \geq 2^{\alpha}m[2^{\alpha}m(2^{\alpha}m+2^{\alpha}-10)/8 + m + 1] + 2^{\alpha-1}m(m-1).$$

Proof. Let E be the square $D \times C$ of order $2^{\alpha}m$, where D comes from the elementary abelian group of order $2^{\alpha-1}$ and C is the square constructed in Theorem 4. Then $I(E) = (2^{\alpha}m)^2(2^{\alpha}m+2^{\alpha}-2)/8$. Now E is essentially formed by substituting squares isomorphic to A and B for the entries in the multiplication table of the elementary abelian group of order 2^{α}, which has a transversal [2, p.170]. Moreover the set of cells $T = \{(i,2-2i): 1 \leq i \leq m\}$ is easily seen to be a transversal in A and B: $\{(i,i): 1 \leq i \leq m\}$ is one in B. So E has a transversal.

Now consider the square F obtained from E by prolongation about the transversal. In this process at most $2^{\alpha}m(2^{\alpha}m-m)$ intercalates are destroyed and at least $2^{\alpha}m + 2^{\alpha}(m(m-1)/2)$ are recovered. So

$$I(F) \geq (2^{\alpha}m)^2(2^{\alpha}m+2^{\alpha}-2)/8 - 2^{\alpha}m(2^{\alpha}m-m) + 2^{\alpha}m + 2^{\alpha-1}m(m-1)$$
$$= 2^{\alpha}m[2^{\alpha}m(2^{\alpha}m+2^{\alpha}-10)/8 + m + 1] + 2^{\alpha-1}m(m-1). \quad \square$$

Theorem 6. *If* $(m,6) = 1$ *then*

$$I(2m+1) \geq m(2m-3)(m-1)/2.$$

Proof. Let C be the square of order 2m, constructed as in Theorem 4. Let T be the transversal in A and B with cells $\{(i,2-2i): 1 \leq i \leq m\}$. For each (i,j) in T, replace a_{ij} by b_{ij} in the upper left copy of A and replace b_{ij} by 2m+1 in both copies of B. Now form an array D by appending a new last row and column: if $(i,j) \in T$ then put $d_{i,2m+1} = d_{2m+1,j} = a_{ij}$ and $d_{i+m,2m+1} = d_{2m+1,j+m} = b_{ij}$, and put $d_{2m+1,2m+1} = 2m+1$. The construction is illustrated in the case $2m+1 = 11$.

1	2	3	4	5	7	8	9	10	6
5	1	2	3	4	8	9	10	6	7
4	5	1	2	3	9	10	6	7	8
3	4	5	1	2	10	6	7	8	9
2	3	4	5	1	6	7	8	9	10
7	8	9	10	6	1	2	3	4	5
8	9	10	6	7	5	1	2	3	4
9	10	6	7	8	4	5	1	2	3
10	6	7	8	9	3	4	5	1	2
6	7	8	9	10	2	3	4	5	1

⟹

1	2	3	4	6	7	8	9	10	11	5
5	1	10	3	4	8	9	11	6	7	2
9	5	1	2	3	11	10	6	7	8	4
3	4	5	8	2	10	6	7	11	9	1
2	7	4	5	1	6	11	8	9	10	3
7	8	9	10	11	1	2	3	4	5	6
8	9	11	6	7	5	1	2	3	4	10
11	10	6	7	8	4	5	1	2	3	9
10	6	7	11	9	3	4	5	1	2	8
6	11	8	9	10	2	3	4	5	1	7
4	3	2	1	5	9	7	10	8	6	11

We now count the intercalates in D. $I(C) \geq m^3$. We must calculate the number of known intercalates destroyed in this process. For $1 \leq i,j \leq m$, each entry a_{ij} of A lay in m intercalates of C: those formed from rows i, m+k and columns j, m+j-i+k. If (i,j) belongs to T, all m intercalates are lost. If not, then two are lost, as there is a case where (i,j-i+k) lies on T and one where (m+k,j) does. So a total of $m^2+2(m^2-m)$ are destroyed. On the other hand, for any i and k, $1 \leq i,k \leq m$, rows m+i and m+k form an intercalate with columns j and 2m+1, where $1 \leq j \leq m$ and $j \equiv 2-k-i \pmod m$, so we have added m(m-1)/2 new intercalates. So

$$I(D) \geq m^3 - m^2 - 2(m^2-m) + (m^2-m)/2$$
$$= m(2m-3)(m-1)/2 \ . \qquad \square$$

We observe that this lower bound is not exact in the case m = 5, because the square of side 11 which we have exhibited has 80 intercalates, while the formula yields 70. This is because of the ten intercalates formed, one for each unordered pair in {1,2,3,4,5}, with the last row. Unfortunately this situation does not generalise (it is easy to show that if n is the equivalent of ½ modulo m, so that 2n = m+1, then $n+1 \equiv -1 \pmod m$ is a necessary condition). However, we have

$$80 \leq I(11) < 220.$$

Corollary 6.1. *If* (m,6) = 1 *then*

$$I(2^\alpha m+1) \geq (2^\alpha m)[(2^\alpha m)(2^\alpha m+2^\alpha-2)-10m+6]/8$$

for $\alpha \geq 2$.

Proof. Let E be the Latin square obtained from the elementary abelian group of order $2^{\alpha-1}$, written so that it has diagonal (1,1,...,1), and let C be defined as before. Then

$$I(E \times C) \geq (2^\alpha m)^2(2^\alpha m+2^\alpha-2)/8.$$

If we carry out the same operation on each of the $2m \times 2m$ diagonal blocks as in the Theorem, we obtain a square D of order $2^\alpha m+1$ in which we have deleted $3m^2-2m$ intercalates $2^{\alpha-1}$ times and added $(m^2-m)/2$ intercalates $2^{\alpha-1}$ times. So

$$I(D) \geq (2^\alpha m)^2(2^\alpha m+2^\alpha-2)/8 - 2^{\alpha-1}(3m^2-2m) + 2^{\alpha-2}(m^2-m)$$
$$= (2^\alpha m)[(2^\alpha m)(2^\alpha m+2^\alpha-2)-10m+6]/8 \ . \qquad \square$$

We observe that this is an improvement on Theorem 5, in the cases where it applies. We can obviously do very slightly better when $m = 5$:

Corollary 6.2. *When* $\alpha \geq 1$, $I(5.2^{\alpha}+1) \geq 5.2^{\alpha-2}[5.2^{\alpha}(3.2^{\alpha}-1)-18]$. □

Theorem 7. *For odd* m, $I(6m+1) \geq m(5m^2+m+3)$.

Proof. We consider the following Latin square of order 6:

$$S = \begin{array}{|cccccc|} \hline 1 & 3 & 5 & 6 & 2 & 4 \\ 4 & 2 & 6 & 5 & 1 & 3 \\ 5 & 4 & 3 & 1 & 6 & 2 \\ 6 & 1 & 2 & 4 & 3 & 5 \\ 2 & 6 & 4 & 3 & 5 & 1 \\ 3 & 5 & 1 & 2 & 4 & 6 \\ \hline \end{array}$$

S has a transversal, shown on the diagonal. It has two intercalates (objects 1 and 4 in rows 3 and 4, objects 2 and 5 in rows 2 and 6) which intersect the transversal, and three others (2 and 3 in rows 5 and 6; 3 and 6 in rows 1 and 5; 5 and 6 in rows 1 and 2) which are all disjoint from the transversal and from the (1,4) intercalate.

Let A and B be the squares of Theorem 4. We define A_k and B_k by adding $(k-1)m$ and $(k-2)m$ respectively to each element of A and of B, so that A_i and B_i are based on $\{(i-1)m+1,\dots,im\}$. Then we form an array of order $6m$ from S by replacing each 1,3,5 by A_1,A_3,A_5 respectively and each 2,4,6 by B_2,B_4,B_6 respectively, except that each diagonal entry k is replaced by B_k. Call this array D. We then form a new array E by prolongation of D about the transversal made up of the transversals T in each B_i on the diagonal; we label the new element $6m+1$.

Each intercalate in S gives rise to a subsquare of size $2m$ in E which contains m^3 intercalates, so we have $5m^3$ intercalates. In the process of prolongation we destroy m^2 intercalates in each of the two which intersected the diagonal. When prolongating we create $6m$ new intercalates involving the symbol $6m+1$. Also, using the transversals as in Theorem 5, we can add a further $m(m-1)/2$ intercalates for each diagonal block. So

$$I(E) \geq 5m^3+6m-2m^2+3m(m-1)$$
$$= m(5m^2+m+3).$$ □

4. ACKNOWLEDGEMENTS

The idea for this research arose in conversation with Donald Preece, who suggested that Latin squares with many intercalates might be useful in statistics.

We wish to thank P.J. Owens and Warren Brisley for useful discussions, and

F.P. Hiner and R.B. Killgrove and Rosemary Bailey for access to their unpublished work.

The first part of this research was carried out when we were both enjoying the hospitality of the University of Surrey; during its completion, Dr. Heinrich was partially supported by a grant from the University of Newcastle's Internal Research Assessment Committee.

REFERENCES

[1] P. Dembowski, *Finite Geometries* (Springer-Verlag, New York, 1968).

[2] J. Dénes and A.D. Keedwell, *Latin Squares and Their Applications* (Akadémiai Kiadó, Budapest, 1974).

[3] H.W. Norton, The 7×7 squares. *Ann. Eugenics* 9 (1939), 269-307.

Department of Mathematics
Simon Fraser University
Burnaby
British Columbia V5A 1S6, Canada

Department of Mathematics
University of Newcastle
New South Wales 2308
Australia

ELEGANT ODD RINGS AND NON-PLANAR GRAPHS

D.A. Holton and C.H.C. Little

We prove that a graph is non-planar if and only if it contains a strict elegant odd ring.

1. INTRODUCTION

Throughout we consider undirected graphs on a finite set of vertices. We denote the vertex set of a graph G by VG and its edge set by EG. If G is a directed graph, C is a directed circuit of G, and a and b are distinct vertices of VC, then we use the notation C(a, b) or $C^{-1}(b, a)$ to mean the directed subpath of C with origin a and terminus b. If a = b, then C(a, b) and $C^{-1}(b, a)$ mean the subpath of C with vertex set $\{a\}$ and empty edge set. Furthermore if P is a path in G with end vertices c, d, then we use IP to denote the set $VP \setminus \{c, d\}$. If a, b $\in$ VP, then P[a, b] denotes the subpath of P with end vertices a and b.

Let S be a collection of circuits of G. If the edges of a graph G can be directed so that every circuit of S is a directed circuit, then we say that S is consistently orientable. The cyclic sequence of circuits $S = (C_0, C_1, \ldots, C_{n-1})$ with $n \geq 3$ is a ring in the graph G, if

(i) S is consistently orientable,

(ii) $EC_i \cap EC_j \neq \emptyset$ if and only if $i = j$, $i \equiv j + 1 \pmod n$ or $i \equiv j - 1 \pmod n$, and

(iii) no edge of G belongs to more than two circuits of G.

We note that (ii) implies (iii) except when n = 3.

The cardinality, $|S|$, of S is the number of circuits in S. If $S = (C_0, C_1, \ldots, C_{n-1})$, then $|S| = n$ and we will refer to S as an n-ring. A ring in which $|S|$ is odd will be called an odd ring.

Let X and Y be distinct paths or circuits of a graph G and suppose that $|VX \cap VY| \geq 2$. Then an $\bar{X}Y$-path is a nondegenerate subpath P of Y, of maximal length, for which $IP \cap VX = \emptyset$ and $EP \cap EX = \emptyset$. An XY-path is a subpath P of Y, of maximal length, for which $EP \subseteq EX \cap EY$.

The ring $S = (C_0, C_1, \ldots, C_{n-1})$ is said to be strict if $|VC_i \cap VC_j| \leq 1$ whenever $EC_i \cap EC_j = \emptyset$. The ring S is elegant if, for each $i = 0, 1, \ldots, n-1$, there is a unique $\bar{C}_iC_{i+1}$ - path. This means that the only vertices that C_i and C_{i+1} have in common are those on the path M_i, where $EC_i \cap EC_{i+1} = EM_i$. We note that here, and throughout this paper, all subscripts are taken as being modulo n.

The purpose of this paper is to provide a combinatorial proof of the following theorem.

Theorem 1.1. A graph is non-planar if and only if it contains a strict elegant odd ring.

The two following results will be of value later.

Lemma 1.2: If $C_i, C_{i+1}, \ldots, C_j$ are consecutive circuits of a ring S, then $\bigcup_{r=i}^{j} C_r$ is 2-connected.

Proof: Clearly $\bigcup_{r=i}^{j} C_r$ is connected. Furthermore the lemma holds for $j = i$. Suppose now that $\bigcup_{r=i}^{j-1} C_r$ is 2-connected. Thus $\bigcup_{r=i}^{j-1} C_r \setminus \{v\}$ is connected for each $v \in \bigcup_{r=i}^{j-1} VC_r$.

Now choose $w \in \bigcup_{r=i}^{j} VC_r$. Then $\bigcup_{r=i}^{j} C_r \setminus \{w\}$ is connected, since $\bigcup_{r=i}^{j-1} C_r \setminus \{w\}$ and $C_j \setminus \{w\}$ are connected and $(VC_j \cap \bigcup_{r=i}^{j-1} VC_r) \setminus \{w\} \neq \emptyset$.

Hence $\bigcup_{r=i}^{j} C_r$ is 2-connected for each fixed i, by induction on j.

Theorem 1.3: (Wagner's Theorem). G is planar if and only if no subgraph of G can be contracted to K_5 or $K_{3,3}$.

Proof: See [4].

Theorem 1.4: (Kuratowski's Theorem). G is planar if and only if no subgraph of G is homeomorphic to K_5 or $K_{3,3}$.

Proof: See [1].

Finally, throughout the paper, whenever S denotes a ring in a graph G, we assume without loss of generality that for each edge $e \in EG$, there exists a circuit C_i in S such that $e \in EC_i$. The justification is that we are attempting to establish the non-planarity of G, and for this purpose the edges that belong to no circuit in S are irrelevant.

2 C_i - AVOIDING PATHS.

If C is a circuit in the graph G, then a C - avoiding path is a path P in G such that $IP \cap VC = \emptyset$. In this section we give two lemmas which guarantee the existence of C_i -avoiding paths, where C_i is a circuit of a ring of G.

Lemma 2.1: Let S be a strict ring and let $C_r, C_s, C_i \in S$, where

$C_i \notin \{C_r, C_{r+1}, \ldots, C_s\}$ and $\{C_r, C_s\} \cap \{C_{i-1}, C_i, C_{i+1}\} = \emptyset$. For any $u \in VC_r$ and $v \in VC_s$, there exists a C_i - avoiding path joining u and v. Furthermore, this path can be chosen so that its edge set is a subset of $\bigcup_{k=r}^{s} EC_k$.

Proof: If r = s, then either $C_s(u, v)$ or $C_s(v, u)$ is a C_i - avoiding path joining u and v, because S is strict and $\{C_s\} \cap \{C_{i-1}, C_i, C_{i+1}\} = \emptyset$, so that $|VC_s \cap VC_i| \leq 1$.

Suppose $S = (C_0, C_1, \ldots, C_{n-1})$. Then we may assume without loss of generality that $i = 0$ and $1 < r < s$. Since S is a strict ring, $EC_r \cap EC_{r+1} \neq \emptyset$ and $|VC_r \cap VC_0| \leq 1$, so that there exists $v_r \in (VC_r \cap VC_{r+1}) \setminus VC_0$. If $IC_r(u, v_r) \cap VC_0 = \emptyset$, then we let $Q_r = C_r(u, v_r)$; otherwise let $Q_r = C_r(v_r, u)$. Since $|VC_r \cap VC_0| \leq 1$ we have $[IQ_r \cup \{v_r\}] \cap VC_0 = \emptyset$.

If $r + 1 \neq s$, we choose $v_{r+1} \in [(VC_{r+1} \cap VC_{r+2}] \setminus VC_0$. If $VC_{r+1}(v_r, v_{r+1}) \cap VC_0 = \emptyset$, let $Q_{r+1} = C_{r+1}(v_r, v_{r+1})$; otherwise let $Q_{r+1} = C_{r+1}(v_{r+1}, v_r)$. It follows by the strictness of S that $VQ_{r+1} \cap VC_0 = \emptyset$.

We may now proceed inductively until Q_{s-1} has been defined. Then let $Q_s = C_s(v_{s-1}, v)$ if $IC_s(v_{s-1}, v) \cap VC_0 = \emptyset$, and let $Q_s = C_s(v, v_{s-1})$ otherwise. Hence $[IQ_s \cup \{v_{s-1}\}] \cap VC_0 = \emptyset$.

If we now let $H = \bigcup_{k=r}^{s} Q_k$, then H is a connected graph containing u and v such that $[VH \setminus \{u, v\}] \cap VC_0 = \emptyset$. Hence there is a path in H which is the required C_0 - avoiding path. That this path consists only of edges of $\bigcup_{k=r}^{s} EC_k$ is clear from the above construction.

We note that the above lemma holds whether or not $u \in VC_r \cap VC_i$ or $v \in VC_s \cap VC_i$. The result can be extended slightly in the case of a strict elegant ring, where we denote the unique C_iC_{i+1} - path by P_i.

Lemma 2.2: Let S be a strict elegant ring with $|S| = n \geq 4$ and let $C_r, C_s, C_i \in S$, where $C_i \notin \{C_r, C_{r+1}, \ldots, C_s\}$. For any $u \in VC_r$ and $v \in VC_s$, there exists a C_i - avoiding path joining u and v provided that

(i) if $r = i - 1$, then $u \notin IP_{i-1}$,

(ii) if $r = i + 1$, then $u \notin IP_i$,

(iii) if $s = i - 1$, then $v \notin IP_{i-1}$,

and (iv) if $s = i + 1$, then $v \notin IP_i$.

Furthermore, this path can be chosen so that the edge set is a subset of $\bigcup_{k=r}^{s} EC_k$.

Proof: If r = s, then the result follows trivially. Therefore once again, without loss of generality, we may assume that $i = 0$ and $r < s$. The result here is precisely that of Lemma 2.1 unless $r = 1$ or $s = n - 1$.

Suppose that $r = 1$ and $s > 1$. By hypothesis $u \notin IP_0$. Therefore for any $u_1 \in VP_1 \setminus VC_0$, there is a C_0 - avoiding subpath of C_1 joining u to u_1 since S is elegant. ($VP_1 \setminus VC_0 \neq \emptyset$ since $|VC_0 \cap VC_2| \leq 1$.) Similarly if $s = n - 1$, then $v \notin IP_{n-1}$ so that there is a C_0 - avoiding subpath of C_{n-1} joining v to a vertex $v_{n-2} \in VP_{n-2} \setminus VC_0$.

If $r = 1$, $s \neq n - 1$, then we may join u to u_1 by the above C_0 - avoiding path and u_1 to v by a C_0 - avoiding path of the type described in Lemma 2.1. Then some subgraph of the union of these two paths must be the path required.

Similarly the lemma holds if $r \neq 1$, $s = n - 1$ and if $r = 1$, $s = n - 1$.

3. INTERLOCKING PATHS.

In this section we investigate some configurations of a pair of paths and a circuit which force a graph to be non-planar.

Let X, Y, Z be circuits in a graph G. Let R_1 be an $\bar{X}Y$-path joining distinct vertices v_1 and v_3 of VX and let R_2 be an $\bar{X}Z$-path joining distinct vertices v_2 and v_4 of VX. Suppose that Q_1 and Q_2 are the two subpaths of X that join v_1 and v_3 and let $v_2 \in IQ_1$ and $v_4 \in IQ_2$. If $IR_1 \cap IR_2 \neq \emptyset$ then we say that Y and Z cross on X, or that R_1 and R_2 cross on X.

The following lemma was proved in [3].

Lemma 3.1: If circuits Y and Z cross on circuit X in a graph G and $|VY \cap VZ| = 1$, then G is non-planar.

The situation of Lemma 3.1 is shown in Figure 3.1.

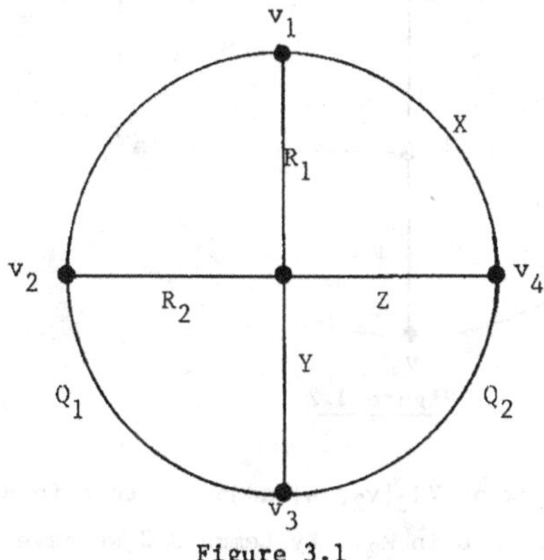

Figure 3.1

Before extending Lemma 3.1 to strict elegant rings via the notion of crossing, we note the following lemma which will be useful throughout our work.

Lemma 3.2: Let S be a strict elegant ring in a graph G with $C_i \in S$ and $|S| > 4$. Then no two vertices of VC_i are joined by an edge $e \in EC_j \setminus EC_i$ for any $C_j \in S$.

Proof: Suppose that for some $C_j \in S$ there exists $e \in EC_j \setminus EC_i$ joining two vertices of VC_i. Since $|VC_j \cap VC_i| > 2$, the strictness of S is contradicted unless $j = i \pm 1$. If $j = i \pm 1$, then the path with edge set $\{e\}$ is the unique $\bar{C}_i C_{i\pm 1}$ - path and so $e \in EC_{i\pm 2}$. Thus we again contradict the strictness of S.

Lemma 3.3: Let S be a strict elegant ring with $|S| > 4$ and let $C_i \in S$ and C be a circuit of G. If C_i and C cross on some circuit A of G, then G is non-planar.

Proof: Since C_i, C cross on the circuit A, we let C_i take the role of Y in the definition of "cross", C take the role of Z, and A the role of X. We may then define v_1, v_2, v_3, v_4, R_1, R_2 as in that definition. Because $IR_1 \cap IR_2 \neq \emptyset$ we can choose a, $b \in IR_1 \cap IR_2$ to minimise $|VR_2[a, v_2]| + |VR_2[b, v_4]|$. (Note that a and b may be the same vertex.) The situation is shown in Figure 3.2.

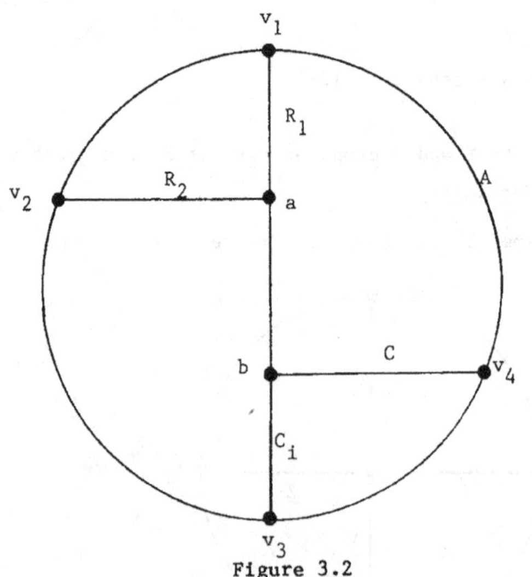

Figure 3.2

Now let u be the vertex of $VR_2[v_2, a]$ adjacent to a in R_2 and let v be the vertex of $VR_2[b, v_4]$ adjacent to b in R_2. By Lemma 3.2 we have u, $v \notin VC_i$, so that

$u, v \notin IP_{i-1} \cup IP_i$. Therefore by Lemma 2.2 there exists a C_i - avoiding path, Q, joining u and v.

We next note that $IR_1[a, b] \cap (VR_2[v_2, a] \cup VR_2[b, v_4] \cup VQ) = \emptyset$. Then let G' be the graph obtained from G by contracting the edges of $R_1[a, b]$ to a single vertex w. We define C_i' and R_1' in a similar fashion, while $R_2' = R_2[v_2, a] \cup R_2[b, v_4] = R_2[v_2, w] \cup R_2[w, v_4]$ and $C' = (C\backslash R_2) \cup R_2'$. For convenience, we let $A_2 = A\setminus\{v_2\}$ and $A_4 = A\setminus\{v_4\}$.

Case 1: Suppose $VQ \cap VA \neq \emptyset$. Choose $s \in VQ \cap VA$ to minimise $|VQ[u, s]|$ and $t \in VQ \cap VA$ to minimise $|VQ[t, v]|$. Further choose $c \in VQ[u, s] \cap VR_2'[v_2, w]$ such that $IQ[c, s] \cap VR_2'[v_2, w] = \emptyset$.

1.1: If $s \in IA_2[v_1, v_3]$ then $A \cup R_1' \cup R_2'[v_2, w] \cup Q[c, s]$ is a subdivision of $K_{3,3}$ and G' is non-planar.

1.2: If $t \in IA_4[v_1, v_3]$, then, similarly, G' is non-planar.

1.3: If $s \in IA_4[v_1, v_3]$ and $t \in IA_2[v_1, v_3]$, then we note that $IQ \cap IR_2'[u, v] = \emptyset$. Hence $Q \cup R_2'[u, v] = T$ is a circuit and $VT \cap VC_i' = \{w\}$. Then G' is non-planar by Lemma 3.1 applied to the circuits $X = A$, $Y = T$, $Z = C_i'$ and paths $Q[s, u] \cup R_2'[u, v] \cup Q[v, t]$ and R_1'.

Case 2: Suppose $VQ \cap VA = \emptyset$, so that $u \neq v_2$ and $v \neq v_4$. Then Q must contain a subpath Q', of minimal length, joining a vertex $u' \in IR_2'[v_2, w]$ to a vertex $v' \in IR_2'[w, v_4]$. Hence $A \cup R_1' \cup Q' \cup R_2'[v_2, w] \cup R_2'[v', v_4]$ is a subdivision of $K_{3,3}$ and so G' is non-planar.

Thus we see that in each case G' is non-planar. Therefore G' can be contracted to K_5 or $K_{3,3}$ by Wagner's Theorem (Theorem 1.3). Since G' is a contraction of G, it follows again from Wagner's Theorem that G is also non-planar.

We note that, throughout the lemma, the fact that C was a circuit was not used. The existence of R_2 is sufficient to obtain the non-planarity of G. Hence we have the following Corollary.

Corollary: Let S be a strict elegant ring with $|S| > 4$ and let $C_i \in S$ and let A be a circuit of G. Suppose there is an $\dot{\bar{A}}C_i$ - path, R_1, joining distinct vertices v_1 and v_3 of VA. Let Q_1, Q_2 be the two subpaths of A joining vertices v_1 and v_3. If $v_2 \in IQ_1$ and $v_4 \in IQ_2$ and there exists an A - avoiding path R_2 joining v_2 and v_4 such that $IR_1 \cap IR_2 \neq \emptyset$, then G is non-planar.

We now extend the result of the previous lemma to circuits which do not intersect.

Suppose A is a circuit and R_1, R_2 are paths which join four distinct vertices v_1, v_3 and v_2, v_4, respectively. Let $(VR_1 \cup VR_2) \cap VA = \{v_1, v_2, v_3, v_4\}$ and $VR_1 \cap VR_2 = \emptyset$. If Q_1 and Q_2 are the two subpaths of A which join v_1 and v_3 and if $v_2 \in IQ_1$ and $v_4 \in IQ_2$, then we say that R_1 and R_2 <u>interlock on A</u>. (See Figure 3.3.)

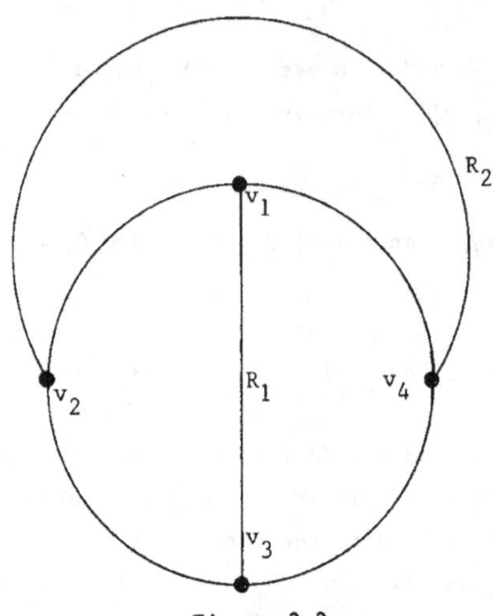

<u>Figure 3.3.</u>

<u>Lemma 3.4</u>: Let $S = (C_0, C_1, \ldots, C_{n-1})$ be a strict elegant ring in the graph G with $n > 4$, and let R_1 and R_2 be interlocking paths on some circuit A of the graph G. If there exists $C_k \in S$ such that

(i) either $IR_1 \cap VC_k \neq \emptyset$ or $ER_1 \cap EC_k \neq \emptyset$,

(ii) either $IR_2 \cap VC_k \neq \emptyset$ or $ER_2 \cap EC_k \neq \emptyset$, and

(iii) there exists at most one non-degenerate AC_k - path,

then G is non-planar.

<u>Proof</u>: Let v_1, v_3 and v_2, v_4 be the ends of R_1, R_2, respectively. By (i) and (ii) there exists a subpath Q of C_k such that $(IR_i \cap VQ) \cup (ER_i \cap EQ) \neq \emptyset$, $i = 1, 2$. By (iii) we may suppose that $EQ \cap EA = \emptyset$. In fact we choose Q to be a path of minimal length satisfying these three conditions. The minimality of the length of Q ensures that Q joins a vertex $u \in VR_1$ to a vertex $v \in VR_2$. We assume that R_1, R_2, A and C_k are chosen to minimise the length of Q. Among the remaining possibilities, we choose R_1, R_2, A and C_k to minimise $|VC_k \cap VA|$.

Case 1: Suppose $|VQ \cap VA| = 0$. This means we must have $u \in IR_1$ and $v \in IR_2$. Hence, by the minimality of $|VQ|$, $A \cup Q \cup R_1 \cup R_2$ is a subdivision of $K_{3,3}$.

Case 2: Suppose $|VQ \cap VA| = 1$. Let $VQ \cap VA = \{w\}$. Define $A_i = A \setminus \{v_i\}$ for $i = 1, 2, 3, 4$.

2.1: If $w \in IA_1[v_2, v_3] \cup IA_1[v_3, v_4] \cup IA_2[v_1, v_4] \cup IA_3[v_1, v_2]$, then again $u \in IR_1$ and $v \in IR_2$. We now consider only the case $w \in IA_2[v_1, v_4]$ (see Figure 3.4) as the others follow by symmetry. In this case, $R_1 \cup A_1[v_3, v_4] \cup R_2 \cup A_3[v_2, v_1]$ is a circuit of the graph, and C_k and A cross on this circuit via paths Q and $A_2[v_1, v_4]$, respectively. Hence, by Lemma 3.3, G is non-planar.

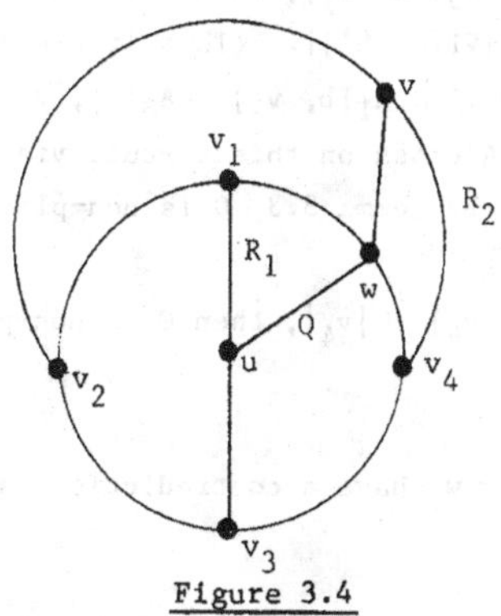

Figure 3.4

2.2: If $w \in \{v_1, v_2, v_3, v_4\}$ then we again note that symmetry allows us to consider only the case $w = v_3$. In this case, $v \in IR_2$ and $u \neq v_1$ since $VQ \cap VA = \{v_3\}$. We then have one of the situations of Figure 3.5.

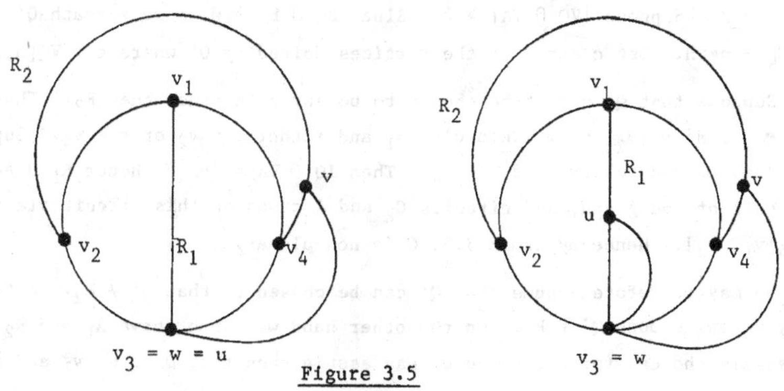

Figure 3.5

2.2.1: $u = v_3$. If $IR_1 \cap VQ \neq \emptyset$, then the minimality of the length of Q is contradicted, since $u \notin IR_1$.

If $IR_1 \cap VQ = \emptyset$, then $ER_1 \cap EQ \neq \emptyset$ by the choice of Q (via hypothesis (i)). Choose $e \in ER_1 \cap EQ$. Since $IR_1 \cap VQ = \emptyset$, e must join v_1 and v_3 and so $v_1, v_3 \in VQ \cap VA$. Hence the assumption that $|VQ \cap VA| = 1$ is contradicted.

2.2.2: $u \in IR_1$. Let T be the shortest subpath of $C_k \setminus \{w\}$ that joins u to a vertex a of $VA \cup IR_2$, and let b be the vertex of $VT \cap IR_1$ that minimises $|VT[a, b]|$. If we define T' to be T[a, b], then $IT' \cap (VA \cup IR_1 \cup IR_2) = \emptyset$.

2.2.2.1: If $a \in IR_2$, then $A \cup R_1 \cup R_2 \cup T'$ is a subdivision of $K_{3,3}$.

2.2.2.2: If $a \in IA_1[v_2, v_3] \cup \{v_2\}$, then let b' be the vertex of $VT \cap VR_1[b, v_1]$ that minimises $|VT[u, b']|$. (Thus if $b \in VR_1[v_3, u]$, then $b' = u$.) Then $R_2 \cup A_1[v_2, a] \cup T' \cup R_1[b, v_1] \cup A_3[v_1, v_4]$ is a circuit of the graph, and the circuits C_k and A cross on this circuit via paths $T[b', u] \cup Q$ and $A_1[a, v_4]$, respectively. Hence by Lemma 3.3, G is non-planar.

2.2.2.3: If $a \in IA_1[v_3, v_4] \cup \{v_4\}$, then G is non-planar by symmetry from Case 2.2.2.2.

2.2.2.4: If $a = v_3$, then we have a contradiction, since by definition T is a subpath of $C_k \setminus \{w\}$.

2.2.2.5: If $a \in IA_3[v_2, v_4]$, then $A_3[v_2, v_4] \cup R_2$ is a circuit of the graph, and circuits C_k and A cross on this circuit via paths $T \cup Q$ and $A_1[v_2, v_4]$. Hence by Lemma 3.3, G is non-planar.

Case 3: Suppose $|VQ \cap VA| \geq 2$. Since $EQ \cap EA = \emptyset$, some subpath Q' of Q must be an $\bar{A}C_k$ - path. Let c and d be the vertices joined by Q' where $c \in VQ[u, d]$.

Suppose that Q' cannot be chosen to be any path other than R_2. Thus $VQ \cap VA = \{c, d\} = \{v_2, v_4\}$. Then $u \in IR_1$ and either $v = v_2$ or $v = v_4$. Suppose without loss of generality that $v = v_4$. Then $IQ \cap VA = \{v_2\}$. Hence $R_1 \cup A_2[v_1, v_3]$ is a circuit of the graph, and circuits C_k and A cross on this circuit via the paths Q and $A_4[v_1, v_3]$. Hence by Lemma 3.3, G is non-planar.

We may therefore assume that Q' can be chosen so that $Q' \neq R_2$. Alternatively we may choose $Q' \neq R_1$. On the other hand we cannot have R_1 and R_2 as the only possible choices for Q', since we may assume then that Q joins v_1 and v_4 and hence some subpath of $Q[v_2, v_3]$ is a possible choice for Q'. It is therefore clear that we may choose $Q' \neq R_1$ and $Q' \neq R_2$.

It also follows that $u \notin VQ'$, for if $u \notin IR_1$, then the edge of EQ incident on u must be the unique edge of ER_1 by the choice of Q. Hence this edge cannot be in EQ'. Similarly $v \notin VQ'$.

We now locate the ends c and d of the path Q'. Suppose that $c \in IA_1[v_2, v_4]$ and $d \in IA_3[v_2, v_4]$, or vice versa. Then there is a subpath $\hat{Q}$ of $Q[c, v]$ which joins a vertex of VQ' to v with the property that $(IQ' \cap V\hat{Q}) \cup (EQ' \cap E\hat{Q}) \neq \emptyset$. Further $|V\hat{Q}| < |VQ|$ since $u \notin VQ'$. Hence Q', R_2 and $\hat{Q}$ can be chosen to replace R_1, R_2 and Q respectively and so contradict the minimality of $|VQ|$. Similarly we cannot have $c \in IA_2[v_1, v_3]$ and $d \in IA_4[v_1, v_3]$, or vice versa. Hence we may assume without loss of generality that $c, d \in VA_1[v_3, v_4]$.

The next step is to establish the existence of a C_k - avoiding path. Let B_1 and B_2 be the two subpaths of A which join the vertices c and d. If $IB_i \subseteq VC_k$ for $i = 1$ or $i = 2$, then by Lemma 3.2 together with the hypothesis that every edge of G belongs to a circuit of S we see that $C_k = Q' \cup B_i$ and so hypothesis (i) is contradicted. Hence we may choose a vertex $x' \in IB_1 \setminus VC_k$ and $y' \in IB_2 \setminus VC_k$. By Lemma 2.2, then, there exists a C_k - avoiding path joining x' and y'. We choose a minimal subpath U of this path which joins a vertex $x \in IB_1$ and a vertex $y \in IB_2$. We note that $IU \cap VA = \emptyset$, and that we may choose $x \in IA_1[c, d]$.

3.1: If $IU \cap IR_1 \neq \emptyset$, then $u \notin \{v_1, v_3\}$. For otherwise, $ER_1 \subset EQ$ and this contradicts the fact that $IU \cap IR_1 \neq \emptyset$.

We now choose $f \in IU \cap IR_1$ so as to minimise $|VU[x, f]|$.

3.1.1: If $IU \cap IR_2 \neq \emptyset$, then $g \in IR_2$ can be chosen to minimise $|VU[f, g]|$. Hence $R_1 \cup R_2 \cup A \cup U[f, g]$ contains a subdivision of $K_{3,3}$ and so G is non-planar.

3.1.2: Suppose $u \in IR_1[v_1, f]$. Then the paths $R_1[v_1, f] \cup U[f, x]$ and Q' interlock on A, and $|VQ[u, d]| < |VQ|$, since $v \notin VQ'$. Hence the choice of Q is contradicted.

3.1.3: We now show that the choice of R_1 and R_2 implies that $VQ[u, d] \cap IA_x[c, d] = \emptyset$, where $A_x = A \setminus \{x\}$. Indeed, suppose the contrary. Let q_1 be the vertex of $VQ[u, d] \cap IA_x[c, d]$ that minimises $|VQ[u, q_1]|$. If $VQ[u, q_1] \cap VA_1[c, d] = \emptyset$, then paths $U[x, f] \cup R_1[f, u] \cup Q[u, q_1]$ and Q' contradict the choice of R_1 and R_2. This is because there is a subpath Q^* of Q which joins u to a vertex of VQ' and satisfies $(IQ' \cap VQ^*) \cup (EQ' \cap EQ^*) \neq \emptyset$ and

$|VQ^*| < |VQ|$, since $v \notin VQ'$. Therefore there exists $q_2 \in VQ[u, q_1] \cap VA_1[c, d]$. But now if q_2 is chosen to minimise $|VQ[q_1, q_2]|$, it follows by similar reasoning that paths Q' and $Q[q_1, q_2]$ contradict the choice of R_1 and R_2. Therefore $VQ[u, d] \cap IA_x[c, d] = \emptyset$.

We next assume that, for a fixed x as defined above, Q' is chosen to minimise $|VQ[u, c]|$. It follows that $VQ[u, c] \cap VA \subset VA_1[c, x]$.

If $v_3 \notin \{c, d\}$, then an argument similar to Case 3.1.2 applies to the paths $R_1[v_3, f] \cup U[f, x]$ and Q', since $u \in IR_1[v_3, f]$.

If $c = v_3$ or $d = v_3$, we first define $K = C_k \setminus \{t\}$, where t is the vertex of VQ adjacent to u in Q. Let z_1 be the vertex of VK $\cap$ (VA $\cup$ VR_2 $\cup$ $IR_1[f, v_1]$) that minimises $|VK[u, z_1]|$. We note that $VK[u, z_1] \cap VU = \emptyset$, since U is C_k - avoiding, and also $VK[u, z_1] \cap VQ' = \emptyset$, since C_k is a circuit. Let z_0 be the vertex of $VK[u, z_1] \cap IR_1[v_3, f]$ that minimises $|VK[z_0, z_1]|$ and let z_2 be the vertex of $VK[u, z_1] \cap VR_1[z_0, f]$ that minimises $|VK[u, z_2]|$. (Thus if $z_0 \in IR_1[v_3, u] \cup \{u\}$, then $z_2 = u$.)

<u>3.1.3.1:</u> If $z_1 \in IR_2$, then $A \cup K[z_0, z_1] \cup R_1 \cup R_2$ is a subdivision of $K_{3,3}$.

<u>3.1.3.2:</u> If $z_1 \in IA_2[x, v_1]$, then $A \cup R_1 \cup U[f, x] \cup K[z_0, z_1]$ is a subdivision of $K_{3,3}$.

<u>3.1.3.3:</u> Suppose $z_1 \in IR_1[f, v_1] \cup \{v_1\} \cup IA_4[v_1, v_2]$. If $c = v_3$, then Lemma 3.3 can be applied to the circuit $R_2 \cup A_1[v_4, x] \cup U[x, f] \cup R_1[f, v_1] \cup A_4[v_1, v_2]$ and the paths $Q[d, u] \cup K[u, z_1]$ and $A_1[v_2, x]$. (Recall that the choice of Q' ensures that $VQ[u, c] \cap VA \subset VA_1[c, x]$.)

On the other hand, suppose that $d = v_3$.

<u>3.1.3.3.1:</u> If $z_1 \in IA_4[v_1, v_2] \cup \{v_1\}$, then Lemma 3.3 can be applied to the circuit $U[x, f] \cup R_1[f, z_0] \cup K[z_0, z_1] \cup A_4[z_1, x]$ and paths $Q[d, u] \cup K[u, z_2]$ and $A_3[x, z_1]$.

<u>3.1.3.3.2:</u> If $z_1 \in IR_1[f, v_1]$, then we apply Lemma 3.3 to the circuit $U[x, f] \cup R_1[f, z_0] \cup K[z_0, z_1] \cup R_1[z_1, v_1] \cup A_4[v_1, x]$ and paths $Q[d, u] \cup K[u,, z_2]$ and $A_3[x, v_1]$.

<u>3.1.3.4</u>: Suppose $z_1 \in IA_1[v_2, v_3] \cup \{v_2\}$. If $c = v_3$, then Lemma 3.3 can be applied to the circuit $U[x, f] \cup R_1[f, z_0] \cup K[z_0, z_1] \cup A_3[z_1, x]$ and paths $Q[d, u] \cup K[u, z_2]$ and $A_1[x, z_1]$. If $d = v_3$, then we argue as in Case 3.1.3.3.1.

<u>3.1.3.5</u>: Suppose $z_1 \in IA_1[x, v_3]$.

<u>3.1.3.5.1</u>: Let $c = v_3$. If $IQ[u, d] \cap IA_1[z_1, x] = \emptyset$, then we may apply Lemma 3.3 to the circuit $C = R_2 \cup A_1[v_4, z_1] \cup K[z_1, z_0] \cup R_1[z_0, v_1] \cup A_4[v_1, v_2]$ and paths $Q[d, u] \cup K[u, z_2]$ and $A_1[z_1, v_2]$.

Therefore we may suppose that there exists a vertex $s \in IQ[u, d] \cap IA_1[z_1, x]$ and we choose s to minimise $|VQ[u, s]|$.

If $IQ[u, s] \cap VA \neq \emptyset$, then $IQ[u, s] \cap VA \subseteq IA_1[v_3, z_1]$ by the choice of Q' and s, so that we may apply Lemma 3.3 to the circuit C and paths $Q[s, u] \cup K[u, z_2]$ and $A_1[z_1, v_2]$.

Therefore we may suppose that $IQ[u, s] \cap VA = \emptyset$. If $z_0 \in IR_1[u, f]$, then $A \cup R_1 \cup K[z_0, z_1] \cup Q[s, u]$ is a subdivision of $K_{3,3}$. We therefore assume that $z_0 \in IR_1[v_3, u] \cup \{u\}$. Since $c \in IQ[s, d]$, we may choose a vertex $s' \in IQ[s, d] \cap (IA_1[v_3, z_1] \cup \{v_3\})$ to minimise $|VG[s, s']|$. Then Lemma 3.3 can be applied to the circuit $U[x, f] \cup R_1[f, z_0] \cup K[z_0, z_1] \cup A_1[z_1, v_2] \cup A_3[v_2, x]$ and paths $Q[u, s']$ and $A_1[z_1, x]$.

<u>3.1.3.5.2</u>: If $d = v_3$, then we apply Lemma 3.3 to the circuit $R_1[v_1, z_0] \cup K[z_0, z_1] \cup A_4[z_1, v_1]$ and paths $Q[d, u] \cup K[u, z_2]$ and $A_2[z_1, v_1]$.

Hence in all cases, G is non-planar.

<u>3.2</u>: If $IU \cap IR_2 \neq \emptyset$, then a similar argument applies.

<u>3.3</u>: Suppose $IU \cap (IR_1 \cup IR_2) = \emptyset$.

<u>3.3.1</u>: If $y \in IA_4[v_1, v_2]$, then $A \cup R_1 \cup R_2 \cup U$ is a subdivision of $K_{3,3}$.

<u>3.3.2</u>: If $y \in IA_1[v_2, v_3] \cup \{v_2\}$, then $R_1 \cup A_1[v_3, x]$ and R_2 are interlocking paths on $U \cup A_3[x, y]$. Suppose $c \in VA_1[v_3, d]$. Since $|VQ[c, v]| < |VQ|$ and $c \in I(R_1 \cup A_1[v_3, x])$, the original choice of R_1, R_2 and Q is contradicted. A similar argument holds if $d \in VA_1[v_3, c]$.

3.3.3: If $y \in IA_3[v_1, v_4] \cup \{v_1\}$, then an argument similar to Case 3.3.2 applies.

3.3.4: Suppose $y \in VA_1[v_3, v_4]$. Without loss of generality, let $y \in VA_1[v_3, x]$. Now R_1 and R_2 are interlocking paths on the circuit $\bar{A} = U \cup A_2[x, v_1] \cup A_4[v_1, y]$. Since $IU \cap VC_k = \emptyset$ and either $c \notin V\bar{A}$ or $d \notin V\bar{A}$, it follows that $|VC_k \cap V\bar{A}| < |VC_k \cap VA|$. Hence the original choice of A, R_1, R_2, Q and C_k is contradicted.

Since all possibilities are exhausted, the lemma follows.

Under slightly stronger conditions the previous lemma generalises to the case where R_1 and R_2 are incident on two distinct circuits of the ring. Hence we have the following lemma.

Lemma 3.5: Let $S = (C_0, C_1, \ldots, C_{n-1})$ be a strict elegant ring in the graph G with $n \geq 4$, and let R_1 and R_2 be interlocking paths on some circuit A of G. If there exist C_j, $C_k \in S$, with $j \leq k$, such that

(i) either $IR_1 \cap VC_j \neq \emptyset$ or $ER_1 \cap EC_j \neq \emptyset$,

(ii) either $IR_2 \cap VC_k \neq \emptyset$ or $ER_2 \cap EC_k \neq \emptyset$,

(iii) $EA \cap \bigcup_{i=j+1}^{k-1} EC_i = \emptyset$ if $j < k - 1$,

(iv) $EA \cap EC_j \cap EC_k = \emptyset$ if $j = k - 1$, and

(v) there exist at most one non-degenerate AC_j - path and at most one non-degenerate AC_k - path,

then G is non-planar.

Proof: We assume that A, R_1, R_2, C_j and C_k are chosen to minimise $k - j$. If $j = k$, then the lemma follows immediately from Lemma 3.4. Therefore we assume that $k - j > 0$. As usual we let v_1, v_3 and v_2, v_4 be the ends of R_1 and R_2, respectively.

Case 1: Suppose $VC_j \cap VA \subseteq \{v_1, v_3\}$. By the minimality of $k - j$, we know that $(IR_1 \cap VC_{j+i}) \cup (ER_1 \cap EC_{j+i}) = \emptyset$ for all i such that $j < j + i \leq k$. Furthermore, if $(IR_2 \cap VC_j) \cup (ER_2 \cap EC_j) \neq \emptyset$, then G is non-planar by Lemma 3.4. Hence we may assume that $VR_2 \cap VC_j = \emptyset$.

1.1: If $|VR_1 \cap VC_j| \geq 2$, then since $ER_1 \cap EC_{j+1} = \emptyset$, there exists a subpath R of C_j, of minimal length, joining two vertices of R_1 and satisfying $ER \cap EC_{j+1} \neq \emptyset$ and $ER \cap ER_1 = \emptyset$. Let a and b be the end vertices of R. We choose a and b so that $b \in IR_1[a, v_3] \cup \{v_3\}$. Because $VR_2 \cap VC_j = \emptyset$, the paths $Q = R_1[v_1, a] \cup R \cup R_1[b, v_3]$ and R_2 have no common vertex. Further since $VC_j \cap VA \subseteq \{v_1, v_3\}$, the paths Q and R_2

interlock on A. But $ER \cap EC_{j+1} \neq \emptyset$, so that $EQ \cap EC_{j+1} \neq \emptyset$, and the minimality of $k - j$ is contradicted.

1.2: If $|VR_1 \cap VC_j| = 1$, then let $VR_1 \cap V\check{C}_j = \{c\}$. By hypothesis (i), it follows that $c \in IR_1$. Let X be a subpath of C_j, of minimal length, joining c to a vertex $d \in VC_j \cap \bigcup_{i=j+1}^{k} VC_i$. Let h be an integer such that $d \in VC_h$ and $j < h \leq k$. Since $\bigcup_{i=h}^{k} C_i$ is connected, for any $f \in VR_2 \cap \bigcup_{i=h}^{k} VC_i$ there exists a path Y joining d and f such that $EY \subset \bigcup_{i=h}^{k} EC_i$. We assume that Y and f are chosen to minimise $|VY|$. Since $EY \subset \bigcup_{i=h}^{k} EC_i$, we have $IR_1 \cap VY = \emptyset$, or else the minimality of $k - j$ is contradicted.

Suppose there exists $e \in EY \cap ER_2$. Then both end vertices of e are in $VC_k \cap VY \cap VR_2$ and so the minimality of $|VY|$ is contradicted, since at least one end vertex must be in IY. Hence $EY \cap ER_2 = \emptyset$.

Similarly if there exists $v \in IY \cap IR_2$, then $v = f$ by the minimality of $|VY|$. Hence $IY \cap IR_2 = \emptyset$.

1.2.1: If $f \in IR_2$, then suppose $VY \cap VA = \emptyset$.

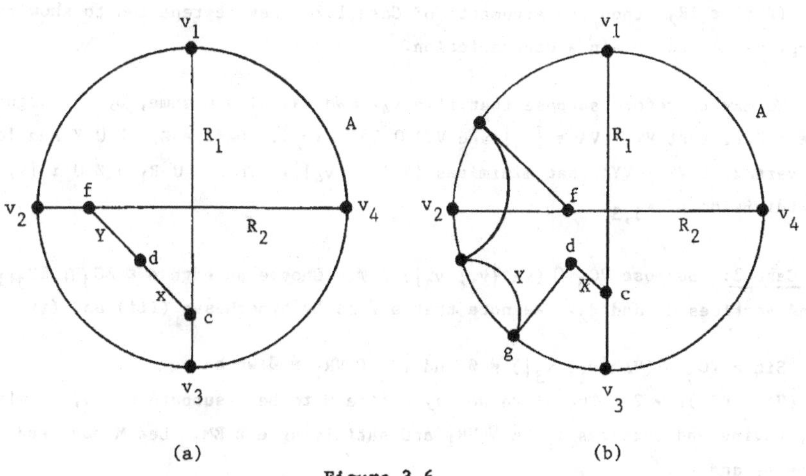

Figure 3.6

This is the situation of Figure 3.6 (a) and we see that $A \cup R_1 \cup R_2 \cup X \cup Y$ is a subdivision of $K_{3,3}$.

We may suppose then that $VY \cap VA \neq \emptyset$. Hence there exists $g \in VY \cap VA$ such that $IY[d, g] \cap VA = \emptyset$. (See Figure 3.6 (b).)

If $g = v_2$ or $g = v_4$, then we contradict the minimality of $|VY|$, since $f \in IR_2$.

If $g \in IA_1[v_2, v_4]$ (we recall that $A_i = A \setminus \{v_i\}$, $i = 1, 2, 3, 4$), then $T = R_1[v_1, c] \cup X \cup Y[d, g]$ is a path since $IR_1 \cap VY = \emptyset$, $IR_1 \cap IX = \emptyset$ and $IX \cap IY = \emptyset$. But T and R_2 interlock on A and, since $d \in VC_h$, they contradict the minimality of $k - j$.

If $g \in IA_3[v_2, v_4]$ a similar argument to the above also contradicts the minimality of $k - j$.

<u>1.2.2</u>: If $f = v_2$ or $f = v_4$, then we can assume without loss of generality that $f = v_2$. By hypothesis (ii), $VC_k \cap VR_2 \neq \{v_2\}$. Now since $\bigcup_{i=h}^{k} C_i$ is 2-connected by Lemma 1.2, there must exist a path Y' joining d to a vertex f' of $(\bigcup_{i=h}^{k} VC_i) \cap (IR_2 \cup \{v_4\})$ such that $VY' \subset \bigcup_{i=h}^{k} VC_i \setminus \{v_2\}$. We now choose Y' and f' to minimise $|VY'|$.

If $f' \in IR_2$, then the arguments of Case 1.2.1 may be repeated to show that G is non-planar or to obtain a contradiction.

We may therefore suppose that $f' = v_4$. We may also assume, by the arguments of Case 1.2.1, that $VY \cap VA = \{v_2\}$ and $VY' \cap VA = \{v_4\}$. Define $Z = X \cup Y$ and let y be the vertex of $VZ \cap VY'$ that minimises $|VY'[y, v_4]|$. Then $A \cup R_1 \cup Z \cup Y'[y, v_4]$ is a subdivision of $K_{3,3}$.

<u>Case 2</u>: Suppose $VC_j \cap (VA \setminus \{v_1, v_3\}) \neq \emptyset$. Choose an edge $e \in EC_j \cap EC_{j+1}$ with end vertices d_1 and d_2. We note that $e \notin EA$ by hypotheses (iii) and (iv).

Since $VC_j \cap (VA \setminus \{v_1, v_3\}) \neq \emptyset$ and $VC_j \cap VR_1 \neq \emptyset$ we have $|VC_j \cap (VA \cup VR_1)| \geq 2$. Therefore we may define M to be a subpath of C_j, of minimal length, having end vertices in $VA \cup VR_1$ and satisfying $e \in EM$. Let M have end vertices p_1 and p_2.

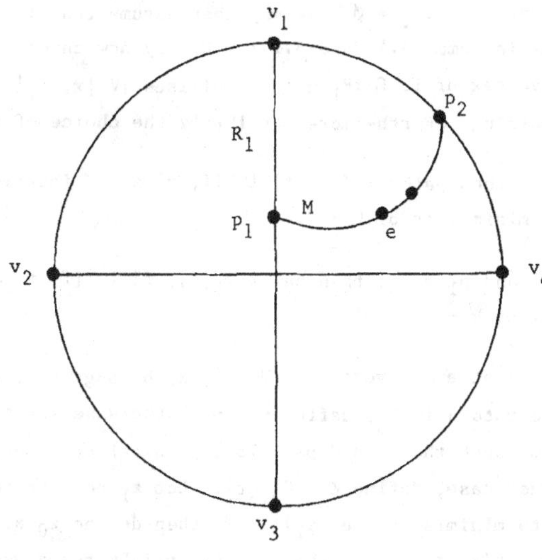

Figure 3.7

2.1: Suppose $p_1 \in IR_1$. Since $e \in EM$ and $IR_1 \cap VC_{j+1} = \emptyset$, it follows that at least one of the end vertices of e belongs to IM. Now we use the arguments of Case 1.2 with c replaced by p_1 and C_j by M.

2.2: If $p_2 \in IR_1$, then we use a similar argument to that of Case 2.1.

2.3: Suppose $p_1, p_2 \in VA$. We assume that A, R_1, R_2 are chosen so as to minimise $|VC_j \cap VA|$. By the minimality of $k - j$ we note that $M \neq R_1$ and $M \neq R_2$.

If $p_1 \in IA_1[v_2, v_4]$ and $p_2 \in IA_3[v_2, v_4]$, then, since $e \in EC_j \cap EC_{j+1} \cap EM$, the circuit A and paths M and R_2 contradict the minimality of $k - j$.

Similarly we cannot have $p_1 \in IA_2[v_1, v_3]$ and $p_2 \in IA_4[v_1, v_3]$. We may therefore assume without loss of generality that $p_1, p_2 \in VA_1[v_3, v_4]$ with $p_1 \in VA_1[v_3, p_2]$.

The arguments now required are analogous to those of Lemma 3.4 Case 3 with M, p_1, p_2 playing the roles of Q', c, d, respectively. We choose u to be any vertex of $VC_j \cap (IR_1 \cup \{v_1\})$. Such a vertex exists by hypothesis (i). We choose $u \in IR_1$ if possible. We then define the path U and vertices x and y as in Lemma 3.4 Case 3, with C_j playing the role of C_k. Here we choose $x \in IA_1[p_1, p_2]$.

<u>2.3.1</u>: Suppose that $IU \cap IR_1 \neq \emptyset$. We may then assume that $IU \cap IR_2 = \emptyset$ for otherwise we may argue as in Lemma 3.4 Case 3.1.1. We may now choose f as in Lemma 3.4 Case 3.1, to be the vertex of $IU \cap IR_1$ which minimises $|VU[x, f]|$. Note that $f \neq u$ since U is C_j - avoiding. Furthermore $u \in IR_1$ by the choice of u.

If $u \in IR_1[v_1, f]$, then paths $R_1[v_1, f] \cup U[f, x]$ and M interlock on A and therefore contradict the minimality of $k - j$.

If $u \in IR_1[v_3, f]$ and $p_1 \neq v_3$, then paths $R_1[v_3, f] \cup U[f, x]$ and M contradict the minimality of $k - j$.

We may now assume that every vertex of $VC_j \cap IR_1$ belongs to $IR_1[v_3, f]$ and $p_1 = v_3$. If p_1 is adjacent to u in C_j, define $t = p_1$; otherwise let t be the vertex of VC_j adjacent to u in C_j such that t and p_2 belong to distinct subpaths of C_j joining u to p_1. In either case, define $K = C_j \setminus \{t\}$. Let z_1 be a vertex of $VK \cap (VA \cup VR_2)$ chosen to minimise $|VK[u, z_1]|$. We then define z_0 as in Lemma 3.4 Case 3.1.3 and so G is non-planar if $z_1 \in IR_2 \cup IA_2[x, v_1]$ by the arguments of Lemma 3.4 Cases 3.1.3.1 and 3.1.3.2.

<u>2.3.1.1</u>: If $z_1 \in IA_4[v_1, v_3] \cup \{v_1\}$, then paths $U[f, x] \cup R_1[f, z_0] \cup K[z_0, z_1]$ and M interlock on A.

<u>2.3.1.2</u>: If $z_1 \in IA_1[v_3, x]$, then paths $R_1[v_1, z_0] \cup K[z_0, z_1]$ and M interlock on A.

In each of the above cases, the minimality of $k - j$ is contradicted.

<u>2.3.2</u>: If $IU \cap IR_2 \neq \emptyset$, then we can repeat the arguments of Case 2.3.1, since $EM \cap EC_{j+1} \neq \emptyset$ and $k - (j + 1) < k - j$.

<u>2.3.3</u>: Suppose $IU \cap (IR_1 \cup IR_2) = \emptyset$.

<u>2.3.3.1</u>: If $y \in IA_4[v_1, v_2]$, then $A \cup R_1 \cup R_2 \cup U$ is a subdivision of $K_{3,3}$.

<u>2.3.3.2</u>: If $y \in IA_4[v_2, p_1] \cup \{v_2\}$, then paths R_2 and either R_1 or $R_1 \cup A_1[v_3, x]$ interlock on the circuit $\bar{A} = U \cup A_1[y, v_2] \cup A_3[v_2, x]$. Since U is C_j - avoiding and $p_1 \notin V\bar{A}$, the minimality of $|VC_j \cap VA|$ is contradicted.

<u>2.3.3.3</u>: If $y \in IA_2[v_1, p_2] \cup \{v_1\}$, then we obtain a contradiction by an argument similar to that in Case 2.3.3.2.

The proof of the lemma is now complete.

4. A SUBSIDIARY LEMMA

In this section we give a lemma which will be useful in the next section. It involves three paths which have common end vertices.

Lemma 4.1: Let $S = (C_0, C_1, \ldots, C_{n-1})$ be a strict elegant ring with $n \geq 4$, in a graph G, and let G be oriented so that every circuit of S is a directed circuit. Let X, Y, Z be three distinct vertex-disjoint paths in G joining vertices a and b. Suppose $C_i \in S$, and let $(VC_i \cap IZ) \cup (EC_i \cap EZ) = \emptyset$ and assume that any subpath of C_i joining a and b contains vertices of $IX \cup IY$ or edges of $EX \cup EY$. Let Q, R be non-degenerate directed subpaths of X, Y, respectively, such that $(EX \cup EY) \cap EC_i = EQ \cup ER$. Let o, t be the origin and terminus, respectively, of Q and o', t', the origin and terminus, respectively, of R.

If (a) $o \in IX[a, t] \cup \{a\}$ and $o' \in IY[a, t'] \cup \{a\}$,

or (b) $t \in IX[a, o] \cup \{a\}$ and $t' \in IY[a, o'] \cup \{a\}$,

then G is non-planar.

Proof: We suppose, without loss of generality, that (a) holds. (See Figure 4.1.)

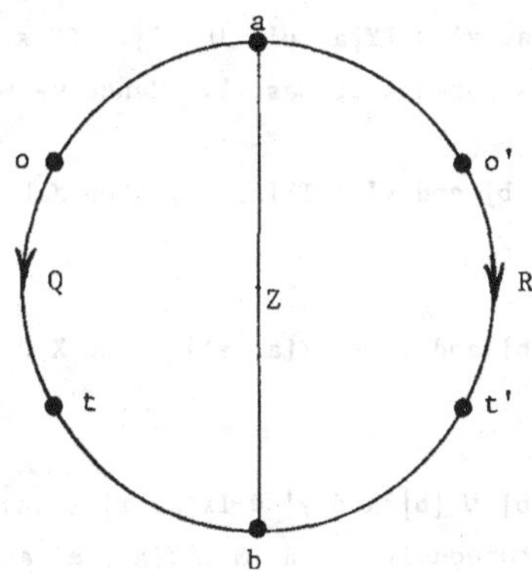

Figure 4.1

Since C_i is a circuit, there must exist a subpath M of C_i, of minimal length, with origin $x \in (VX[t, b] \setminus \{t'\}) \cup IY[b, t']$ and terminus $x' \in IX[o, a] \cup (VY[a, o'] \setminus \{o\})$. Similarly there exists a subpath N of C_i, of minimal length, with origin $y \in IA[x, x']$ and terminus $y' \in IB[x, x']$, where $A = (X \cup Y) \setminus \{o\}$ and $B = (X \cup Y) \setminus \{t'\}$.

We now consider the various possibilities for x, x', y, y'. By the symmetry between X and Y, we note that we may suppose without loss of generality that $x \in VX[t, b]$.

<u>Case 1</u>: Suppose $x' \in IX[o, a] \cup \{a\}$.

<u>1.1</u>: If $y \in IX[x, b] \cup \{b\}$, then $x \neq b$ and the paths M and N interlock on the circuit $X \cup Z$. Then G is non-planar by Lemma 3.4.

<u>1.2</u>: If $y \in IY[b, t'] \cup \{t'\}$ and $x \neq b$, then the paths M and $N \cup Y[y, b]$ interlock on the circuit $X \cup Z$ and G is non-planar by Lemma 3.4.

<u>1.3</u>: If $y \in IY[b, t'] \cup \{t'\}$ and $x' \neq a$, then the paths M and $N \cup Y[y, a]$ interlock on the circuit $X \cup Z$ and G is non-planar by Lemma 3.4.

<u>1.4</u>: If $x = b$ and $x' = a$, the hypotheses of the lemma are contradicted.

<u>1.5</u>: If $y \in IX[x', a] \cup IY[a, o'] \cup \{a\}$, then arguments similar to those above can be applied to give G non-planar.

<u>Case 2</u>: Suppose that $x' \in IY[a, o'] \cup \{o'\}$. If $x = b$, then by the symmetry of X and Y we may use the arguments of Case 1. Hence we may assume that $x \neq b$.

<u>2.1</u>: If $y \in IY[x', b]$ and $y' \in IX[x, a]$, then $X \cup Y \cup Z \cup M \cup N$ is a subdivision of $K_{3,3}$.

<u>2.2</u>: If $y \in IX[x, b]$ and $y' \in IY[a, x']$, then $X \cup Y \cup Z \cup M \cup N$ is a subdivision of $K_{3,3}$.

<u>2.3</u>: If $y \in IX[x, b] \cup \{b\}$ and $y' \in IX[a, x] \cup \{a\}$, then since $y = b$ and $y' = a$ cannot happen simultaneously, paths $M \cup Y[x', a]$ and N or paths $M \cup Y[x', b]$ and N interlock on $X \cup Z$. Hence G is non-planar by Lemma 3.4.

<u>2.4</u>: If $y \in IY[b, x'] \cup \{b\}$ and $y' \in IY[x', a] \cup \{a\}$, then the result follows from Case 2.3 by the symmetry of X and Y.

5. THE MAIN LEMMA

We now come to the main lemma of this paper.

Lemma 5.1: Let $S = (C_0, C_1 \ldots, C_{n-1})$ be a strict odd elegant ring, with $n \geq 5$, in a graph G. If there exists $C_k \in S$ such that $VC_k \cap VC_j \neq \emptyset$ for some $C_j \in S \setminus \{C_{k-1}, C_k, C_{k+1}\}$, then G is non-planar.

Proof: Since S is consistently orientable, we may assume that G is oriented so that all the circuits of S are directed.

For all i, we let the unique $\bar{C}_i C_{i+1}$ - path have origin v_i and terminus u_i. Then $C_i(u_i, v_i)$ is a $C_i C_{i+1}$ - path which we denote by P_i.

Since S is strict, there exists a vertex v such that $VC_k \cap VC_j = \{v\}$. Suppose that $v \in VC_i$ for all i. Hence $v \in VP_i$ for all i. In fact v is an end vertex of P_i since $EC_{i-1} \cap EC_{i+1} = \emptyset$. Moreover, since C_{i+1} is a directed circuit, $v = u_i$ if and only if $v = v_{i+1}$. Thus the oddness of S is contradicted and there exists h such that $v \notin VC_h$.

By reordering S cyclically if necessary, we may assume that $0 \leq j < h < k$. We may also assume that C_k and C_j are chosen to minimise $k - j$. If follows that $v \notin VC_h$ for all h satisfying $j < h < k$. In particular $v \neq u_{k-1}$ and $v \neq v_{k-1}$.

We next show that if $\ell, m \in \{j, j+1, \ldots, k-1\}$ and $\ell \leq m - 2$ then $VC_\ell \cap VC_m \subset \{u_{m-1}, v_{m-1}\}$. Indeed, choose $u \in VC_\ell \cap VC_m$. Then $u \in VC_h$ for all $h \in \{\ell, \ell+1, \ldots, m\}$, for otherwise the minimality of $k - j$ is contradicted, since $m - \ell \leq k - 1 - j$. Since $VC_m \cap VC_{m-1} = VP_{m-1}$, it follows that $u \in VP_{m-1}$. Similarly $u \in VP_{m-2}$, so that u is an end vertex of P_{m-1}. In fact, it follows that u is an end vertex of P_h for all $h \in \{\ell, \ell+1, \ldots, m-1\}$. A similar argument yields the same conclusion if $\ell, m \in \{j+1, j+2, \ldots, k\}$ and $\ell \leq m - 2$.

We now define

$$A = \begin{cases} C_j(v_j, v) \cup \left[\bigcup_{i=0}^{\frac{1}{2}(k-j-2)} C_{j+2i+1}(v_{j+2i}, u_{j+2i+1}) \right] \\ \quad \cup \left[\bigcup_{i=1}^{\frac{1}{2}(k-j-2)} C_{j+2i}(v_{j+2i}, u_{j+2i-1}) \right] & \text{if } k-j \text{ is even} \\ C_j(v, u_j) \cup \left\{ \bigcup_{i=0}^{\frac{1}{2}(k-j-3)} \left[C_{j+2i+1}(v_{j+2i+1}, u_{j+2i}) \right. \right. \\ \quad \left. \left. \cup \; C_{j+2i+2}(v_{j+2i+1}, u_{j+2i+2}) \right] \right\} & \text{if } k-j \text{ is odd.} \end{cases}$$

The conclusion of the preceding paragraph shows that A is a path. In either case, define B to be the path with edge set $\bigcup_{i=j}^{k-1} EC_i \setminus [EA \cup \bigcup_{i=j}^{k-1} EP_i]$. The choice of C_j and C_k ensures that $VC_k \cap (IA \cup IB) = \emptyset$.

Case 1: Suppose that $v \in IP_k$. This possibility leads to a number of cases which are dismissed either by using elementary properties of the ring or by discovering a subdivision of $K_{3,3}$. Hence we will show that in all such possible cases, G is non-planar.

If $C_j = C_{k+2}$, then there exists a unique C_jC_{k+1} - path, P_{k+1}. We have $EP_{k+1} \neq \emptyset$ since $EC_j \cap EC_{k+1} \neq \emptyset$. But the path with vertex set $\{v\}$ is also a C_jC_{k+1} - path, since $EP_k \subset EC_{k+1}$. Thus the elegance of the ring S is contradicted. We conclude that $C_j \neq C_{k+2}$, so that $EC_j \cap EC_{k+1} = \emptyset$ and $|VC_j \cap VC_{k+1}| \leq 1$.

We note that $IC_{k+1}(v_k, u_k) \cap VC_k = \emptyset$ for otherwise there exists more than one $\bar{C}_kC_{k+1}$ - path. We also note that we cannot have both $v_k = u_{k-1}$ and $u_k = v_{k-1}$ for otherwise $|VC_{k-1} \cap VC_{k+1}| \geq 2$. By the symmetry of A and B we may therefore assume without loss of generality, that $v_k \neq u_{k-1}$. The situation of this case is shown in Figure 5.1.

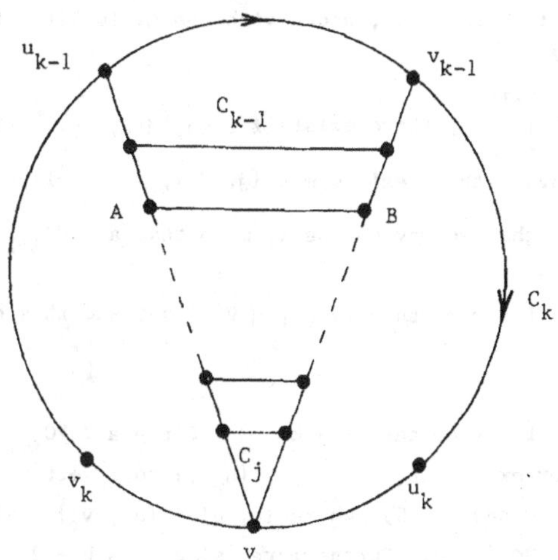

Figure 5.1

1.1: Suppose $VC_{k+1}(v_k, u_k) \cap VA = \emptyset$. We note that some subgraph of $C_{k+1}(v_k, u_k)$ is the $C_{k+1}\ C_{k+2}$ - path, P_{k+1}, with origin u_{k+1} and terminus v_{k+1}. Since $|VC_k \cap VC_{k+2}| \leq 1$, either $u_{k+1} \in IC_{k+1}(v_k, u_k)$ or $v_{k+1} \in IC_{k+1}(v_k, u_k)$. Hence $IC_{k+1}(v_k, u_k) \cap VC_{k+2} \neq \emptyset$. Choose $w \in IC_{k+1}(v_k, u_k) \cap VC_{k+2}$.

1.1.1: If $u_j \in VA$, then, by the minimality of $k - j$, $v \notin VC_{j+1}$ and so $u_j \neq v$. Further $VC_j \cap VC_k = \{v\}$ and so $u_j \neq u_{k-1}$. Hence $u_j \in IA$. Thus, by Lemma 2.1, there is a C_k - avoiding path R' joining w and u_j. Let R be a shortest subpath of R' joining a vertex of $IC_{k+1}(v_k, u_k)$ to a vertex of IA. Then $C_k \cup C_{k+1} \cup A \cup R$ is a subdivision of $K_{3,3}$ and so G is non-planar.

1.1.2: If $v_j \in VA$, then we use a similar argument to that of Case 1.1.1.

1.2: Suppose $VC_{k+1}(v_k, u_k) \cap VA \neq \emptyset$. Since $v_k \neq u_{k-1}$, we see that $u_{k-1} \notin VC_{k+1}$, for otherwise there would be more than one $\tilde{C}_k C_{k+1}$ - path. Further $v \notin VC_{k+1}(v_k, u_k)$ and so, from the hypothesis, $VC_{k+1}(v_k, u_k) \cap IA \neq \emptyset$.

1.2.1: If $VC_{k+1}(v_k, u_k) \cap IB \neq \emptyset$, then there exists a subpath Q of $VC_{k+1}(v_k, u_k)$, of minimal length, joining vertices $a \in IA$ and $b \in IB$. Suppose $v_{k-1} \in IQ$. Then $u_k = v_{k-1}$ since there is only one $C_k C_{k+1}$ - path. Hence v_{k-1} is an end vertex of Q, in contradiction to the assumption that $b \in IB$. Hence we may suppose that $v_{k-1} \notin IQ$ and this, along with the minimality of Q, implies that $IQ \cap (VA \cup VB) = \emptyset$.

Since $EA \subset \bigcup_{i=j}^{k-1} EC_i$, there exists $\ell \in \{j, j+1, \ldots, k-1\}$ such that $a \in VC_\ell$ and similarly, there exists $m \in \{j, j+1, \ldots, k-1\}$ such that $b \in VC_m$. But a, $b \notin VC_k$, so that we may choose ℓ, m so that $a \notin VC_{\ell+1}$ and $b \notin VC_{m+1}$.

1.2.1.1: If $\ell = m$, then $|VC_{k+1} \cap VC_\ell| \geq 2$ and this contradicts the fact that S is strict.

1.2.1.2: If $\ell < m$, then $\ell \leq k - 2$. Since $a \notin VC_{\ell+1}$, it follows that P_ℓ joins a vertex $p \in IA[u_{k-1}, a] \cup \{u_{k-1}\}$ to a vertex $q \in IB[b, v]$ (since $b \notin VC_\ell$ by the strictness of S), where $\{p, q\} = \{u_\ell, v_\ell\}$. Also $IP_\ell \cap IQ = \emptyset$, for otherwise $|VC_{k+1} \cap VC_\ell| \geq 2$. Furthermore, since $\ell \leq k - 2$, we have $IP_\ell \cap VC_k = \emptyset$ since any vertex of $VC_\ell \cap VC_k$ must be an end vertex of P_ℓ. It follows that $A \cup B \cup P_\ell \cup Q \cup C_k(u_{k-1}, v)$ is a subdivision of $K_{3,3}$. Hence G is non-planar.

1.2.1.3: If $m < \ell$, then $A \cup B \cup P_\ell \cup Q \cup C_k(v, v_{k-1})$ is a subdivision of $K_{3,3}$ and so G is again non-planar.

1.2.2: If $VC_{k+1}(v_k, u_k) \cap IB = \emptyset$, then there are two cases to consider.

1.2.2.1: If $u_k = v_{k-1}$ then there is a subpath Q of $C_{k+1}(v_k, u_k)$, of minimal length, joining v_{k-1} to a vertex $a \in IA$. Then $a \notin VC_{k-1}$, because otherwise $|VC_{k-1} \cap VC_{k+1}| \geq 2$. Therefore $a \in IA[v, v_{k-2}]$. Also $IQ \cap VC_{k-1} = \emptyset$ by the strictness of S.

1.2.2.1.1: Suppose that $v_{k-2} \neq u_{k-1}$ and $u_{k-2} \neq v_{k-1}$. Then $Q \cup P_{k-2} \cup A \cup B \cup C_k(v, v_{k-1})$ is a subdivision of $K_{3,3}$.

1.2.2.1.2: Suppose next that $u_{k-2} = v_{k-1}$. We note that $u_j \neq v_{k-1}$ and $v_j \neq v_{k-1}$ since $VC_k \cap VC_j = \{v\}$. Therefore we may let r be the largest integer in $\{j, j+1, \ldots, k-1\}$ such that $u_r \neq v_{k-1}$ and $v_r \neq v_{k-1}$. Hence $r < k - 2$. By the choice of r, we have $v_{k-1} \in VP_h$ for all $h \in \{r+1, r+2, \ldots, k\}$, so that $a \notin VC_h$ by the strictness of S. Therefore if p' and q' are the vertices of VA and VB respectively joined by P_r, it follows that $p' \in IA[a, u_{k-1}] \cup \{u_{k-1}\}$ and $q' \in IB$. If $p' = u_{k-1}$, then $u_{k-1}, v_{k-1} \in VC_{r+1} \cap VC_k$ in contradiction to the strictness of S, since $r + 1 \leq k - 2$. Therefore $p' \in IA[a, u_{k-1}]$, so that $Q \cup P_r \cup A \cup B \cup C_k(v, v_{k-1})$ is a subdivision of $K_{3,3}$.

1.2.2.1.3: Finally, suppose that $u_{k-2} \neq v_{k-1}$ but $v_{k-2} = u_{k-1}$. Thus $j < k - 2$, since $u_{k-1} \notin \{u_j, v_j\}$. Without loss of generality, let $u_j \in IA$. Upon contraction of the edges of $EA[u_j, a] \cup EB[v_j, u_{k-2}]$, the graph $C_k \cup A \cup B \cup Q \cup P_{k-2} \cup P_j$ then yields a subdivision of K_5, so that G is non-planar by Wagner's Theorem.

1.2.2.2: If $u_k \in IC_k(v_{k-1}, v)$, then we let Q be the subpath of $C_{k+1}(v_k, u_k)$, of minimal length, joining u_k to a vertex $a \in IA$.

Now $a \notin VC_j$, since otherwise $|VC_{k+1} \cap VC_j| > 2$. Similarly $IQ \cap VP_j = \emptyset$. Hence $C_k \cup A \cup P_j \cup B[x, v_{k-1}] \cup Q$ is a subdivision of $K_{3,3}$, where $\{x\} = VP_j \cap VB$.

Case 2: Suppose without loss of generality that $v \in VC_k(v_{k-1}, u_k)$. Hence $u_k, v_k \in VC_k(v, u_{k-1})$. Without loss of generality we will assume that $u_j \in IA$ and $v_j \in IB$.

2.1: If $(IB[v_{k-1}, v_j] \cup IP_j \cup \{v_j\}) \cap \bigcup_{i=k+1}^{j-1} VC_i \neq \emptyset$ (see Figure 5.2), then paths $C_k(v, u_{k-1})$ and $B(v_{k-1}, v_j) \cup P_j$ interlock on the circuit $A \cup C_k(u_{k-1}, v)$. The hypotheses of Lemma 3.5 are satisfied and so G is non-planar.

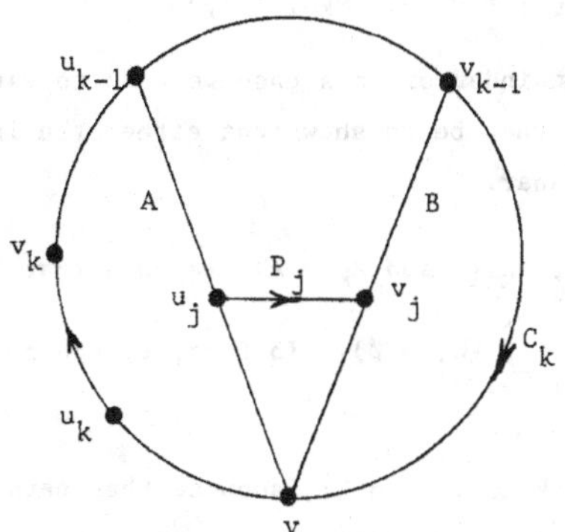

Figure 5.2

2.2: If $(IB[v_{k-1}, v_j] \cup IP_j \cup \{v_j\}) \cap \bigcup_{i=k+1}^{j-1} VC_i = \emptyset$ but $IB[v, v_j] \cap \bigcup_{i=k+1}^{j-1} VC_i \neq \emptyset$, then by Lemma 2.2, there exists a C_k - avoiding path joining v_k to a vertex of $IB[v, v_j]$, such that the edges of this path are in

$\bigcup_{i=k+1}^{j-1} EC_i$. Let P be a subpath of this C_k - avoiding path which is of minimal length and which joins a vertex $x \in \{v_k\} \cup IA \cup VP_j \cup IB[v_{k-1}, v_j]$ to a vertex $y \in IB[v, v_j]$.

Now since $(IB[v_{k-1}, v_j] \cup IP_j \cup \{v_j\}) \cap \bigcup_{i=k+1}^{j-1} VC_i = \emptyset$, we have $x \notin IB[v_{k-1}, v_j] \cup IP_j \cup \{v_j\}$.

If $x \in IA$, then paths $C_k(v, u_{k-1})$ and $B[v_{k-1}, y] \cup P$ interlock on $C_k(u_{k-1}, v) \cup A$. Hence by Lemma 3.5, G is non-planar.

Finally, if $x = v_k$ then $x \in IC_k(v, u_{k-1}) \cup \{u_{k-1}\}$ and $C_k(x, v) \cup A \cup B \cup P_j \cup P$ is a subdivision of $K_{3,3}$, and so G is non-planar.

<u>2.3</u>: We may now suppose that $(IB \cup IP_j) \cap \bigcup_{i=k+1}^{j-1} VC_i = \emptyset$.

Now if, for some $\ell \in \{j, j+1, \ldots, k-2\}$ and $i \in \{k+1, k+2, \ldots, j-1\}$, we have $IP_\ell \cap VC_i \neq \emptyset$, $u_{k-1} \neq u_\ell$, and $u_{k-1} \neq v_\ell$, then the paths $C_k(v, u_{k-1})$ and $B[v_{k-1}, q] \cup P_\ell$ interlock on $C_k(u_{k-1}, v) \cup A$, where $q = v_\ell$ if $u_\ell \in IA$ and $q = u_\ell$ if $v_\ell \in IA$. Hence by Lemma 3.5, G is non-planar. Henceforward we may assume that $IP_\ell \cap VC_i = \emptyset$ whenever $\ell \in \{j, j+1, \ldots, k-2\}$, $i \in \{k, k+1, \ldots, j-1\}$, $u_{k-1} \neq u_\ell$ and $u_{k-1} \neq v_\ell$.

To settle the remainder of this case we need to establish an inductive procedure. Our aim will then be to show that either the inductive step may be made or that G is non-planar.

Define $F_k = C_k(v, v_{k-1})$ and $A_k = A$. We note that $EP_k \subsetneq F_k(v, u_{k-1})$ and $EP_{j-1} \subset EA_k$ (since $IB \cap \bigcup_{i=k+1}^{j-1} VC_i = \emptyset$). In fact, by the properties of the ring, $EP_{j-1} \subsetneq EA[v, u_j]$.

For some $i \in \{k, k+1, \ldots, j-2\}$, suppose that paths F_i and A_i have been defined with the following properties:

(i) F_i joins v and v_{k-1};
(ii) A_i joins two distinct vertices a_i and a_i' of VF_i, where $a_i' \in VF_i[a_i, v_{k-1}] \cap VA[a_i, u_{k-1}]$;
(iii) $(IA_i \cap VF_i) \cup (EA_i \cap EF_i) = \emptyset$;
(iv) $EF_i[v, a_i] \subset_i EA$;
(v) $EF_i \subset EA \cup \bigcup_{h=k} EC_h$ and $\emptyset \subset EF_i \cap EC_{i+1} \subsetneq EF_i[a_i, a_i']$;

(vi) $EA \cap EC_{j-1} \subseteq EA_i \subseteq EA$.

We also assume that F_i satisfies a seventh property, to be described shortly.

By hypothesis (v) together with the hypotheses governing this present Case 2.3, we observe that $VF_i \cap (IB \cup IP_j) = \emptyset$. We also note that F_k and A_k satisfy the properties listed above.

In order to describe the seventh property of F_i we need a definition. For all i, let $N_i = F_i \cap C_{i+1}$ and let N_i be a path with origin o_i and terminus t_i. If either $t_i \in VF_i[o_i, a_i]$ and $t_{i-1} \in VF_{i-1}[o_{i-1}, a'_{i-1}]$ or $t_i \in VF_i[o_i, a'_i]$ and $t_{i-1} \in VF_{i-1}[o_{i-1}, a_{i-1}]$, then F_i and F_{i-1} are said to alternate. The final property of F_i is

(vii) F_i and F_{i-1} alternate for all $i > k$.

In what follows we show that given F_i and A_i, we can either construct F_{i+1} and A_{i+1} to satisfy (i)-(vii) or demonstrate that G is non-planar.

2.3.1: Suppose that $i \neq j - 2$.

2.3.1.1: Assume that $VC_{i+1} \cap [(VF_i[v, a_i] \cup VF_i[a'_i, v_{k-1}]) \setminus \{a_i, a'_i\}] \neq \emptyset$.

2.3.1.1.1: If $a'_i \in VC_j$, then $a_i \in VC_j$. The situation, in its simplest form, is given in Figure 5.3. Now by the strictness of S, we cannot have both $a_i \in VC_{i+1}$ and $a'_i \in VC_{i+1}$. Since $EF_i[a_i, a'_i] \cap EC_{i+1} \neq \emptyset$, $VC_{i+1} \cap [(VF_i[v, a_i] \cup VF_i[a'_i, v_{k-1}]) \setminus \{a_i, a'_i\}] \neq \emptyset$ and $|VC_{i+1} \cap VC_j| \leq 1$, there must be a subpath X of C_{i+1}, of minimal length, joining a vertex of $IF_i[a_i, a'_i]$ to a vertex of $IF_i[a'_i, v_{k-1}] \cup \{v_{k-1}\} \cup [VF_i[v, a_i] \setminus \{a_i\}]$ and satisfying $[IX \cap (VF_i \cup IA_i)] \cup (EX \cap EF_i) = \emptyset$. Then the paths X and A_i interlock

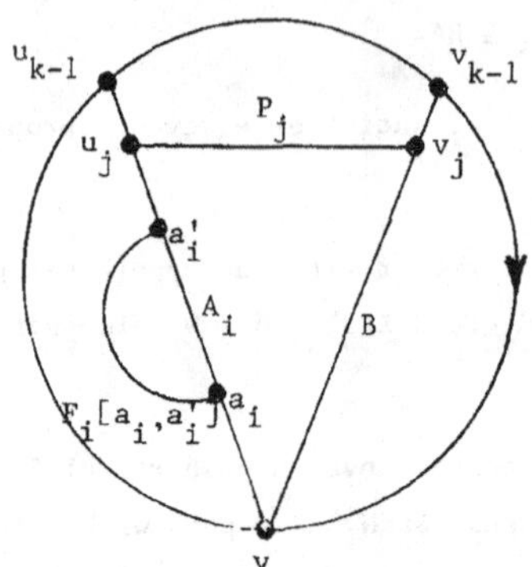

Figure 5.3

on $F_i \cup B$. Furthermore, $EX \cap EC_{i+1} \neq \emptyset$ and $EA_i \cap EC_{j-1} \neq \emptyset$. Also, if $i \neq j - 3$, then $\bigcup_{h=i+2}^{j-2} EC_h \cap (EF_i \cap EB) = \emptyset$ since $EF_i \subset \bigcup_{h=k}^{i} EC_h \cup EA$ and $EB \subset \bigcup_{h=j}^{k-1} EC_h$. In addition $(EF_i \cup EB) \cap EC_{i+1} \cap EC_{j-1} = \emptyset$ because $i \in \{k, k+1, \ldots, j-3\}$ so that $EC_{i+1} \cap EC_j = \emptyset$, $EC_i \cap EC_{j-1} = \emptyset$ and $(EC_{i+1} \cup EC_{j-1}) \cap \bigcup_{h=j+1}^{i-1} EC_h = \emptyset$. Finally there exists at most one non-degenerate F_iC_{i+1} - path since $EF_i \subset \bigcup_{h=k}^{i} EC_h \cup EA$, and no non-degenerate BC_{i+1} -, F_iC_{j-1} - or BC_{j-1} - paths. Thus G is non-planar by Lemma 3.5.

<u>2.3.1.1.2:</u> Suppose $a_i' \notin VC_j$. Since $EA_i \subseteq EA$ and $EA_i \cap EC_{j-1} \neq \emptyset$, then $a_i \in IA[v, u_j] \cup \{v\}$. Further, since $a_i \notin IP_{j-1}$, then $EF_i \cap EC_{j-1} = \emptyset$.

Now choose any vertex $x \in VC_{i+1} \cap [(VF_i[v, a_i] \cup VF_i[a_i', v_{k-1}]) \setminus \{a_i, a_i'\}]$, and let $X = F_i[v, a_i'] \cup A_i[a_i', u_j] \cup P_j \cup B[v_j, v]$. Then X is a circuit, since $(IB \cup IP_j) \cap \bigcup_{h=k}^{j-1} VC_h = \emptyset$.

<u>2.3.1.1.2.1:</u> If $x \in VF_i[a_i', v_{k-1}] \setminus \{a_i'\}$, then we observe that $EX \cap \bigcup_{h=i+2}^{j-2} EC_h = \emptyset$ for all $i \neq j-3$ and since $EC_{j-1} \cap EX = \emptyset$, then $EC_{i+1} \cap EC_{j-1} \cap EX = \emptyset$. Because there exists at most one non-degenerate XC_{i+1} - path and no non-degenerate XC_{j-1} - path, we conclude that G is non-planar by Lemma 3.5 applied to the circuit X and the paths $A_i[a_i, u_j]$ and $F_i[a_i', v_{k-1}] \cup B[v_{k-1}, v_j]$.

2.3.1.1.2.2: If $x \in VF_i[v, a_i] \setminus \{a_i\}$, then since $x \in VC_{i+1}$, we must have $VC_j \cap VC_{i+1} = \{x\}$ by the strictness of S. Since $EF_i[a_i, a_i'] \cap EC_{i+1} \neq \emptyset$, there exists an $\bar{X}C_{i+1}$ - path, Q, joining x to a vertex of $VF_i[a_i, a_i'] \cup IA[a_i', u_j]$. $IQ \cap VC_j = \emptyset$ as was noted by the strictness of S. Hence Q and $A_i[a_i, u_j]$ interlock on X and so by Lemma 3.5, G is non-planar.

2.3.1.2: We may now assume that $VC_{i+1} \cap [(VF_i[v, a_i] \cup VF_i[a_i', v_{k-1}]) \setminus \{a_i, a_i'\}] = \emptyset$. There are now two types of argument. One is to show, as we have done before, that G is non-planar, while the other is to establish the existence of F_{i+1} and A_{i+1} and so continue the induction.

By the hypotheses of this sub-case we know that there exists a subpath Y of C_{i+1}, of minimal length, joining two vertices of $VA_i \cup VF_i[a_i, a_i']$ and satisfying $EY \cap EC_{i+2} \neq \emptyset$. Let y, y' be the end vertices of Y. Then, by hypotheses, $\{y, y'\} \subseteq VA_i \cup VF_i[a_i, a_i']$ and $IY \cap [VA_i \cup VF_i[a_i, a_i']] = \emptyset$.

We observe here that if we let $C = A_i \cup F_i[a_i, a_i']$, then by Lemma 3.4 we may assume that there are no two $\bar{C}C_{i+1}$ - paths which interlock on C.

2.3.1.2.1: Suppose $y, y' \in VF_i[a_i, a_i']$. Without loss of generality, we assume $y' \in VF_i[y, a_i']$. The situation is shown in Figure 5.4 although a_i' may be in VC_j.

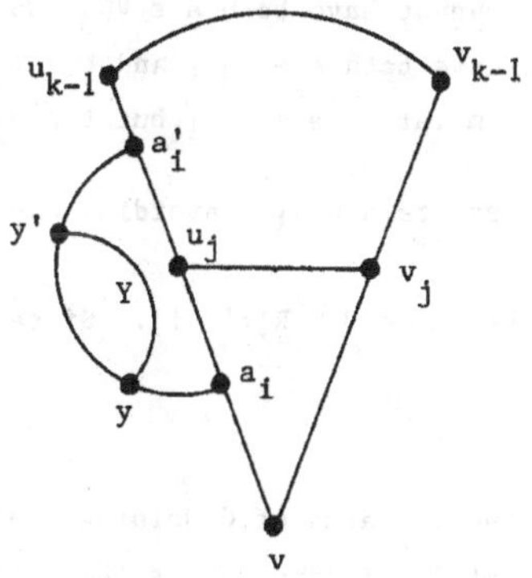

Figure 5.4

Let $A_{i+1} = A_i$, $a_{i+1} = a_i$, $a_{i+1}' = a_i'$ and $F_{i+1} = F_i[v, y] \cup Y \cup F_i[y', v_{k-1}]$. We now establish that G is non-planar or A_{i+1} and F_{i+1} satisfy the required properties (i)-(vii). Clearly properties (i)-(vi) are straightforward. We must show that F_i and F_{i+1} alternate.

First we note that $IB \cap \bigcup_{h=k+1}^{j-1} VC_h = \emptyset$ by the hypothesis governing Case 2.3 and $VC_{i+1} \cap [(VF_i[v, a_i] \cup VF_i[a_i', v_{k-1}]) \setminus \{a_i, a_i'\}] = \emptyset$, by the hypothesis of the present subcase. Hence if we define $B_i = F_i[a_i, v] \cup B \cup F_i[v_{k-1}, a_i']$, then $VC_{i+1} \cap IB_i = \emptyset$.

<u>2.3.1.2.1.1</u>: Suppose that there exists $e \in EF_i[y', a_i'] \cap EC_{i+1}$. We note that $e \in EC_i \cap EC_{i+1}$. Let e join the vertices c and d, where $c \in VF_i[a_i', d]$.

Since $EC_{i+2} \cap EY \neq \emptyset$, $EC_i \cap EC_{i+1} \cap EC_{i+2} = \emptyset$, and no two $\bar{C}C_{i+1}$ - paths interlock on C, it follows that P_{i+1} has an end vertex $f' \in VC_{i+1}'[d, y]$, where $C_{i+1}' = C_{i+1} \setminus \{c\}$. If $f' \notin VC$, let R be the subpath of C_{i+1} such that $f' \in IR$, $IR \cap VC = \emptyset$, and R joins two vertices of VC. Let f and g be the vertices of VC joined by R. If $f' \in VC$, let $f = g = f'$ and let R be the path with vertex set $\{f'\}$. Since no two $\bar{C}C_{i+1}$ - paths interlock on C it is clear that $f, g \in VF_i[d, y]$. Without loss of generality, let $f \in VF_i[d, g]$.

Define $D = C \setminus \{e\}$. Since $y \in VD[d, u_{j-1}] \cap VC_{i+1}$, $c \in VD[c, u_{j-1}] \cap VC_{i+1}$ and $EC_{i+1} \cap EC_j = \emptyset$, there must exist a $\bar{C}C_{i+1}$ - path, L, joining a vertex $a \in VD[y, u_{j-1}]$ to a vertex $b \in ID[u_{j-1}, c] \cup \{c\}$.

We now note that we cannot have both $a \in VC_j$ and $b \in VC_j$ by the strictness of S. Therefore, we cannot have both $a = u_{j-1}$ and $b = v_{j-1}$. We thus assume that $a \neq u_{j-1}$. The argument is similar if $a = u_{j-1}$ but $b \neq v_{j-1}$.

By Lemma 2.2, there exists a C_{i+1} - avoiding path U joining f' to u_{j-1} such that $VU \subset \bigcup_{h=i+2}^{j-1} VC_h$. Let $K' = U \cup R[f', f]$. Since U is C_{i+1} - avoiding, then K' is a path.

Let X and Z be the two subpaths of C joining a and b. For the sake of the discussion we will assume that $f \in IX$ and $u_{j-1} \in IZ$. Since $a, b \notin VR[f', f] \cup \{u_{j-1}\} \supseteq VK' \cap VC_{i+1}$ and K' joins a vertex in IX to a vertex in IZ, there must be a subpath K of K' joining a vertex of IX to one of IZ and satisfying $(IK \cap VC) \cup (EK \cap EC) = \emptyset$.

Then K and L interlock on C, so that G is non-planar by Lemma 3.5.

2.3.1.2.1.2: Suppose that $EF_i[a_i, y] \cap EC_{i+1} \neq \emptyset$. A similar argument to that of Case 2.3.1.2.1.1 shows that G is non-planar.

2.3.1.2.1.3: Suppose that $EF_i[y, y'] \cap EC_{i+1} \neq \emptyset$.

We apply Lemma 4.1 to show that F_i and F_{i+1} must alternate. Now $F_i[y, y']$, Y and $W = F_i[y', a_i'] \cup A_i \cup F_i[a_i, y]$ are three vertex-disjoint paths in G joining the vertices y and y', since no two $\bar{C}C_{i+1}$ - paths interlock on C, $VC_{i+1} \cap IW = \emptyset$. Further there exists no subpath T of C_{i+1}, other than Y, joining y and y' and containing no edges of $EF_i[y, y']$, since then $T \cup Y$ would be C_{i+1}. Let Q be the non-degenerate directed subpath of $F_i[y, y']$ such that $(EF_i[y, y'] \cup EY) \cap EC_{i+1} = EQ \cup ER$ where $R = Y$. Further let o, t be the origin and terminus, respectively, of Q and o', t' be the origin and terminus, respectively, of R. Then, by lemma 4.1 if

(a) $o \in IF_i[y, t] \cup \{y\}$ and $o' \in IY[y, t'] \cup \{y\}$ or

(b) $t \in IF_i[y, o] \cup \{y\}$ and $t' \in IY[y, o'] \cup \{y\}$, then G is non-planar.

Hence we may assume that F_i and F_{i+1} alternate.

2.3.1.2.2: Suppose Y joins a vertex of $IF_i[a_i, a_i']$ to one of IA_i. Without loss of generality, we may assume that $y \in VF_i$ and $y' \in VA_i$. Since $EA_i \cap EC_{j-1} \neq \emptyset$ and $a_i' \notin IP_{j-1}$, we have $v_{j-1} \in IA[u_{j-1}, a_i'] \cup \{a_i'\}$. There are three cases to consider.

2.3.1.2.2.1: Suppose $y' \in IA_i[u_{j-1}, v_{j-1}]$. We show that G is non-planar in this case. First we observe that without loss of generality, we may assume that $(IF_i[a_i', y] \cup \{a_i'\}) \cap VC_{i+1} \neq \emptyset$.

Since no two $\bar{C}C_{i+1}$ - paths can interlock on C, then we must have $[IF_i[a_i, y] \cup IA_i[a_i, y'] \cup \{a_i\}] \cap VC_{i+1} = \emptyset$. Hence there must exist a $\bar{C}C_{i+1}$ - path H joining y' to a vertex $h \in IA_i[y', a_i'] \cup \{a_i'\} \cup IF_i[a_i', y]$. Then we also have $ID[y', h] \cap VC_{i+1} = \emptyset$ where $D = C \setminus \{a_i\}$, because no two $\bar{C}C_{i+1}$ - paths can interlock on C.

Since $EY \cap EC_{i+2} \neq \emptyset$, $EC_{i+1} \cap EF_i \neq \emptyset$ and $EC_i \cap EC_{i+1} \cap EC_{i+2} = \emptyset$, then u_{i+1} and v_{i+1} are not both in VH. We assume, without loss of generality, that $u_{i+1} \notin VH$. We also note that by Lemma 2.2, there exists a C_{i+1} - avoiding path, K', joining u_{i+1} to v_{j-1} such that $EK' \subseteq \bigcup_{r=i+2}^{j-1} EC_r$.

We note that $u_{i+1} \notin ID[y', h]$ because otherwise there exists a $\bar{C}C_{i+1}$ - path that interlocks with H on C. Thus $u_{i+1} \notin VH \cup ID[y', h]$. But $v_{j-1} \in ID[y', h]$ and so there must exist a subpath K of K', of minimal length, joining a vertex $s \in ID[y', h]$ to a vertex $t \in (VC \setminus VD[y', h]) \cup \{u_{i+1}\}$. We now define

$$J = \begin{cases} K & \text{if } t \in VC \\ K \cup C^*_{i+1}[t, t'] & \text{otherwise,} \end{cases}$$

where $C^*_{i+1} = C_{i+1} \setminus \{h\}$ and t' is a vertex of $(VC^*_{i+1} \cap VC) \setminus \{y'\}$ that minimises $|VC^*_{i+1}[t, t']|$. Then paths H and J interlock on the circuit C and so G is non-planar by Lemma 3.5.

<u>2.3.1.2.2.2:</u> Suppose $y' \in IA_i[a_i^!, v_{j-1}] \cup \{v_{j-1}\}$. If $VC_{i+1} \cap (IF_i[y, a_i] \cup \{a_i\} \cup IA_i[a_i, y']) \neq \emptyset$, then there exists a subpath H of C_{i+1} of minimal length joining a vertex of $VA_i[v_{j-1}, y']$ to a vertex of $IF_i[y, a_i] \cup VA_i[a_i, u_{j-1}]$ since no two $\bar{C}C_{i+1}$ - paths interlock on C, and S is strict. Since S is strict, H does not join v_{j-1} to u_{j-1} and so an analogous argument to that of the previous subcase can be used to show that G is non-planar. Hence we may assume that $VC_{i+1} \cap (IF_i[y, a_i] \cup \{a_i\} \cup IA_i[a_i, y']) = \emptyset$.

We now define $F_{i+1} = F_i[v, y] \cup Y \cup A_i[y', a_i^!] \cup F_i[a_i^!, v_{k-1}]$ and $A_{i+1} = A_i[a_i, y']$. The induction hypotheses (i) through (vi) are now readily verified. We next prove that F_i and F_{i+1} are alternating.

Let $X = A_i[y', a_i^!] \cup F_i[a_i^!, y]$, $Z = A_i[a_i, y'] \cup F_i[a_i, y]$, $a = y'$ and $b = y$. Then X, Y, Z are three vertex-disjoint paths in G joining vertices a and b. Furthermore, $VC_{i+1} \cap IZ = \emptyset$ and there is no subpath T of C_{i+1}, other than Y, joining a and b and containing no vertices of $IX \cup IY$ for otherwise $T \cup Y$ would be C_{i+1}. If Q, R are the non-degenerate directed subpaths of X, Y, respectively, such that $(EX \cup EY) \cap EC_{i+1} = EQ \cup ER$, we see that $EQ = EF_i \cap EC_{i+1}$ and $ER = EY$. Then Lemma 4.1 guarantees that F_{i+1} and F_i alternate unless G is non-planar.

<u>2.3.1.2.2.3:</u> Suppose that $y' \in IA_i[a_i, u_{j-1}] \cup \{u_{j-1}\}$. Then we define $F_{i+1} = F_i[v, a_i] \cup A_i[a_i, y'] \cup Y \cup F_i[y, v_{k-1}]$ and $A_{i+1} = A_i[y', a_i^!]$, and an argument similar to that of Case 2.3.1.2.2.2 shows that F_{i+1} and A_{i+1} satisfy all the inductive hypotheses.

<u>2.3.1.2.3:</u> Suppose $y, y' \in VA_i$. Without loss of generality, we assume that $y' \in IA_i[y, a_i^!] \cup \{a_i^!\}$. Since $|VC_j \cap VC_{i+1}| \leq 1$ we see that we cannot have

$y' \in IP_{j-1}$. Again we will show that either G is non-planar, or the induction can be continued.

2.3.1.2.3.1: Suppose that $y \in IP_{j-1}$. If $i = j - 3$, then C_{i+1} and C_{j-1} are consecutive circuits. Since there exists a unique $C_{i+1}C_{j-1}$ - path, P_{j-2}, y is an end vertex of P_{j-2}. Therefore y is an end vertex of some edge $e \in EC_{j-1} \cap EA_i \cap EC_{i+1}$. But $e \in EP_{j-1}$, so that $e \in EC_{j-1} \cap EC_j \cap EC_{i+1}$, in contradiction to the fact that S is a ring.

If $i \neq j - 3$, then $VC_{j-1} \cap VC_{i+1} = \{y\}$. Since $u_{j-1} \in IA_i[a_i, y] \cup \{a_i\}$ and $v_{j-1} \in IA_i[y, y']$ and $y' \notin VC_{j-1}$, there exists a subpath J of $C_{j-1}(v_{j-1}, u_{j-1})$, of minimal length, joining a vertex of $IA_i[y, y']$ to a vertex of $IA_i[y', a_i'] \cup VF_i[a_i, a_i'] \cup IA_i[a_i, y]$. Clearly $VJ \cap VC_{i+1} = \emptyset$ since $VC_{j-1} \cap VC_{i+1} = \{y\}$. Hence $VJ \cap VY = \emptyset$. Then J and Y interlock on C and hence G is non-planar by Lemma 3.5.

2.3.1.2.3.2: Suppose that $u_{j-1}, v_{j-1} \in VA_i[y, y']$. We then define $F_{i+1} = F_i[v, a_i] \cup A_i[a_i, y] \cup Y \cup A_i[y', a_i'] \cup F_i[a_i', v_{k-1}]$ and $A_{i+1} = A_i[y, y']$. Again properties (i) through (vi) of the induction step are readily checked. We must now show that F_{i+1} and F_i alternate. Again we use Lemma 4.1, with $X = A_i[y', a_i'] \cup F_i[a_i', a_i] \cup A_i[a_i, y]$ and $Z = A_i[y, y']$, and $a = y'$, $b = y$.

We note that $VC_{i+1} \cap IZ = \emptyset$ since otherwise two $\bar{C}C_{i+1}$ - paths interlock on C. Hence by Lemma 4.1, F_i and F_{i+1} must alternate unless G is non-planar.

2.3.1.2.3.3: Suppose that $y' \in VA_i[a_i, v_{j-1}]$. Then $\{y, y'\} \subseteq VC_{i+1} \cap VC_j$ and the strictness of S is contradicted.

2.3.1.2.3.4: The final possibility is that $y, y' \in VA_i[v_{j-1}, a_i']$. Now $IA_i[y, y'] \cap VC_{i+1} = \emptyset$, since otherwise two $\bar{C}C_{i+1}$ - paths interlock on C. Since $EC_{i+1} \cap EC_j = \emptyset$ and $EC_{i+1} \cap EF_i \neq \emptyset$, there must therefore exist a subpath R of C_{i+1}, of minimal length, joining a vertex r of $VA_i[v_{j-1}, y]$ to a vertex r' of $IA_i[a_i, v_{j-1}] \cup IF_i[a_i, a_i'] \cup \{a_i\}$.

If $r' \in IA_i[a_i, v_{j-1}] \cup \{a_i\}$, then by the strictness of S, there exist

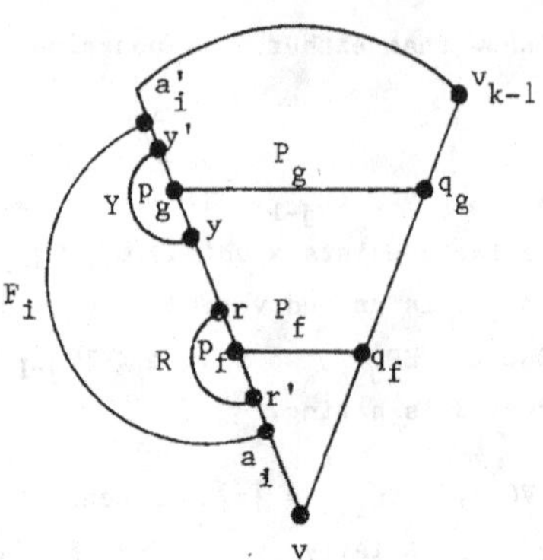

Figure 5.5

$f, g \in \{j, j+1, \ldots, k-2\}$ such that P_f and P_g have an end vertex in $IA_i[r, r']$ and $IA_i[y, y']$, respectively, and $(IP_f \cup IP_g) \cap VC_{i+1} = \emptyset$. Suppose $p_f \in VP_f \cap IA_i[r, r']$, $p_g \in VP_g \cap IA_i[y, y']$ and q_f, q_g are the other end vertices of P_f, P_g, respectively. (See Figure 5.5.) Hence $p_f \in IA_i[v_{j-1}, r] \cup \{v_{j-1}\}$. Furthermore, if $IP_g \cap VC_h = \emptyset$ for some $h \in \{k, k+1, \ldots, j-1\}$, then P_g and Y interlock on the circuit $A \cup B \cup C_k(u_{k-1}, v_{k-1})$, and so G is non-planar by Lemma 3.5. Hence we may assume that $IP_g \cap VF_i = \emptyset$, and similarly $IP_f \cap VF_i = \emptyset$. Thus paths $A_i[r', p_f] \cup P_f \cup B[q_f, q_g]$ and $Y \cup A_i[y', a_i']$ interlock on $F_i[z, v_{k-1}] \cup B[v_{k-1}, q_g] \cup P_g \cup A_i[p_g, r] \cup D[r, z]$, where $D = C_{i+1} \setminus \{y'\}$ and z is the vertex of $VF_i[a_i, a_i'] \cap VC_{i+1}$ that minimises $|VD[r, z]|$. Hence G is non-planar by Lemma 3.5.

If $r' \in IF_i[a_i, a_i']$, then by the strictness of S, there exists $\ell \in \{j, j+1, \ldots, k-2\}$, such that P_ℓ has an end vertex p_ℓ, say, in $IA_i[y, y']$. Let q_ℓ be the other end vertex of P_ℓ. Then paths $A_i[a_i', y'] \cup Y$ and $B[q_\ell, v] \cup A[v, a_i] \cup F_i[a_i, r']$ interlock on $F_i[r', v_{k-1}] \cup B[v_{k-1}, q_\ell] \cup P_\ell \cup A_i[p_\ell, r] \cup R$. Hence G is non-planar by Lemma 3.5.

2.3.2: Suppose that $i = j - 2$.

Define $N = P_{j-2} \cap F_{j-2}$ and let N be a path with origin o and terminus t. We have already shown that G is non-planar unless F_i and F_{i+1} alternate for all $i \in \{k, k+1, \ldots, j-3\}$. Since $u_j \in VA$ and $|S|$ is odd we may assume that $o \in VF_{j-2}[a_{j-2}, t]$.

If $o = a_{j-2}$, then since there is only one $\bar{C}_{j-1}C_j$ - path, we must have $u_{j-1} = a_{j-2}$, but then the fact that C_{j-1} is a directed circuit is contradicted.

It follows that $IF_{j-2}[a_{j-2}, a'_{j-2}] \cap VC_{j-1} \neq \emptyset$.

Suppose now that $IB_{j-2} \cap VC_{j-1} \neq \emptyset$. There are two cases to consider.

<u>2.3.2.1:</u> Assume that $u_j \notin IA_{j-2}$. We note that $VA_{j-2} \subset VC_j$, $IB_{j-2} \cap VC_{j-1} \neq \emptyset$ and $IF_{j-2}[a_{j-2}, a'_{j-2}] \cap VC_{j-1} \neq \emptyset$. Since there exists a unique $\bar{C}_{j-1}C_j$ - path, it follows that there exists a subpath T of C_{j-1}, of minimal length, joining a vertex of IB_{j-2} to a vertex of $IF_{j-2}[a_{j-2}, a'_{j-2}]$ and satisfying $VT \cap VA_{j-2} = \emptyset$. Then paths T and A_{j-2} interlock on the circuit $F_{j-2} \cup B$. Hence G is non-planar by Lemma 3.4.

<u>2.3.2.2:</u> Assume that $u_j \in IA_{j-2}$.

<u>2.3.2.2.1:</u> If $IB_{j-2}[v_j, a'_{j-2}] \cap VC_{j-1} \neq \emptyset$, then paths $B_{j-2}[v_j, a'_{j-2}]$ and $A_{j-2}[a_{j-2}, u_j]$ interlock on the circuit $F_{j-2}[a_{j-2}, a'_{j-2}] \cup A_{j-2}[a'_{j-2}, u_j] \cup P_j \cup B_{j-2}[v_j, a_{j-2}]$. Hence G is non-planar by Lemma 3.4.

<u>2.3.2.2.2:</u> If $[IB_{j-2}[a_{j-2}, v_j] \cup \{v_j\}] \cap VC_{j-1} \neq \emptyset$, then since $IB_{j-2}[a_{j-2}, v_j] \subset VC_j$, we force the contradiction that $u_{j-1} \notin VA_{j-2}$.

We must now consider the case $IB_{j-2} \cap VC_{j-1} = \emptyset$. If there is a subpath U of C_{j-1} joining a_{j-2} and a'_{j-2} that has no vertices or edges in common with $F_{j-2}[a_{j-2}, a'_{j-2}]$ or A_{j-2}, then there must exist another subpath W of C_{j-1}, of minimal length, joining a vertex $p \in IA_{j-2}$ and a vertex $q \in IF_{j-2}[a_{j-2}, a'_{j-2}]$. Then U and $W \cup A_{j-2}[p, u_j] \cup P_j$ interlock on $F_{j-2} \cup B$. Hence G is non-planar by Lemma 3.4. Therefore we may assume that no such path U exists. Then we may apply Lemma 4.1 with $a = a_{j-2}$, $b = a'_{j-2}$, $X = F_{j-2}[a_{j-2}, a'_{j-2}]$, $Y = A_{j-2}$, $Z = B_{j-2}$, $Q = N$ and $R = P_{j-1}$. Hence G is non-planar.

6. <u>THE MAIN THEOREM.</u>

We are now able to give the main result of this paper.

<u>Theorem 6.1:</u> A graph is non-planar if and only if it contains a strict odd elegant ring.

<u>Proof:</u> Suppose G is a non-planar graph. Then by Kuratowski's theorem it suffices to exhibit a strict odd elegant ring in K_5 and $K_{3,3}$. Strict odd rings in

these graphs have been found in [2], and their elegance is obvious.

Let S be a strict odd elegant ring. Suppose $|S| = 3$. Let $S = (C_0, C_1, C_2)$, and let u and v be the origin and terminus respectively of the unique C_0C_1-path P_0. Let e_1 be the edge of P_0 incident on v, let e_2 be the other edge of C_0 incident on v and let e_3 be the other edge of C_1 incident on v. Thus $e_1 \notin EC_2$, and e_2 and e_3 cannot both belong to EC_2. Thus if $v \in VC_2$ then the degenerate path with vertex set $\{v\}$ is either a C_0C_2 - path or a C_1C_2 - path. Since there must be a non-degenerate such path, the elegance of S is contradicted. Thus $v \notin VC_2$ and similarly $u \notin VC_2$. It is now immediate that $C_0 \cup C_1 \cup C_2$ is a subdivision of $K_{3,3}$.

Suppose therefore that $|S| \geq 5$. By Lemma 5.1 we may assume that $VC_k \cap VC_j \neq \emptyset$ if and only if $j \in \{k - 1, k, k + 1\}$. Then, if $S = (C_0, C_1, \ldots, C_{n-1})$, the graph

$$\bigcup_{k=0}^{n-2} [C_{k+1}(v_k, u_{k+1}) \cup C_{k+1}(v_{k+1}, u_k)] \cup C_0(v_{n-1}, u_0) \cup C_0(v_0, u_{n-1}) \cup P_0 \cup P_1 \cup P_2$$

is a subdivision of $K_{3,3}$, where u_i, v_i, P_i are defined for all i as in the proof of Lemma 5.1. Hence G is non-planar.

REFERENCES

[1] K. Kuratowski, Sur le probleme des courbes gauches en topologie, Fund. Math. 15 (1930), 271-283.

[2] C.H.C. Little, A Conjecture About Circuits In Planar Graphs, Combinatorial Mathematics III, Lecture Notes in Mathematics, Springer, New York 452 (1975), 171-175.

[3] C.H.C. Little, A Theorem On Planar Graphs, Combinatorial Mathematics IV, Lecture Notes in Mathematics, Springer, New York 560 (1976), 136-141.

[4] K. Wagner, Ueber eine Eigenschaft der ebenen Komplexe, Math. Ann. 114 (1937), 570-590.

Department of Mathematics
University of Melbourne
Parkville Victoria 3052

Department of Mathematics
Royal Melbourne Institute of Tech.
G.P.O. Box 2476V
Melbourne Victoria 3001

ON CRITICAL SETS OF EDGES IN GRAPHS

05 C 38
05 C 99

Mordechai Lewin

Let $G = (V,E)$ be a graph. Let α denote the minimum number of vertices that cover (are incident with) all the edges of G. Let β denote the maximum number of mutually nonadjacent (independent) edges of G. For any graph G we have $\beta(G) \leq \alpha(G)$. A basis of G is an independent set of β edges in G. The subset $F \subset E$ is α-critical if $\alpha(G\setminus F) < \alpha(G)$. F is β-critical if $\beta(G\setminus F) < \beta(G)$. A set of mutually coinciding edges is a star. For $v \in V$, S_v is the set of all edges of G incident with v.

The well known König's Theorem [4] states that for a bipartite graph G we have $\alpha(G) = \beta(G)$. But there are other graphs for which $\alpha = \beta$, as for example the (4,5)-graph [2, p.215]. We shall call such graphs k-perfect. We here suggest a non-constructive characterization of k-perfect graphs by means of critical sets of edges.

Theorem 1. *The graph* G *is* k-*perfect if and only if every* α-*critical star in* G *is also* β-*critical*.

Remark. A star may be α-critical but not β-critical, it may be β-critical but not α-critical, it may be both and it may be none of the two, so the theorem is really meaningful.

Proof. Let $S \subset E$ be an α-critical star in G, which is not β-critical. Then $\beta(G) = \beta(G\setminus S) \leq \alpha(G\setminus S) < \alpha(G)$ and hence G is not k-perfect.

Now let G be such that $\beta(G) < \alpha(G)$. Let $X = \{x_1, x_2, \ldots, x_\alpha\} \subset V$ be a minimum cover of E (line-cover of G), and let B be a basis of G. Let H be the subgraph of G spanned by X.

Case 1. $H \cap B \neq \emptyset$. Without loss of generality we may assume $(x_1,x_2) \in B$. Put $S_{x_1}\setminus(x_1,x_2) = S'$. Since X is a cover of E, $X\setminus x_1$ is a cover of $E\setminus S'$. But X is a minimum cover and hence $S' \neq \emptyset$. Therefore S' is α-critical. On the other hand $S' \cap B = \emptyset$ and hence S' is not β-critical.

Case 2. $H \cap B = \emptyset$. Then all β edges of B emanate from β vertices of X. Since $\alpha > \beta$, there is a vertex y of X which is not incident with any edge of B. The set X is a minimum cover and hence y is not isolated. Then S_y is an α-critical star, but is not β-critical. This completes Case 2 and proves the theorem.

From now on we shall not use the term β-critical and hence we shall write critical instead of α-critical. A graph is critical if all its edges are critical.

In [1] Beineke, Harary and Plummer prove the following result.

Theorem BHP. *Two adjacent critical edges lie on an odd cycle.*

This interesting theorem receives another proof by Jeurissen [3]. Jeurissen in fact strengthens the result in that he replaces odd cycle by chordless odd cycle.

Let (x,y) be a critical edge of a graph G and let {x,y} be contained in some minimum cover of G. We shall refer to such a doubly covered edge as d-<u>edge</u>.

In [3] Jeurissen also proves

<u>Theorem J</u>. *Every* d-*edge of a graph* G *belongs to a chordless odd cycle.*

We shall here supply among other results an independent proof of Theorem J. First a lemma.

<u>Lemma 1</u>. *Two adjacent critical edges of a graph* G *are both* d-*edges.*

<u>Proof</u>. Let (x,y) and (y,z) be two distinct critical edges in G. Consider $G\setminus(x,y)$. Since (x,y) is critical a minimum cover S' of $G\setminus(x,y)$ does not contain x or y and hence it contains z. Then $S'\cup y = S$ is a minimum cover of G and so (y,z) is a d-edge in G. Likewise (x,y) is a d-edge in G, proving the lemma.

We now prove Theorem J.

The smallest graph with a d-edge is K_3, which satisfies the theorem. Now assume that the theorem is true for all graphs of smaller size than G.

Suppose y is a cutvertex. Let X,Y be subgraphs of G such that $x\in X, X\cap Y = y$, $X\cup Y = G$, and X is connected. No minimum cover of Y contains y and hence (x,y) is a d-edge in X. By the induction hypothesis (x,y) belongs to a chordless odd cycle. We therefore assume G to be a block.

Let S' be a minimum cover of $G\setminus(x,y)$. Then clearly $\{x,y\}\cap S' = \phi$. Let S be a minimum cover of G.

<u>Case 1</u>. $S\cap S' \neq \phi$. Let $z\in S\cap S'$. Since $z\in S'$ we have $\alpha(G\setminus(x,y)\setminus z)=\alpha(G)-2$, and $\alpha(G\setminus z) = \alpha(G)-1$, so that (x,y) is critical in $G\setminus z$, and $S\setminus z$ is a cover and hence a minimum cover of $G\setminus z$. Clearly $z\notin\{x,y\}$ and so $\{x,y\}\subset S\setminus z$. $G\setminus z$ has less edges than G and hence by the induction hypothesis (x,y) lies on a chordless odd cycle.

<u>Case 2</u>. $S\cap S' = \phi$. By passing from x to y along a cycle (in fact any arbitrary cycle) containing (x,y) we have to pass vertices of S' and vertices of S alternately. This is only possible if the cycle is odd. By choosing a minimal such cycle we ensure that it is chordless. This completes case 2 and proves the theorem.

<u>Corollary 1</u>. *A bipartite graph has no* d-*edges.*

Considering that part of the proof of Theorem J which reflects upon the possibility of y being a critical vertex in G we have in fact shown:

<u>Corollary 2</u>. <u>(Th. 1 in [1])</u>. *Two adjacent critical edges lie on a cycle.*

It follows from the proof of case 2 of the last theorem that in this particular case (x,y) belongs only to odd cycles. This brings us to the following:

<u>Corollary 3</u>. *Let* $G = (V,E)$ *be a graph. Let* (x,y) *be a critical edge of* G

lying on an even cycle. Let S *be a minimum cover of* G *containing* {x,y}. *Then any minimum cover of* $G \setminus (x,y)$ *meets* S.

We conclude with an extremely short proof of König's Theorem for bipartite graphs.

For $G = K_2$, the theorem is clearly true, so assume that the theorem holds for bipartite graphs with less edges than G. We prove it for G. We may assume G to be connected. If G contains an edge which is not critical, delete it and use induction. Otherwise G is critical and connected and hence by Lemma 1 and Corollary 1 we have $G = K_2$ and therefore k-perfect, proving the theorem.

REFERENCES

[1] L. W. Beineke, F. Harary and M. D. Plummer, On the critical lines of a graph, *Pacific J. Math.*, 22 (1967), 205-212.

[2] F. Harary, *Graph Theory*. (Addison Wesley, Reading, Mass., 1972).

[3] R. H. Jeurissen, Covers, Matchings and odd cycles of a graph, *Discrete Math.*, 13 (1975), 251-260.

[4] D. König, Graphen und Matrizen, *Mat. Fiz. Lapok*, 38 (1931), 116-119.

Department of Mathematics
Technion, Israel Institute of Technology
Haifa

05C60

FURTHER EVIDENCE FOR A CONJECTURE ON TWO-POINT DELETED SUBGRAPHS OF CARTESIAN PRODUCTS

K.L. McAVANEY

Another theorem is proved that supports the conjecture : a connected composite graph $G \times H$ with G and H on more than two points is uniquely determined by each of its two-point deleted subgraphs.

This paper is a sequal to [1]. We refer the reader to [1] for the necessary preliminaries especially properties P1, P2, P3, and P6 of connected composite graphs and Lemmas L1, L2, L3, and L4. Regrettably there is an error in Figure 4.13 of [1] and the remarks relating to it : point 31 is not necessarily adjacent to 32. Instead we observe that there is a point $3\ell \sim 33$ and x (L2, P6) and hence $3\ell \sim 23$ (L4) which is impossible.

It should be pointed out that throughout [1] the condition "J is isomorphic to $G \times H$" can obviously be weakened to "J is isomorphic to $K \times L$ where K and L are connected graphs each with more than two points". Indeed we need to strengthen our results in this way if the Conjecture in [1] is to be established.

The following theorem further supports the conjecture. We recall our approach is to try to show that if $G \times H - u - v + w + x \cong K \times L$ then, but for a few minor exceptions, $Nw = Nu$ (or Nv) and $Nx = Nv$ (or Nu respectively). Indeed this paper, [1], and some more work that will appear elsewhere achieves this result for the case $u = 11$ and $v = 22$. Present work is on the other case $u = 11$ and $v = 21$, which is complicated by a greater number of exceptions. Known exceptions to the neighbourhood property do not contradict the conjecture.

The theorem below uses the following terms. An *end point* is a point of degree 1. A *penultimate point* is a point of degree 2 that is adjacent to an end point. An *antipenultimate point* v is a point that is adjacent to a penultimate point which in turn is adjacent to an end point distinct from v (an antipenultimate point may be an end point itself).

Theorem. *If* G, H, K, and L *are connected graphs each with more than two points and* $J = G \times H - 11 - 22 + w + x \cong K \times L$ *then for all* gi~g1 *and* hj~h1, *with* i *and* j *not both* 2, *if* w~1j *then (1)* w~i1 *or (2)* x~1j *and* i1 *unless*

(a) g1 *and* h2 *are antipenultimate points with* g2 *and* h1 *the corresponding end points or*

(b) g2 *and* h1 *are antipenultimate points with* g1 *and* h2 *the corresponding end points or*

(c) G = o(1)—o(2)—o *and* H = o(1)—o—o(2)—o *or*

(d) G = o—o(1)—o—o(2) *and* H = o—o(1)—o(2) *or*

(e) g1 *and* h2 *are antipenultimate points with* g2 *and* h1 *the corresponding penultimate points or*

(f) G = o—o(1)—o(2) = H.

The remaining seven variations of the statement above, obtained by interchanging G *and* H *or* 1 *and* 2 *or* x *and* w *are also true.*

Proof. Suppose neither (1) nor (2) is true. Then (1j, ij) and (ij, i1) are in the same section set, E say, of J (P3). By P1 there is an edge $(ij,y) \in \bar{E}$. See Figure 1. Then y = ik, kj, or x (P6).

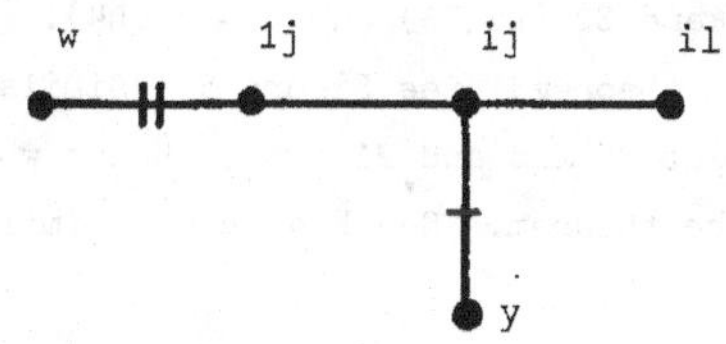

Figure 1

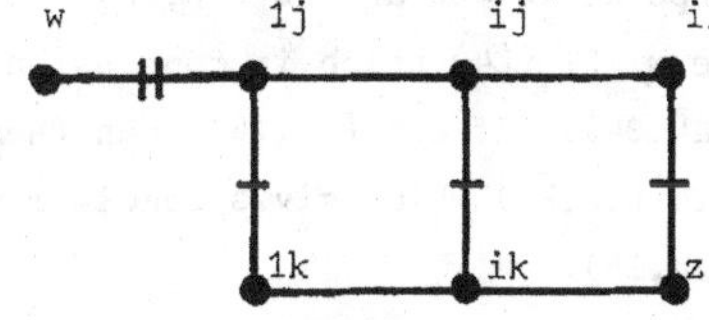

Figure 2

Case 1. y = ik.

There is a point z ~ ik and i1 (P3). See Figure 2. Now z = iℓ or x (P6).

Case 1.1. z = iℓ.

By P3 either w ~ 1ℓ and i1 which gives conclusion (1) of the theorem, or x ~ 1ℓ and i1 and hence x ~ 1j (L4) which gives conclusion (2) of the theorem. See Figure 3.

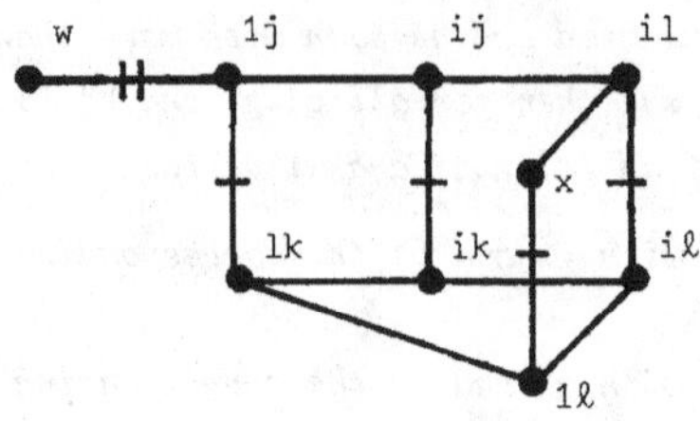

Figure 3

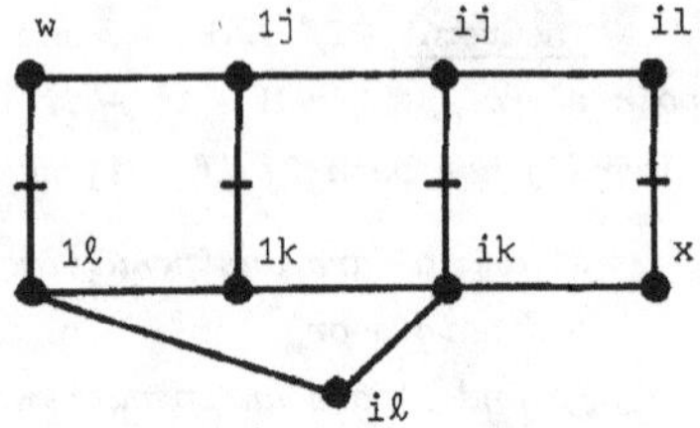

Figure 4

Case 1.2. z = x.

Case 1.2.1. (1j, w) ε E.

Either there is a point 1ℓ ~ w and 1k or j = 2 and 2k ~ w and 1k (L2,P6).

Case 1.2.1.1. 1ℓ ~ w and 1k.

If iℓ ≠ 22 then x or i1 ~ ij and iℓ (L1) which contradicts P6. See Figure 4.

So iℓ = 22. Set j = 3 and k = 4. Because J $\cong$ K x L and both K and L are connected and have more than one edge, there is an edge e ε $\overline{E}$ incident to 13 or 14. Thus e = (13,33), (14,34), (13,15), or (14,15) (P6). If e = (13,33) there is a point 43 ~ 33 and 23 (P3,P6). Also 31 ~ 33 and 32 (L2,P6) and 31 ~ w (L4). Hence w ~ 21 (L4) which is conclusion (1) of the theorem. See Figure 5. Similarly e ≠ (14,34). If e = (13,15) then there is a point 26 ~ 25 and 21 (P6). Hence w ~ 16 and 21 (P3,P6) which gives conclusion (1) of the theorem. See Figure 6. Similarly e ≠ (14,15).

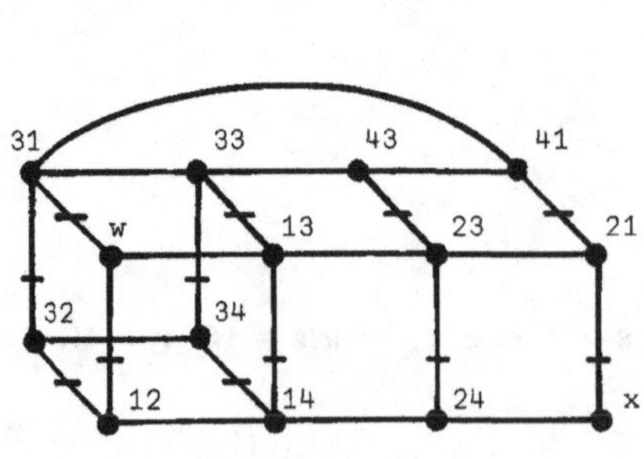

Figure 5

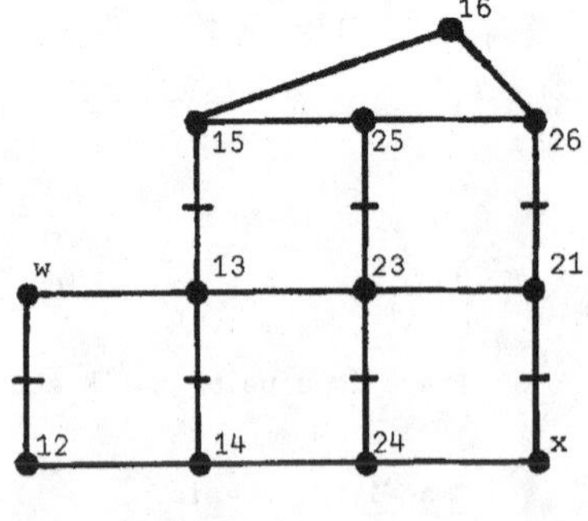

Figure 6

<u>Case 1.2.1.2.</u> $j = 2$ and $2k \sim w$ and $1k$.

Set $i = 3$ and $k = 3$. Again there is an edge $e \in \overline{E}$ incident with 12 or 13. Thus $e = (12,42)$, $(13,43)$, $(12,14)$, or $(13,14)$ (P6). If $e = (12,42)$ there is a point $52 \sim 42$ and 32 (P3,P6). Hence $w \sim 53$ and 51 (L1) which contradicts P6. See Figure 7. Similarly $e \neq (13,43)$. If $e = (12,14)$ there is a point $35 \sim 34$ and 31 (P3,P6). Then $w \sim 15$ and 31 (P3,P6) which gives conclusion (1) of the theorem. See Figure 8. If $e = (13,14)$ there is a point $35 \sim 34$ and x (L2,P6). Then $w \sim 13$ and 15 (L1) which contradicts P6.

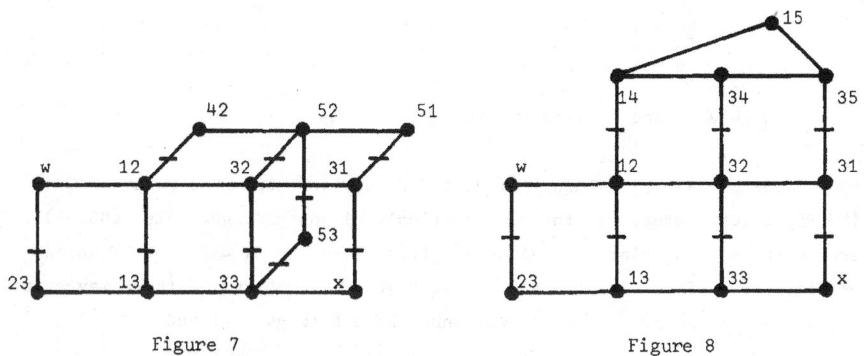

Figure 7 Figure 8

<u>Case 1.2.2.</u> $(1j,w) \in \overline{E}$.

There is a point $\ell j \sim \omega$ and ij (L2,P6). Suppose $\ell k \neq 22$. Then $w \sim \ell k$ and $1k$ if $i \neq 2$ or $x \sim \ell k$ and $1k$ if $j \neq 2$ (L1). This contradicts P6. See Figure 9.

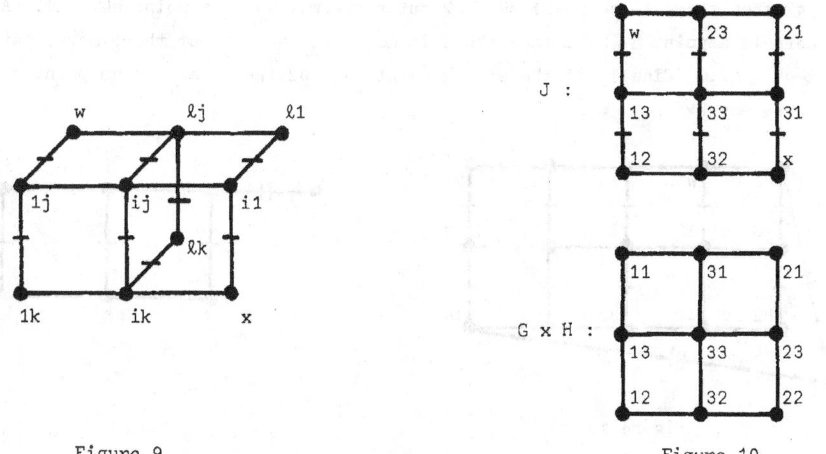

Figure 9 Figure 10

So $\ell k = 22$. Set $j = 3 = i$. Then $G = \overset{1}{\circ}\!-\!\!-\!\overset{3}{\circ}\!-\!\!-\!\overset{2}{\circ} = H$ or there is a point $g4 \sim g1, g2$, or $g3$ or there is a point $h4 \sim h1, h2$, or $h3$. See Figure 10.

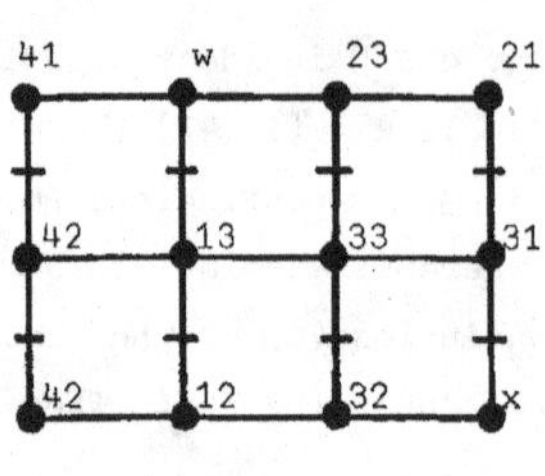

Figure 11

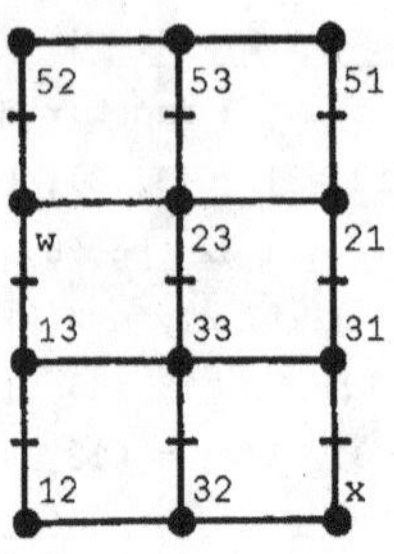

Figure 12

Now g4 $\not\sim$ g3 and h4 $\not\sim$ h3 (L1,P6).

Suppose g4 $\sim$ g1. Then (13,43) ε E otherwise there is a point 53 $\sim$ 43 and 33 (P3,P6) which contradicts the fact that only g1 and g2 $\sim$ g3. Also (43,41) ε $\overline{E}$ otherwise there is a point 44 $\sim$ 41 and 42 (P3,P6) and so 34 $\sim$ 31 and 32 which contradicts P6. Then $\omega \sim$ 41 (P3,P6). See Figure 11. Likewise there may exist g5 $\sim$ g2. See Figure 12. However we cannot have both g4 $\sim$ g1 and g5 $\sim$ g2 (L3).

Similarly there may exist h4 $\sim$ h1 or h5 $\sim$ h2 but not both.

Moreover if there is a point g4 $\sim$ g1 then there is no point h4 $\sim$ h1 (P6, see Figure 13) but there may exist a point h4 $\sim$ h2. Likewise if there is a point g4 $\sim$ g2 then there is no point h4 $\sim$ h2 but there may exist a point h4 $\sim$ h1. Also if there is a point h4 $\sim$ h1 then there is no point g4 $\sim$ g1 but there may exist a point g4 $\sim$ g2. Finally if there is a point h4 $\sim$ h2 then there is no point g4 $\sim$ g2 but there may exist g4 $\sim$ g1.

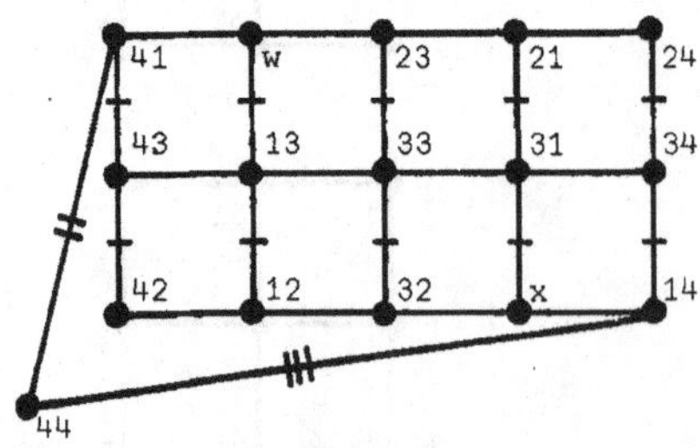

Figure 13

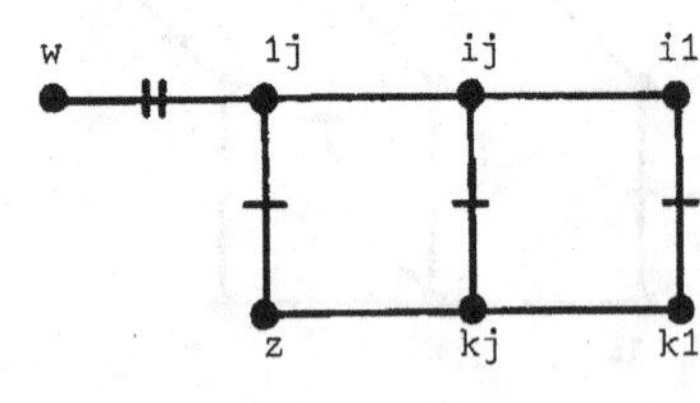

Figure 14

Thus, because g1 $\not\sim$ g2 and h1 $\not\sim$ h2 (P6), g1 and h2 are antipenultimate points with g2 and h1 the corresponding end points or g2 and h1 are antipenultimate points with g1 and h2 the corresponding end points.

<u>Case 2.</u> y = kj.

There is a point $z \sim 1j$ and kj (P3). See Figure 14. Now $z = \ell j$, x, or w.

<u>Case 2.1.</u> $z = \ell j$.

Either x or $w \sim 1j$ and $\ell 1$ (P3) and hence x or w respectively $\sim i1$ (L4.) This gives conclusion (1) or (2) of the theorem.

<u>Case 2.2.</u> $z = x$.

Because H is connected and has more than two points, there is a point $h\ell \sim hj$ or $h1$ ($\ell \neq j,1$).

<u>Case 2.2.1.</u> $h\ell \sim hj$.

Suppose $i\ell \neq 22$ then $(ij, i\ell) \in E$ otherwise we have Case 1. If $k\ell \neq 22$ then $j = 2$ and $2\ell \sim 1\ell$ and 4ℓ (L1,P6). So $(k1,21) \in E$ (P3,P6) and hence $w \sim i1$ and and 21 (P3,P6) which gives conclusion (1) of the theorem. See Figure 15. If $k\ell = 22$ then w or $x \sim i2$ and $2j$ (P3) which contradicts P6. See Figure 16.

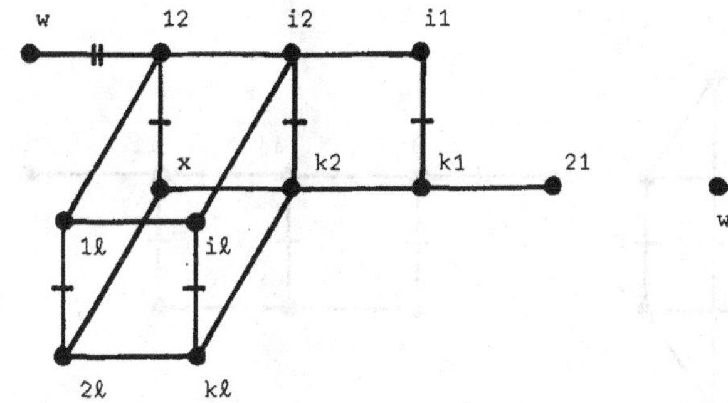

Figure 15

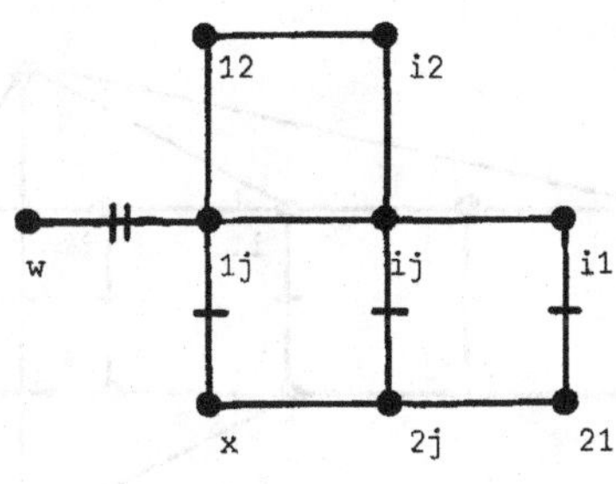

Figure 16

So $i\ell = 22$. Set $j = 3 = k$. Suppose $(33,32) \in \overline{E}$. Then there is a point $34 \sim 32$ and 31 (P3,P6) and x or $w \sim 32$ and 24 (P3) which contradicts P6. See Figure 17. So $(33,32) \in E$. Then $w \sim 23$ and 32 (P3,P6). Now $(13,12) \in E$ (P3,P6). Hence there is a point $14 \sim 12$ and x (L2,P6) and $(14,24) \in \overline{E}$ (P3,P6). Therefore $34 \sim x$ (L2,L3,P6). So $(32,33) \in \overline{E}$ (P2), a contradiction. See Figure 18.

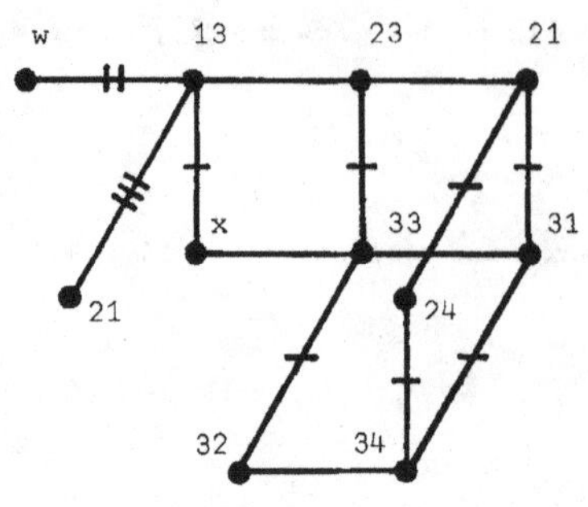

Figure 17

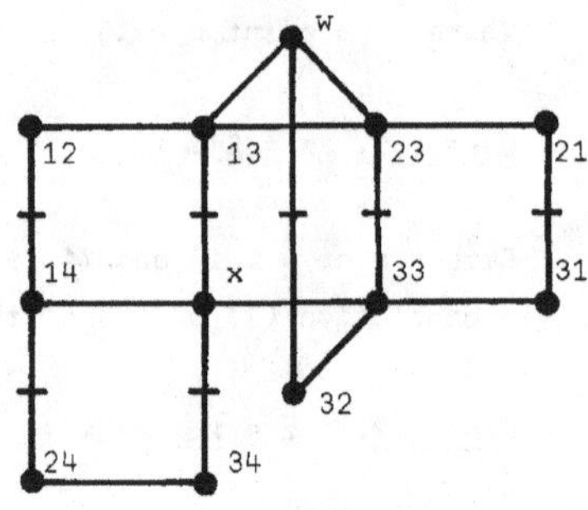

Figure 18

Case 2.2.2. hℓ ~ h1.

Suppose iℓ ≠ 22. Then (i1,iℓ) ε E, otherwise there is a point im ~ ij and iℓ (P3,P6) and we have Case 1. Hence (iℓ,1ℓ) ε E (P3,P6). If kℓ ≠ 22, there is a point mℓ ~ kℓ and 1ℓ (P3,P6). So w ~ m1 and 1ℓ (P3,P6) and hence w ~ i1 (L4) which is conclusion (1) of the theorem. See Figure 19. If kℓ = 22 then x or w ~ i2 and 21 (P3) which contradicts P6. See Figure 20.

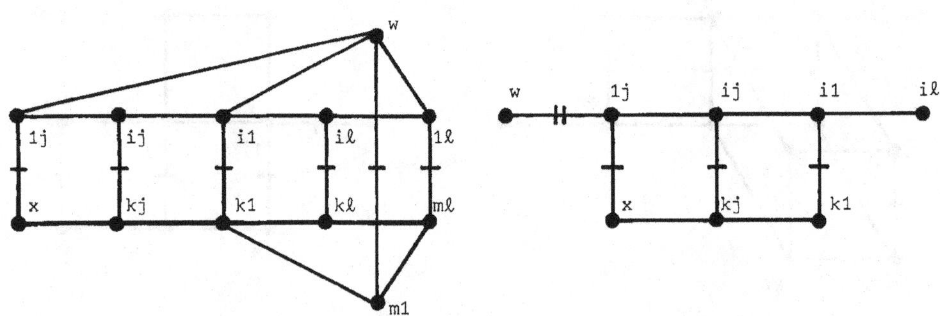

Figure 19

Figure 20

So iℓ = 22. Set k = 3 = j. If (31,32) ε E then w ~ 21 and 32 (P3,P6) which gives conclusion (1) of the theorem. See Figure 21. If (31,32) ε $\overline{E}$ then there is a point 34 ~ 33 and 32 (P3,P6). So w ~ 24 and 32 (P3,P6) and hence w ~ 21 (L4) which is conclusion (1) of the theorem. See Figure 22.

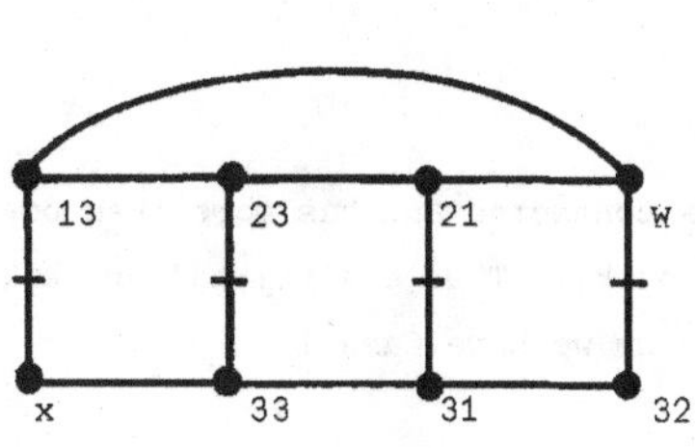

Figure 21

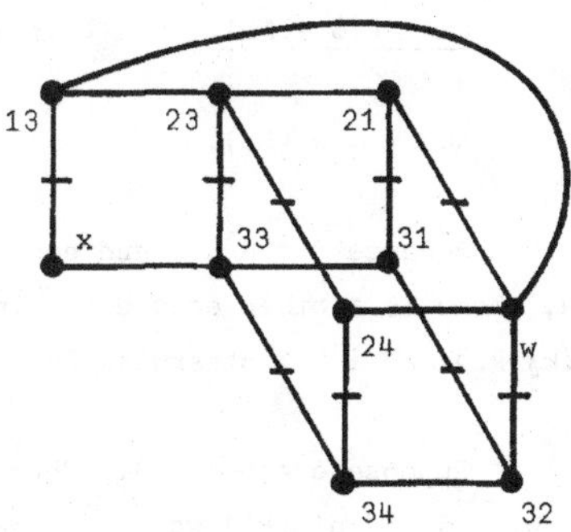

Figure 22

Case 2.3. z = w.

Again there is a point hℓ ~ hj or h1 (ℓ ≠ j,1).

Case 2.3.1. hℓ ~ hj.

Case 2.3.1.1. iℓ ≠ 22.

Now (ij,iℓ) ε E otherwise we have Case 1. Suppose kℓ = 22. Set i = 3 = j. Then x ~ 32 and 23 (P3,P6) and there is a point 14 ~ 12 and w (L2,P6). So 34 ~ 14 and 32 which contradicts P6. See Figure 23.

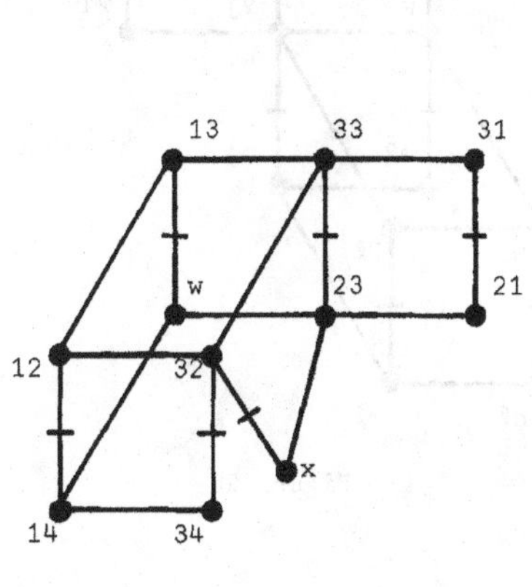

Figure 23

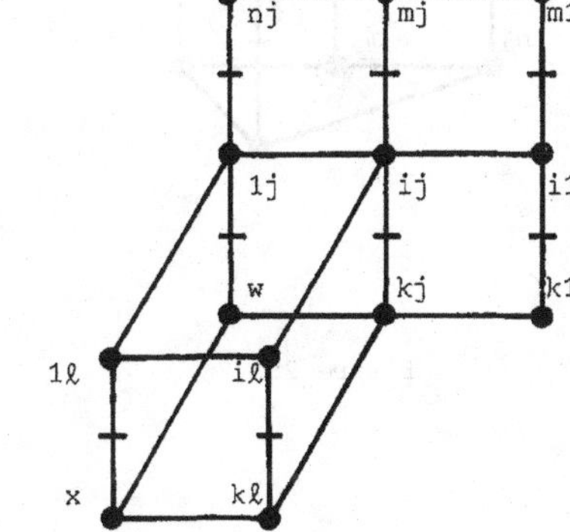

Figure 24

So kℓ ≠ 22. By L1, either x ~ 1ℓ and kℓ or j = 2 and 2ℓ ~ 1ℓ and kℓ.

<u>Case 2.3.1.1.1.</u> $x \sim 1\ell$ and $k\ell$.

Now $x \sim w$ (L4).

Because $J \cong K \times L$ and each of K and L is connected and has more than one edge, there is another edge $e \in \bar{E}$ incident to ij or kj. Thus e = (ij,mj) or (kj,mj) or (kj,k2) and i = 2 otherwise P6 is contradicted or we have Case 1.

Suppose e = (ij,mj). Then there is a point nj ~ 1j and mj (P3,P6). Hence x or w ~ 1j and n1 (P3) which contradicts P6. See Figure 24.

If e = (kj,mj) then either there is a point nj ~ w and mj or k = 2 and m2 ~ w and mj (L2,P6). In the former case x or 21 ~ n1 and k1 (L1). Then, respectively, 1ℓ ~ i1 (L4) which is impossible, or 21 ~ w (L4) and x ~ 1j and 21 (P3,j = 2) and hence x ~ i1 (L4) which gives conclusion (2) of the theorem. See Figure 25. In the latter case there is a point $n\ell$ ~ x and $m\ell$ (L2,P6) and hence $n\ell$ ~ m2 (L4) which is impossible. See Figure 26.

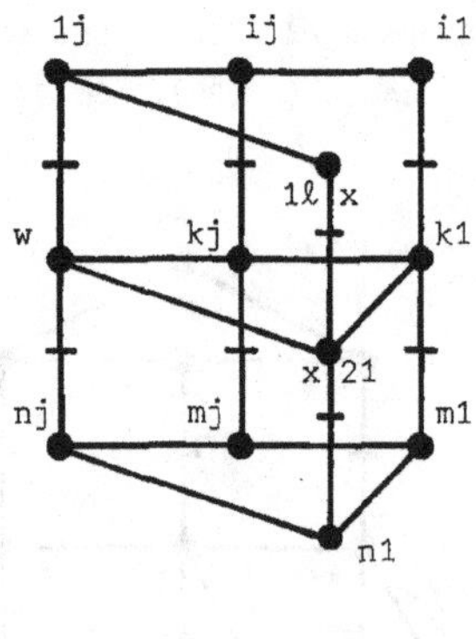

Figure 25

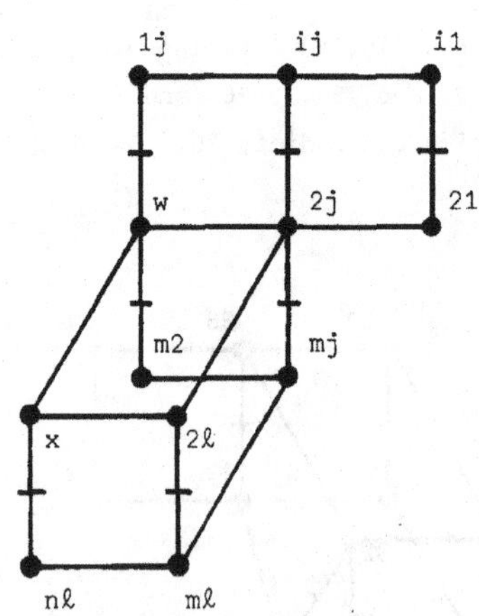

Figure 26

So i = 2 and e = (kj,k2). Then there is a point km ~ k2 and k1 (P3,P6). Hence x or w ~ k2 and 2m (P3) which contradicts P6. See Figure 27.

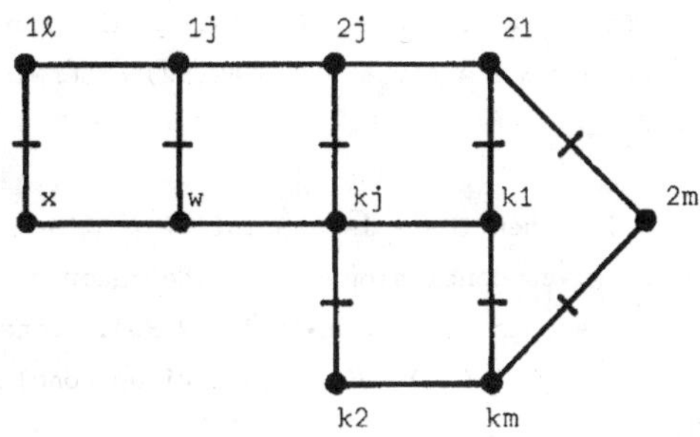

Figure 27

<u>Case 2.3.1.1.2.</u> j = 2 and 2ℓ ~ 1ℓ and kℓ.

Now 2ℓ ~ w (L4). Set i = 3 = ℓ and k = 4.

Suppose (41,21) ε $\bar{E}$. Then x ~ 42 and 21 (P3,P6). Hence there is a point 44 ~ 43 and x (L2,P6). Consequently there is a point 54 ~ 34 and x (L2,P6). Then 54 ~ 32 (L4) which is impossible. See Figure 28.

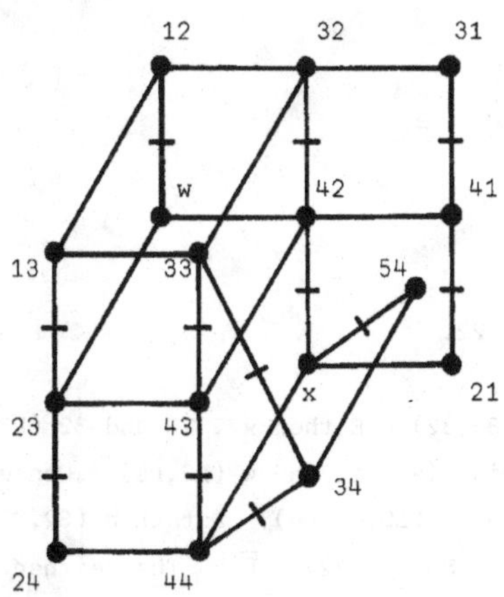

Figure 28

So (41,21) ε E. Then x ~ 31 and 21 (P3,P6). Again there is an edge e ε $\overline{E}$ incident to 32 or 42. Here e = (32,52) or (42,52) otherwise P6 is contradicted or we have Case 1.

Suppose e = (32,52). Then there is a point 62 ~ 52 and 12 (P3,P6). Hence x ~ 12 and 61 (P3,P6) which gives conclusion (2) of the theorem. See Figure 29. If e = (42,52) then there is a point 62 ~ w and 52 (L2,P6). Consequently 21 ~ 61 (L1,P6) and hence 21 ~ w and x ~ 12 (L4) which again gives conclusion (2) of the theorem. See Figure 30.

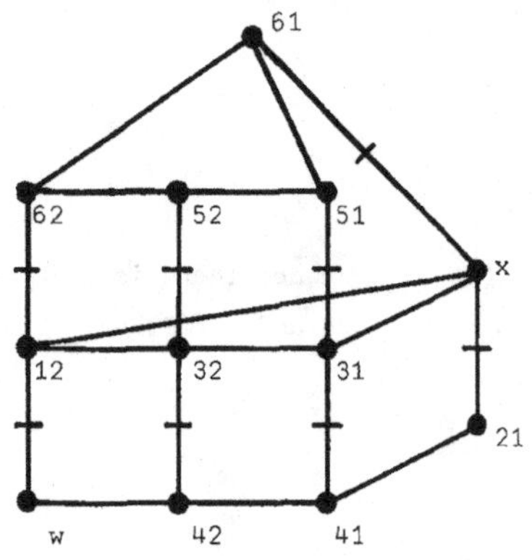

Figure 29

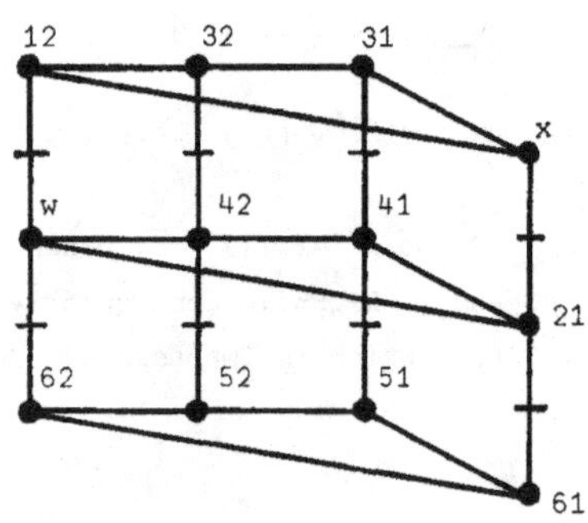

Figure 30

Case 2.3.1.2. iℓ = 22.

Set j = 3 = k. If (33,32) ε E then x ~ 23 and 32 (P3,P6). Now (13,12) ε E (P3,P6). Hence there is a point 14 ~ 12 and w (L2,P6). Consequently (14,24) ε $\overline{E}$ (P3,P6). Therefore 34 ~ 24 and w (L2,L3,P6). But then (32,33) ε $\overline{E}$ (P2), a contradiction. See Figure 31. So (33,32) ε $\overline{E}$. Then either x ~ 32 and 31 or there is a point 34 ~ 32 and 31 (P3). In the latter case x ~ 32 and 24 (P3,P6). Hence there is a point 35 ~ w and 32 (L2,P6). Therefore 25 ~ 35 and x or there is a point 36 ~ 35 and x (L2,P6). Consequently 25 or 36 respectively ~ 13 (L4) which is impossible. See Figure 32.

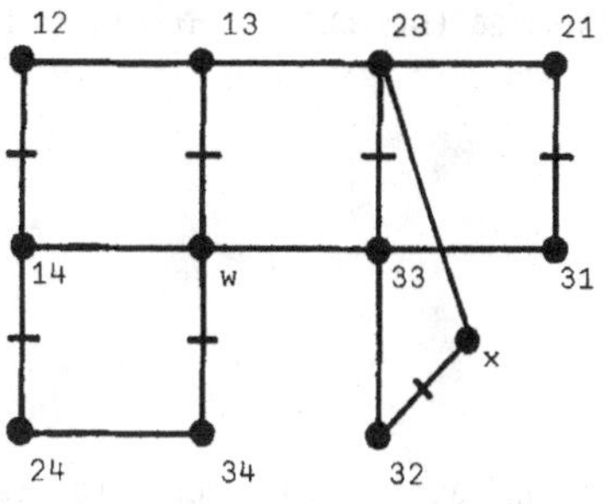

Figure 31

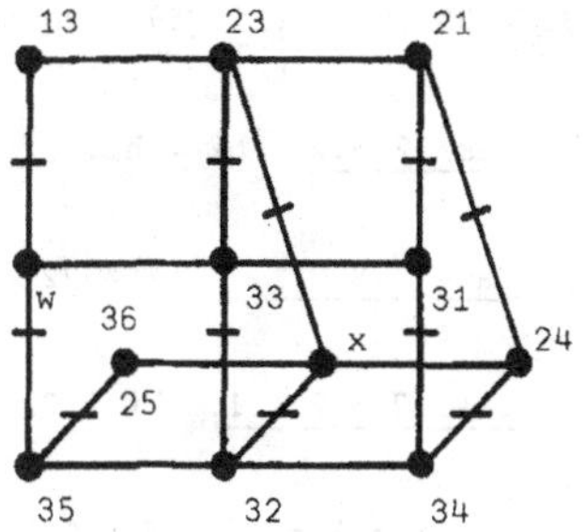

Figure 32

So $x \sim 32$ and 31. Then $(13,12) \in E$ (P3,P6). Hence there is a point $14 \sim 12$ and w (L2,P6) and therefore $34 \sim w$ (L2,L3,P6). Thus $G = \overset{1}{\circ}\!-\!\overset{2}{\circ}\!-\!\overset{3}{\circ}$ and $H = \overset{1}{\circ}\!-\!\overset{3}{\circ}\!-\!\overset{2}{\circ}\!-\!\overset{4}{\circ}$, or there is a point $g4 \sim g1$, $g2$, or $g3$, or there is a point $h5 \sim h1$, $h2$, $h3$, or $h4$. See Figure 33.

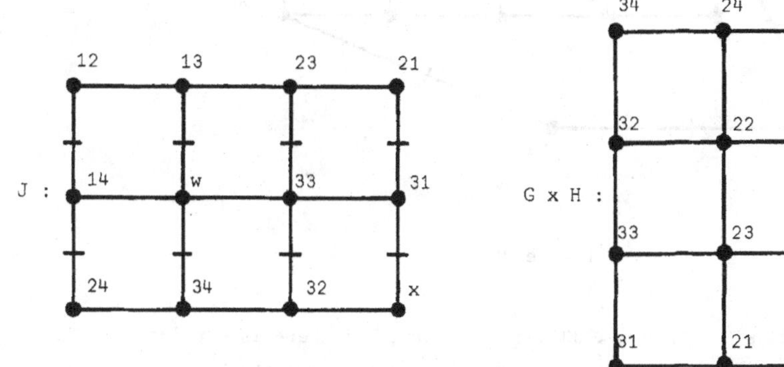

Figure 33

If $g4 \sim g1$ then $x \sim 44$ and 43 (L1,P6) which contradicts P6. Similarly $g4 \not\sim g3$, $h5 \not\sim h3$, $h5 \not\sim h4$. Suppose $g4 \sim g2$. If $(24,44) \in \overline{E}$ then there is a point $54 \sim 44$ and 34 (P3,P6) which contradicts the fact that only $g2 \sim g3$. But a similar contradiction results if $(24,44) \in E$. So $g4 \not\sim g2$. Likewise $h5 \not\sim h2$. Finally suppose $h5 \sim h1$. Then $(21,25) \in E$ otherwise there is a point $26 \sim 23$ and 25 (P3,P6) which contradicts the fact that only $h1$ and $h2 \sim h3$. Hence there is a point $36 \sim 35$ and x (L2,P6). Consequently $w \sim 21$ and 26 (L1) which contradicts P6.

<u>Case 2.3.2.</u> $h\ell \sim h1$.

<u>Case 2.3.2.1.</u> $i\ell \neq 22$.

<u>Case 2.3.2.1.1.</u> $k\ell = 22$.

Set $i = 3 = j$. If $(31,32) \in E$ then $x \sim 21$ and 32 (P3,P6). Hence $(32,12) \in E$ (P3,P6) and there is a point $42 \sim x$ and 12 (L2,P6). Therefore $(42,41) \in \overline{E}$ (P3,P6). Consequently (41,43) and (43,13) are in the same section set (P3,P6). This contradicts P6. See Figure 34.

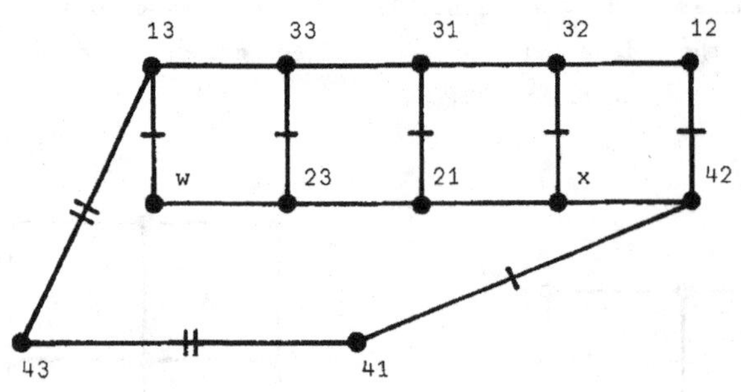

Figure 34

So $(31,32) \in \overline{E}$. Then $x \sim 33$ and 32 (P3,P6, if there is a point $34 \sim 33$ and 32 we have Case 1). Hence there is a point $43 \sim 13$ and x (L2,P6). Also $(43,41) \in \overline{E}$ (P3,P6). Therefore $42 \sim x$ (L2,L3,P6). Thus $G = \overset{4}{\circ}\!-\!\overset{1}{\circ}\!-\!\overset{3}{\circ}\!-\!\overset{2}{\circ}$ and $H = \overset{3}{\circ}\!-\!\overset{1}{\circ}\!-\!\overset{2}{\circ}$ or there is a point $g5 \sim g1$, g2, g3, or g4 or a point $h4 \sim h1$, h2, or h3. See Figure 35. These possible extra points yield contradictions in the same way as in Case 2.3.1.2.

<u>Case 2.3.2.1.2.</u> $k\ell \neq 22$.

If $(i1,i\ell) \in \overline{E}$ then, by P3 and P6, there is a point $im \sim ij$ and $i\ell$ and we have Case 1, or $x \sim ij$ and $i\ell$ and then $i = 2$ and k2 kj and $k\ell$ (L1). In the latter case $(2\ell,1\ell) \in \overline{E}$ (P3,P6) and then $x \sim 12$ (P3,P6). Hence $(1j,2j) \in \overline{E}$ (P2), a contradiction. See Figure 36. So $(i1,i\ell) \in E$. $(i\ell,1\ell)$ $\overline{E}$ then $x \sim i1$ and 1ℓ (P3,P6) and hence there is a point $im \sim ij$ and x (L2,P6) an have Case 1.

See Figure 37. So $(i\ell, 1\ell) \in E$.

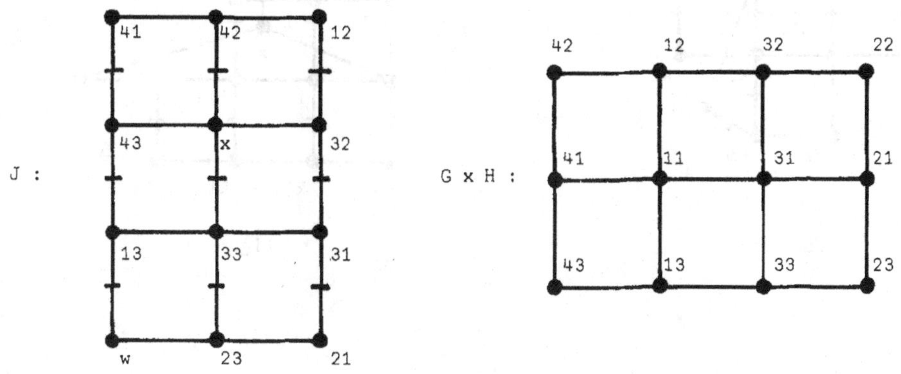

Figure 35

Then $x \sim k\ell$ and 1ℓ or $j = 2$ and $2\ell \sim k\ell$ and 1ℓ (P3,P6). In the latter case $x \sim 21$ and 1ℓ (P3,P6). So in both cases $x \sim 1\ell$. Now, as before, there is an edge $e \in \overline{E}$ incident to ij or kj. Thus $e = (ij,mj)$ or $e = (kj,mj)$ or $i = 2$ and $e = (kj,k2)$ otherwise P6 is contradicted or we have Case 1.

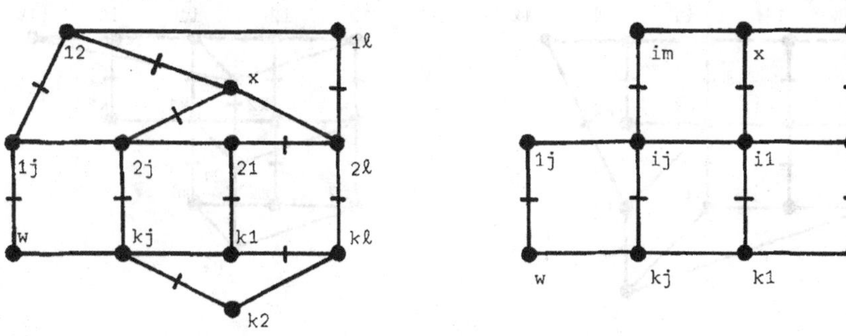

Figure 36

Figure 37

Suppose $i = 2$ and $e = (kj,k2)$. Then there is a point $km \sim k2$ and $k1$ (P3,P6). Hence x or $w \sim k2$ and $2m$ (P3) which contradicts P6. See Figure 38. Suppose $e = (ij,mj)$. Then there is a point $nj \sim 1j$ and mj (P3,P6). Hence $x \sim 1j$ and $n1$ (P3,P6). Therefore $x \sim i1$ (L4) which gives conclusion (2) of the theorem. See Figure 39.

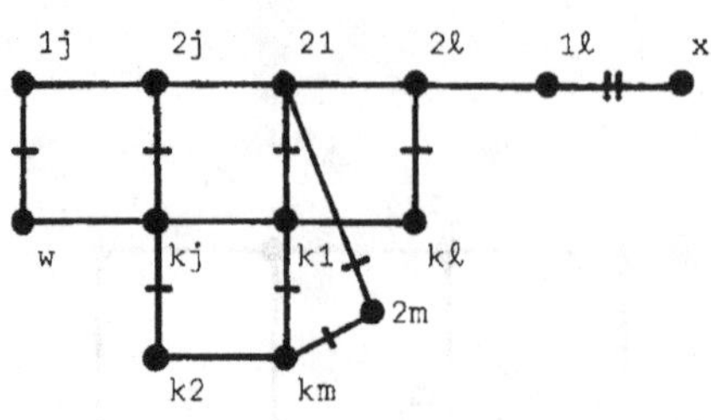

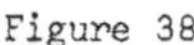

Figure 38

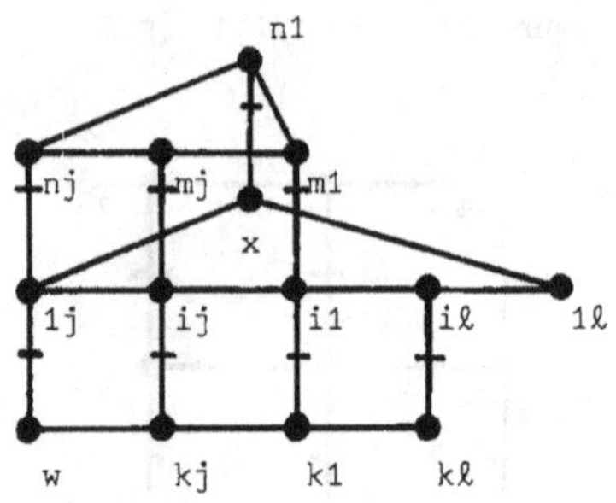

Figure 39

So $e = (kj,mj)$. Either there is a point $nj \sim mj$ and w or $k = 2$ and $m2 \sim mj$ and w (L2,P6). In the latter case $(ij,i2) \in E$ (otherwise we have Case 1) and then x or $w \sim i2$ and $2j$ (P3) which contradicts P6. In the former case x or $21 \sim n1$ and $k1$ (L1). If $x \sim n1$ and $k1$ then $(x,1\ell) \in \overline{E}$ (P6) and hence w or $k\ell \sim k1$ and 1ℓ (P3) which contradicts P6. See Figure 40. If $21 \sim n1$ and $k1$ then $21 \sim w$ (L4) and hence $x \sim 12$ and 21 (P3,$j = 2$). Then $x \sim i1$ (L4) which gives conclusion (2) of the theorem. See Figure 41.

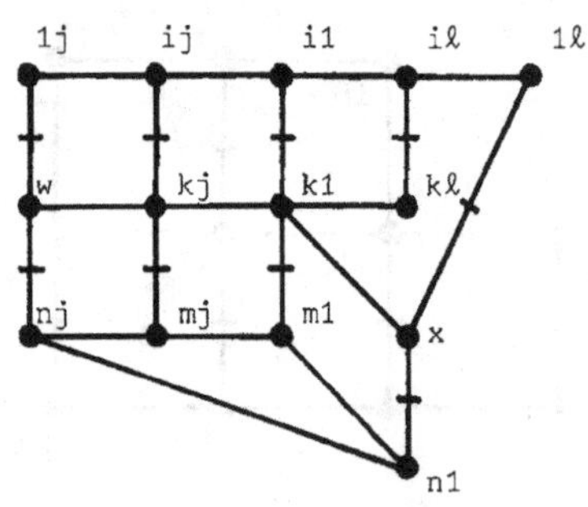

Figure 40

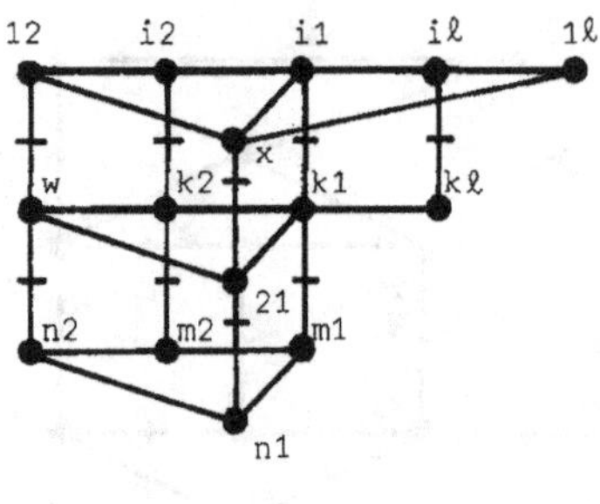

Figure 41

<u>Case 2.3.2.2.</u> $i\ell = 22$.

Set $j = 3 = k$. If $(31,32) \in E$ then $x \sim 21$ and 32 (P3,P6). Again there is an edge $e \in \overline{E}$ incident with 23 or 33 and $e = (23,43)$ which give conclusion (2) of the theorem, or e = (33,43) which gives a contradiction. See Figure 42.

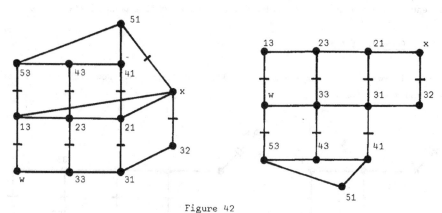

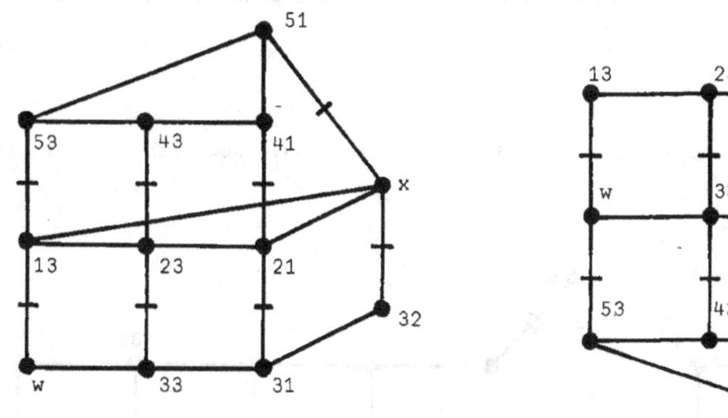

Figure 42

So $(31,32) \in \overline{E}$. If there is a point 34 ~ 33 and 32 then 24 ~ 23 and we have Case 1. So x ~ 33 and 32 (P3,P6). Thus 12 ~ w and x and G = o—o—o (1 2 3) and H = o—o—o (2 1 3) or there is a point g4 ~ g1, g2, or g3 or there is a point h4 ~ h1, h2, or h3. See Figure 43.

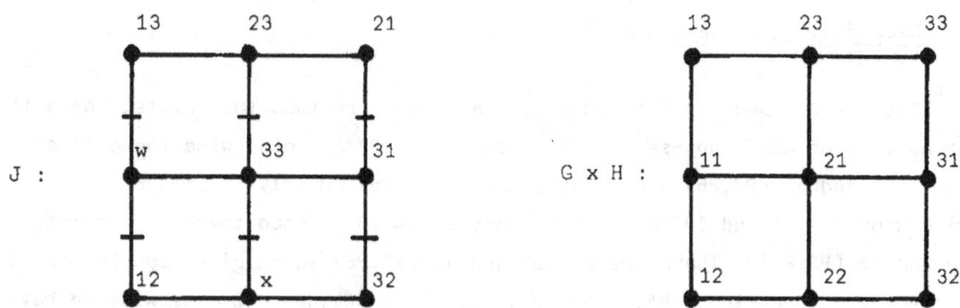

Figure 43

If g4 ~ g3 then w ~ 43 and 42 (L1) which contradicts P6. Similarly h4 $\not\sim$ h3.

Suppose g4 ~ g2. Then $(21,41) \in \overline{E}$ otherwise there is a point 51 ~ 31 and 41 (P3,P6) which contradicts the fact that only g2 ~ g3. Hence there is a point 53 ~ 13 and 43. Consequently x or w ~ 13 and 51 (P3) which contradicts P6. So g4 $\not\sim$ g2. Similarly h4 $\not\sim$ h1.

However there may exist g4 ~ g1 or h4 ~ h2. Thus g1 and h2 are antipenultimate points and g2 and h1 are the corresponding penultimate points.

Case 3. y = x.

There is a point z ~ x and i1 (P3). See Figure 44. Now either z = ik or j = 2 and z = 21 (L2,P6).

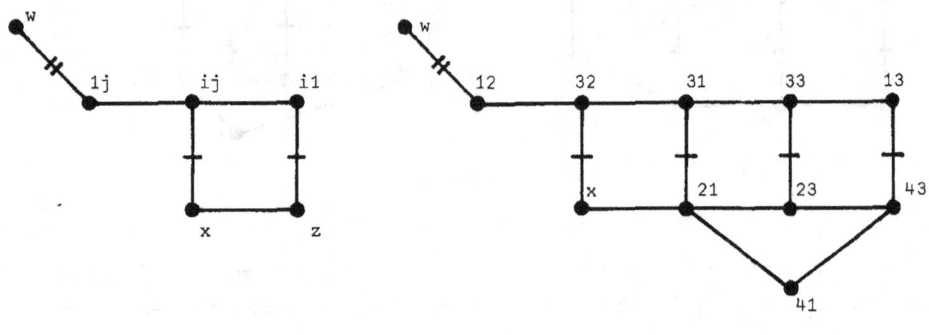

Figure 44

Figure 45

Case 3.1. j = 2 and z = 21.

Set i = 3. Because H is connected and has more than two points, there is a point h3 ~ h1 or h2. Suppose h3 ~ h1. Then (31,33) ε E otherwise there is a point 34 ~ 33 and 32 (P3,P6) and we have Case 1. Consequently (33,13) ε E otherwise x or w ~ 31 and 13 (P3) which contradicts P6. Hence there is a point 43 ~ 13 and 23 (P3,P6). Therefore w ~ 31 and 41 (P3,P6) which gives conclusion (1) of the theorem. See Figure 45. So h3 ~ h2. Then (32,33) ε E otherwise we have Case 1.

As before there is an edge e ε $\overline{E}$ incident to 31 or 21. Here e = (21,41) (P6, only h2 ~ h1, or we have Case 2). Suppose (41,42) ε E. Then x ~ 42 (P3,P6). Either (33,23) or (23,43) ε $\overline{E}$ (P6) hence x ~ 23 (P3,P6). But there is a point 53 ~ 13 and 23 (P3,P6). Hence 52 ~ x (L4). Therefore w ~ 51 and 31 (P3,P6) which is conclusion (1) of the theorem. See Figure 46. So (41,42) ε $\overline{E}$. If (33,23) ε E then (23,43) ε $\overline{E}$ and (43,42) ε E (P6) and hence x or w ~ 42 and 23 (P3) which contadicts P6. If (33,23) ε $\overline{E}$ then x ~ 23 (P3,P6) and the same contradiction results. See Figure 47.

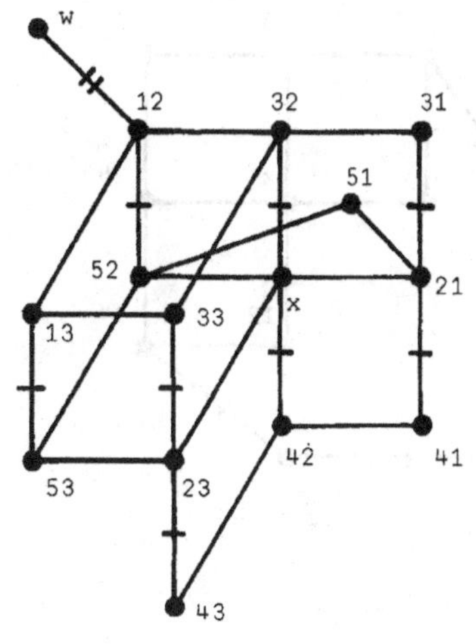

Figure 46

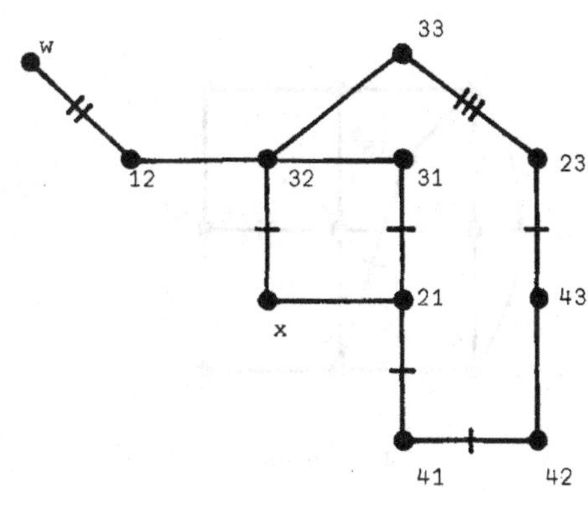

Figure 47

Case 3.2. $z = ik$.

If $(ik,1k) \in E$ then $w \sim i1$ and $1k$ (P3,P6) and we have conclusion (1) of the theorem. Suppose $(ik,1k) \in \overline{E}$. Then either $\ell k \sim x$ and $1k$ or $i = 2$ and $12 \sim x$ and $1k$.

Case 3.2.1. $i = 2$ and $12 \sim x$ and $1k$.

Set $j = 3$ and $k = 4$. Now either w or a point $43 \sim 13$ and x (P3,P6). If $43 \sim 13$ and x then $w \sim 43$ and 12 (L3). Hence there is a point $15 \sim 13$ and 12 (P3,P6) which contradicts P6. See Figure 48. So $w \sim 13$ and x. Now there is a point $g3 \sim g2$ or $g1$. If $g3 \sim g2$ then $w \sim 33$ and 34 (L1) which contradicts P6. So $g3 \sim g1$. Then $(14,34) \in \overline{E}$ otherwise $w \sim 24$ and 34 (P3,P6) which contradicts L3. Therefore $(34,31) \in \overline{E}$ (P3,P6) and hence $(31,33) \in E$ and $(33,13) \in \overline{E}$ (P6). So w or $x \sim 13$ and 31 (P3) which contradicts P6. See Figure 49.

Case 3.2.2. $\ell k \sim x$ and $1k$.

Now $(\ell k,\ell 1) \in E$ (P3,P6).

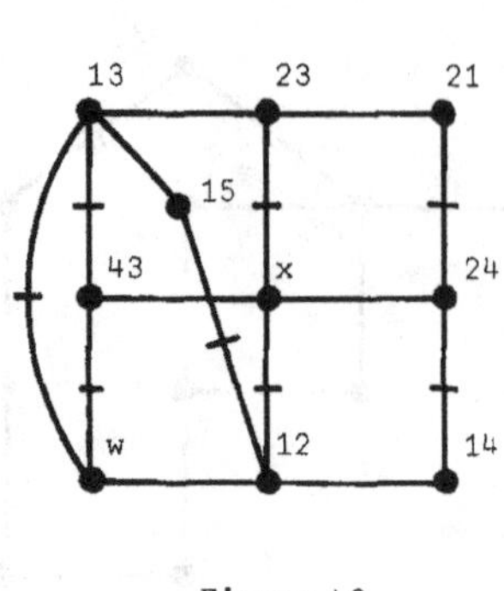

Figure 48

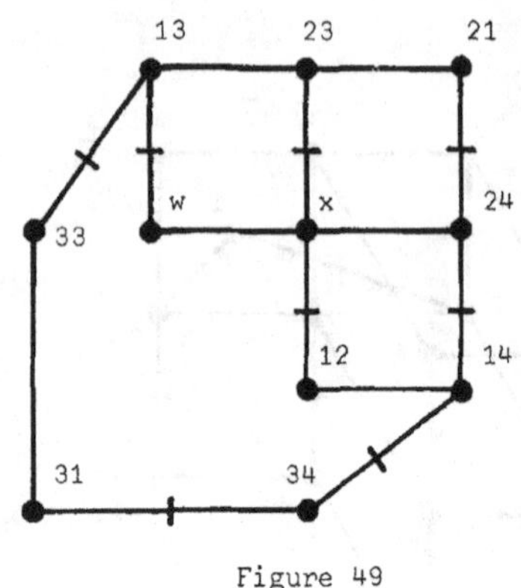

Figure 49

<u>Case 3.2.2.1.</u> $\ell j = 22$.

Set i = 3 = k. Now either w or a point 24 ~ 21 and x. If 24 ~ 21 and x then (31,34) ε E otherwise there is a point 35 ~ 32 and 34 (P3,P6) and we have Case 1. Hence there is a point 35 ~ 33 and 34 (P3,P6). But then 25 ~ 23 and 24 which contradicts P6. So w ~ 21 and x. Thus G = $\overset{2}{\circ}\!-\!\overset{1}{\circ}\!-\!\overset{3}{\circ}$ = H or there is a point g4 ~ g1, g2, or g3 or a point h4 ~ h1, h2, or h3. See Figure 50.

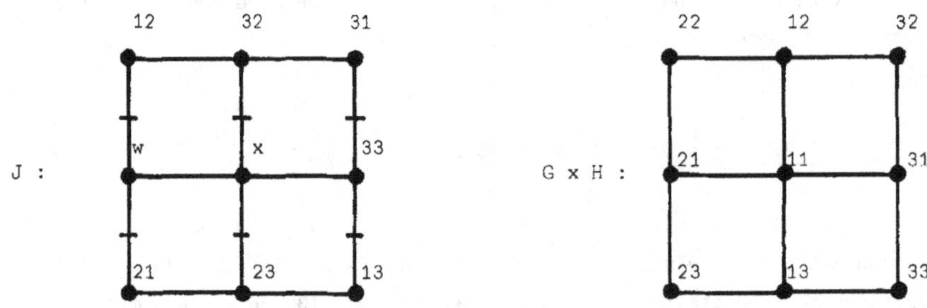

Figure 50

Now g4 $\not\sim$ g3 and h4 $\not\sim$ h3 (L1,P6). Suppose g4 ~ g1. Then (12,42) ε E otherwise there is a point 52 ~ 32 and 42 (P3,P6) which contradicts the fact that only g1 ~ g3. Similarly (13,43) ε $\overline{E}$. Hence (43,41) ε E (P6). But then x or w ~ 4 and 13 (P3) which contradicts P6. See Figure 51. So g4 $\not\sim$ g1. Similarly h4 $\not\sim$ h1. Now suppose g4 ~ g2. Then (23,43) ε E otherwise there is a point 53 ~ 43 and 13 (P3,P6) which contradicts the fact that only g2 and g3 ~ g1. Consequently there is a point 53 ~ x and 43 (L2,P6). Hence 51 ~ w (L4). But then 52 ~ 12 (P3,P6) which again contradicts the fact that only g2 and g3 ~ g1. See Figure 52. So g4 $\not\sim$ g2. Similarly h4 $\not\sim$ h2.

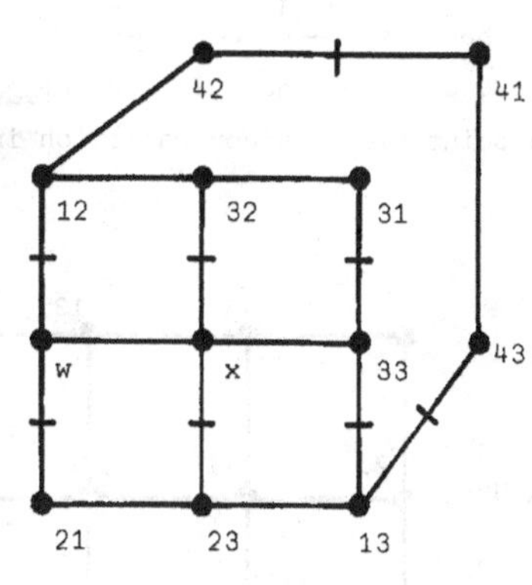

Figure 51

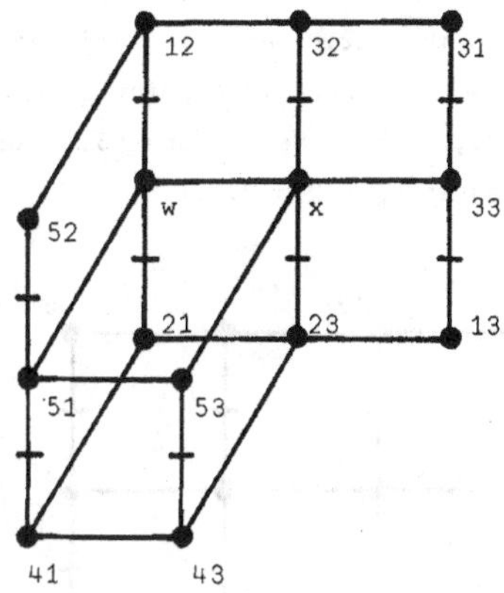

Figure 52

<u>Case 3.2.2.2</u>. $\ell j \neq 22$.

Suppose $x \not\sim \ell j$. Then $(1j,\ell j) \in E$ otherwise there is a point $mj \sim \ell j$ and ij (P3,P6) and hence $m1 \sim mj$ and $\ell 1$ which contradicts P6. See Figure 53. Similarly $(\ell 1,\ell j) \in E$. This contradicts P6.

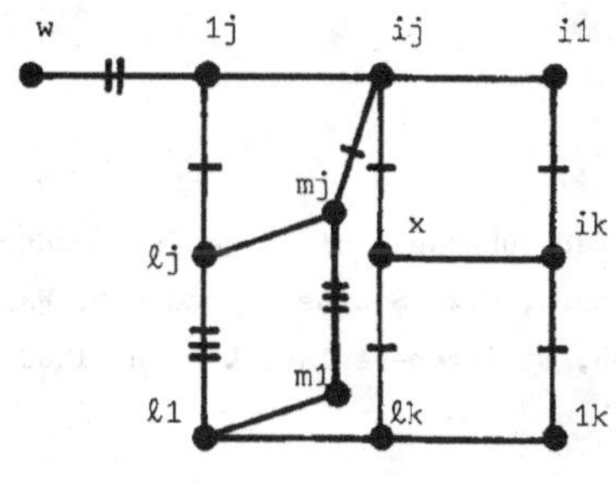

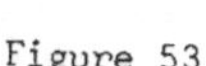
Figure 53

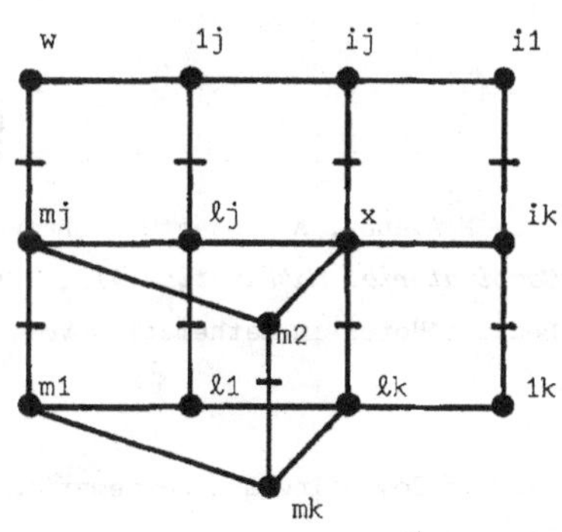

Figure 54

So $x \sim \ell j$. If $(1j,w) \in \overline{E}$ then there is a point $mj \sim w$ and ij (L2,P6) and we have Case 2. If $(1j,w) \in E$ then there is a point $mj \sim w$ and ℓj (L2,P6). If $mk \neq 22$ then $\ell = 2$ and $m2 \sim mj$ and mk (L1,P6). Hence $m2 \sim x$ (L4) which contradicts L3. See Figure 54.

So $mk = 22$. Set $i = 3 = j$ and $\ell = 4$. Then $G = \overset{3}{\circ}\!-\!\overset{1}{\circ}\!-\!\overset{4}{\circ}\!-\!\overset{2}{\circ}$ and $H = \overset{3}{\circ}\!-\!\overset{1}{\circ}\!-\!\overset{2}{\circ}$ or there is a point $g5 \sim g1, g2, g3$, or $g4$ or there is a point $h4 \sim h1, h2$, or $h3$. See Figure 55. Each extra point yields a contradiction in the same way as in Case 2.3.2.1.1.

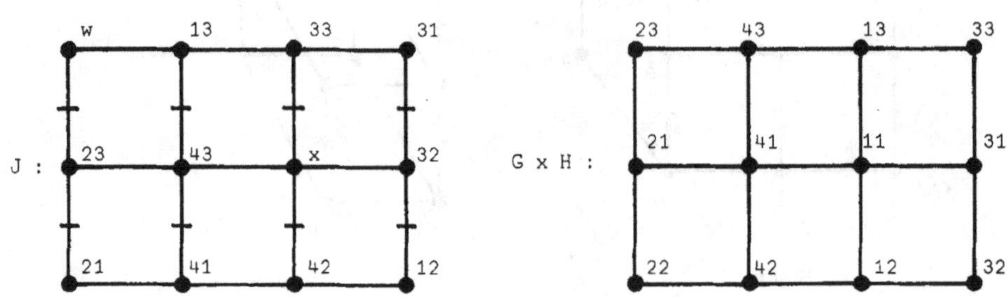

Figure 55

This completes the proof of Case 3. The remaining seven variations of the statement in the theorem, obtained by interchanging G and H or 1 and 2 or x and w, follow by symmetry.

REFERENCE

[1] K.L. McAvaney, A conjecture on two-point deleted subgraphs of cartesian products, *Combinatorial Mathematics VII*, Eds. R.W. Robinson, G.W. Southern, and W.D. Wallis, Lecture Notes in Mathematics Vol. 829, 172-185, Springer-Verlag, Berlin, 1980.

Division of Computing and Mathematics
Deakin University
Victoria 3217

DEQUES, TREES AND LATTICE PATHS

D.G. ROGERS AND L.W. SHAPIRO

A double ended queue or deque *is a linear list for which all insertions and deletions occur at the ends of the list. We give a direct, 'pictorial' proof of a result of Knuth on the enumeration of permutations obtainable from output restricted deques. This approach readily identifies the numbers of these permutations as* Schröder numbers *and leads naturally to correspondences with other equinumerous sets of trees and lattice paths. We also gather together other references to occurrences of the Schröder numbers.*

1. INTRODUCTION.

The *Schröder* numbers r_n and s_n are given by

$$r_n = \sum_{i=0}^{n} \binom{2n-i}{i} C_{n-i} \quad , n \geq 0 , \tag{1}$$

and

$$s_0 = 1 \ ; \ s_n = \tfrac{1}{2} r_n \quad , n \geq 1 , \tag{2}$$

where, in turn, C_n is the nth *Catalan* number,

$$C_n = \frac{1}{n+1} \binom{2n}{n} \quad , n \geq 0 . \tag{3}$$

The first few values of these sequences are given in Table 1.

n	0	1	2	3	4	5	6	7	8	9
C_n	1	1	2	5	14	42	132	429	1,430	4,862
r_n	1	2	6	22	90	394	1,806	8,558	41,586	206,098
s_n	1	1	3	11	45	197	903	4,279	20,793	103,049

TABLE 1

The Catalan numbers occur in a wide variety of enumeration problems, involving, among other things, stack permutations, rooted planar trees and restricted lattice paths. It is a challenge to find both correspondences between such equinumerous sets and particularly simple ways of enumerating them. The Schröder numbers often arise in variants of the Catalan problems. We explore here connections between variants associated with the three examples mentioned above, namely deque permutations, marked rooted planar trees and restricted lattice paths with diagonal steps, with a view to giving a 'pictorial' proof of a result of Knuth on the first of these (see [6; §2.2.1,

Exercise 11, pp. 239,534]).

We note, in an appendix, a form of Lagrange's inversion theorem which is particularly useful in enumeration problems of the sort considered here. We also draw attention to its use in the enumeration of Davenport-Schinzel sequences (see [12]), giving a further occurrence of the Schröder numbers. Additional references to the Schröder numbers are gathered in the list at the end of the paper.

This paper is a sequel to [17]and [20], these papers forming a companion to the bibliography [5] of the Catalan numbers (see also [26; §3.1]) and the review [2] of occurrences of the Motzkin numbers. The Schröder numbers s_n, $n \geq 0$, appear as Sequence 1163 and 1170 (correcting a misprint) of [22]. The Schröder numbers are so called after the occurrence of the numbers s_n, $n \geq 0$, in a problem considered by Schröder in [21].

2. DEQUES.

Following Knuth [6; §2.2.1, esp. p. 235], a *d*ouble *e*nded *que*ue or *deque* is a linear list for which all insertions and deletions occur at the ends of the list: one representation is as the railway shunting network in Figure 1. A deque is *output-restricted* if deletions take place only at one end, say, the left, in which case the bottom track in Figure 1 is closed. (An *input-restricted* deque is defined similarly and is obtained when an output-restricted deque is run backwards.)

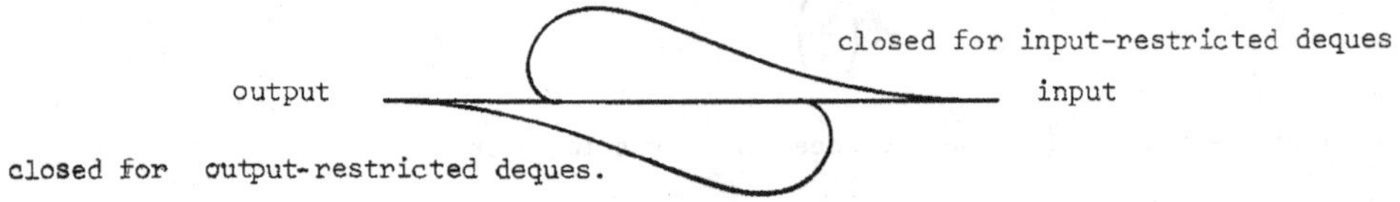

FIGURE 1: A deque as a railway shunting network

A deque, like other linear lists or shunting networks, may be viewed as a device for permuting the inputs. Knuth develops a generating function solution to the problem of enumerating the set $\mathcal{D}_n$ of permutations obtainable on an output - restricted deque with n inputs, $n \geq 1$ (see [6; §2.2.1, Exercises 10,11, pp. 239,533-4]). It may be deduced from this that the number $d(n)$ of permutations in $\mathcal{D}_n$ is the Schröder number r_{n-1}, $n \geq 1$. Although this deduction is not made in [6], Knuth does later note, in discussing a tree enumeration problem, an equivalence with Schröder's problem in [21] and raises the question of finding a connection between these and the problem of enumerating $\mathcal{D}_n$ (see [6; § 2.3.4.4, Exercise 31, pp. 398,587]). We present an enumeration of $\mathcal{D}_n$ which is more combinatorial in spirit, leading explicitly to (1) and providing correspondences with other objects enumerated by the Schröder numbers.

As Knuth observes (see [6; §2.2.1, Exercise 10, pp. 239,533]), the handling of any set of n inputs, $n \geq 1$, and so also the permutation achieved in the process, may be encoded as a word of length 2n in the letters Q (insertion at right), S (insertion at left) and X (deletion at left). The code word is uniquely determined if we adopt the 'priority' rules: -

(i) *the word begins with a Q ;* and

(ii) *Q never follows X .*

Code words are then characterized by (i) and (ii) and the further 'balancing' property

(iii) *the number of Q's and S's in any proper initial segment always exceeds the number of X's with equality for the word as a whole, there being n X's in all.*

For $n \geq 2$, if we delete the initial Q and final X, we obtain a word of length 2(n-1) satisfying (ii) and

(iv) *the number of X's in any initial segment never exceeds the number of Q's and S's with equality for the word as a whole (and possibly elsewhere), there being n-1 X's in all.*

Now similar codes using two letters (for instance, up-down codes) are familiar in connection with rooted planar trees (see, for example, [26; p.24]) and they may also be used in the analysis of stacks (see [5; §2.2.1, Exercises 3,4,pp. 238,531-2]). As a modification of this, we now describe a three letter code for a family of marked rooted planar trees which satisfies conditions (ii) and (iv), showing thereby that these trees are in one-to-one correspondence with output-restricted deque permutations.

3. TREES.

Using the conventional drawing of a rooted planar tree (see Figure 2), the left most, upward branch (if any) at a vertex is called the *eldest branch* (see [20, §5]). Consider the set $T_n^{(k)}$, $n \geq 1$, of rooted planar trees with n edges in which eldest branches may be marked independently of each other in k different ways (k=1, unmarked; k=2, marked or unmarked). For k=2, the up-down code (case k=1) may be modified to give a unique code word of length 2n in the letters D (down), U (up an unmarked edge) and V (up a marked edge) for each tree in $T_n^{(2)}$, $n \geq 1$. An illustration is shown in Figure 2.

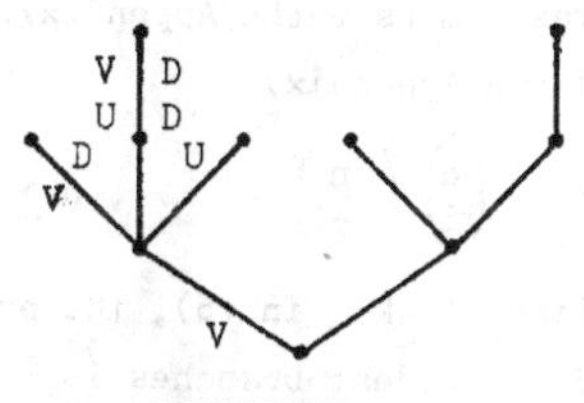

FIGURE 2: Encoding a marked tree

The code words produced in this way are characterized by the properties (compare (i) and (iv) above): -

(a) V *never follows* D *; and*

(b) the number of D*'s in any initial sequence never exceeds the number of* U*'s and* V*'s with equality for the word as a whole (and possibly elsewhere), there being n* D*'s in all.*

This establishes the correspondence between $\mathcal{D}_{n+1}$ and $\mathcal{T}_n^{(2)}$, $n \geq 1$. Note that $\mathcal{D}_1$ and $\mathcal{T}_0^{(k)}$ are singleton sets. (Similarly $\mathcal{T}_n^{(k)}$ may be put in correspondence with permutations arising from the deque in which there are $k-1$ mutually exclusive ways of making an insertion at the right, $k=1$ giving the stack.)

Now, $\mathcal{T}_n^{(k)}$ may be enumerated by considering the valence of the roots of the trees as shown in the pictorial scheme in Figure 3 in which $t_k(n)$ is the number of trees in $\mathcal{T}_n^{(k)}$ and $T_k(x)$ is the associated generating function.

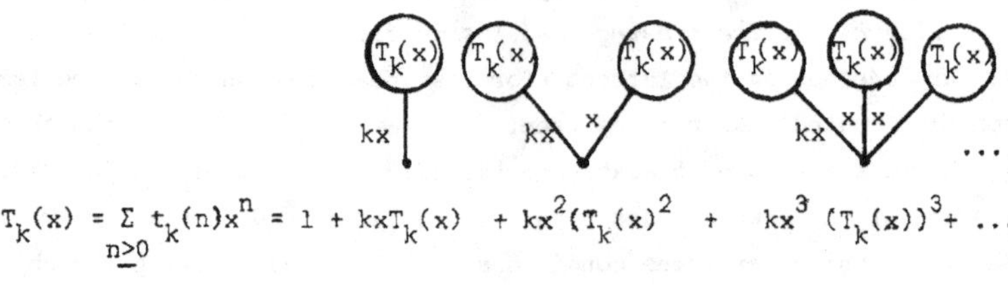

$$T_k(x) = \sum_{n\geq 0} t_k(n)x^n = 1 + kxT_k(x) + kx^2(T_k(x))^2 + kx^3(T_k(x))^3 + \ldots$$

FIGURE 3

It follows that $T_k(x)$ satisfies the functional equation

$$T_k(x) = 1 + xT_k(x)(k + T_k(x) - 1) . \tag{4}$$

For $k=1$ and 2, (4) gives well known functional equations for the generating functions for the Catalan numbers C_n, $n \geq 0$, and the Schröder numbers r_n, $n \geq 0$, respectively, so that

$$t_1(n) = C_n \; ; \; t_2(n) = r_n \quad , \quad n \geq 0 .$$

Indeed, the explicit expressions (1) and (3) may be deduced from (4) in these cases by using Lagrange's inversion theorem (see the Appendix). More generally, the same technique shows that (again see the Appendix)

$$t_k(n) = \sum_{i=1}^{n} \frac{1}{n}\binom{n}{i}\binom{n}{i-1} k^i \quad , \; n \geq 1 . \tag{5}$$

Hence, considering the coefficient of k^i in (5), the number $t(n,i)$ of rooted planar trees with n edges and i eldest branches is

$$t(n,i) = \frac{1}{n}\binom{n}{i}\binom{n}{i-1} , \; 1 \leq i \leq n . \tag{6}$$

Note that it follows from (6) that

$$t(n,i) = t(n,n+1-i) \quad , 1 \leq i \leq n . \tag{7}$$

Now a rooted planar tree with n edges, $n \geq 1$, and i eldest branches, $1 \leq i \leq n$, has i non-terminal vertices (including the root) and so $n+1-i$ endpoints (terminal vertices). So $t(n,i)$, $1 \leq i \leq n$, is also the number of rooted planar trees with n edges and $n+1-i$ endpoints or, in view of (7), n edges and i endpoints (compare [14] and [15; p. 428]). For a 'pictorial' proof of this after the above manner, consider the set $\mathcal{E}_n^{(k)}$ of rooted planar trees with n edges in which endpoints (other than the root) are coloured, independently of each other, with any of k colours. The scheme in Figure 4 shows that the associated generating function in $E_k(x)$ satisfies the same functional equation (4) as $T_k(x)$, so the result follows.

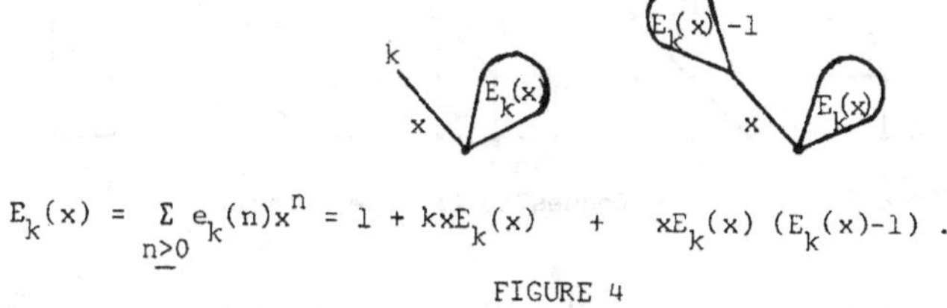

$$E_k(x) = \sum_{n \geq 0} e_k(n)x^n = 1 + kxE_k(x) + xE_k(x)\,(E_k(x)-1) .$$

FIGURE 4

Both rules (iv) and (b) above mention the possibility of a balance in the number of occurrences of the code letters. This happens every time one item is left on the deque or a return is made to the root vertex. The correspondence shows that the number $d(n,m)$, $1 \leq m < n$, of permutations in $\mathcal{D}_n$ leaving a single item on the deque on m occasions (after the first insertion) is the same as the number of trees in $\mathcal{T}_{n-1}^{(2)}$ whose roots have valence m and, from Figure 3, this is the coefficient of x^{n-1} in $2x^m(T_2(x))^m$. This number is later identified as being also the number of certain restricted lattice paths (see(14)) and an explicit expression for it may be found using the results in the Appendix.

Although the correspondence between $\mathcal{D}_n$ and $\mathcal{T}_{n-1}^{(2)}$, $n \geq 1$, readily leads to the determination of $d(n)$ as the Schröder number r_{n-1}, $n \geq 1$, much of the finer detail just described still uses generating function methods. A more completely combinatorial approach is obtained by looking instead at restricted lattice paths with diagonal steps. However, in order to be in a position to do this, we need first to transform the code words used in Section 2.

4. LATTICE PATHS.

In a code word of length $2(n-1), n \geq 2$, satisfying rules (ii) and (iv) of

Section 2, each occurrence of the letter Q begins a string of letters which also satisfies these rules and is minimal subject to this condition. Thus to each Q there is associated a unique X , namely that at the end of the minimal string begun by the Q . We now delete the Q's and replace their associated X's by P's (see Figure 5). The result is a code word characterized by the following property:

> (α) *the number of occurrences of X in any initial segment never exceeds the number of S's with equality for the word as a whole, the number of P's and X's together being n-1 .*

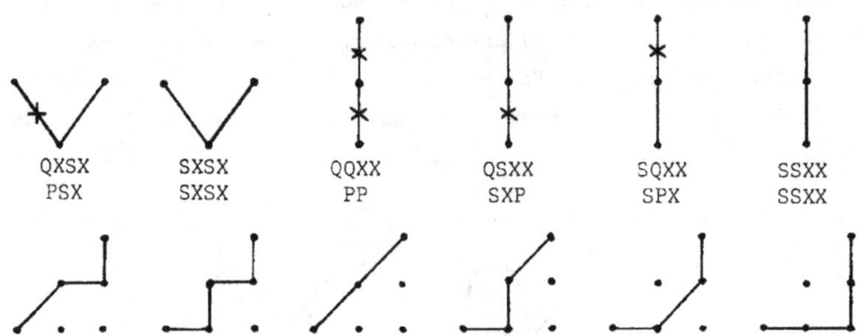

FIGURE 5: Trees and Deques to Lattice Paths.

Now a code word of this sort may be reinterpreted as a lattice path on the integral square lattice by taking S,P and X to stand for a step to the right, diagonally and up, respectively (compare [20, §5]). A path on the integral square lattice, allowing these types of steps, which, starting from the origin, remains in the non-negative quadrant, on or below the main diagonal, is called a *restricted lattice path.* So, if we start from the origin, the path traced out by a code word with property (α) is a restricted lattice path terminating at the point (n-1,n-1), $n \geq 2$. Thus, if $R_{n,m}$, $0 \leq m \leq n$, is the set of restricted lattice paths terminating at the point (n,m), then the foregoing establishes a correspondence between $\mathcal{D}_n$ or $T^{(2)}_{n-1}$ and $R_{n-1,n-1}$, $n \geq 2$. Note, again, that $\mathcal{D}_1$ and $R_{0,0}$ are singleton sets. (The correspondence extends to $T^{(k)}_{n-1}$ by using diagonal steps with multiplicity k-1, see also (13).)

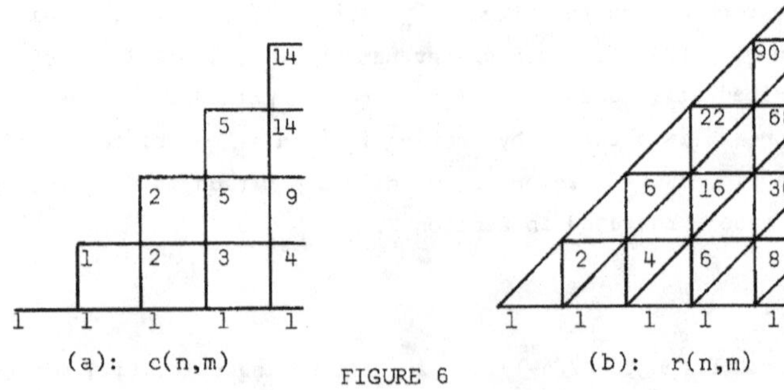

FIGURE 6

If $c(n,m)$, $0 \leq m \leq n$, is the number of paths in $\mathcal{R}_{n,m}$, without diagonal steps (see Figure 6a), then it is well known that $c(n,m)$ is the *ballot number* given by

$$c(n,m) = \frac{n-m+1}{n+1}\binom{n+m}{n} \quad , \ 0 \leq m \leq n \ . \tag{8}$$

In particular, taking $n=m$ and comparing (3),

$$c(n,n) = C_n \quad , \ n \geq 0 \ . \tag{9}$$

Now a path in $\mathcal{R}_{n-i,m-i}$ without diagonal steps can be extended to one in $\mathcal{R}_{n,m}$ by the insertion of i diagonal steps in $\binom{n+m-i}{i}$ ways, $0 \leq i \leq \min(n,m)$. So, if $r(n,m)$, $0 \leq m \leq n$, is the number of paths in $\mathcal{R}_{n,m}$ (see Figure 6b), then

$$r(n,m) = \sum_{i=0}^{\min(n,m)} \binom{n+m-i}{i} c(n-i,m-i) \quad , \ n \geq 0 \ . \tag{10}$$

Taking $m=n$ and noting (1) and (11), we see that

$$d(n+1) = t_2(n) = r(n,n) = r_n \quad , \ n \geq 0 \ ,$$

which completes a purely combinatorial enumeration of the sets $\mathcal{D}_n, n \geq 1$. (Note that this also gives an interpretation of the terms in the sum in (1) as the number of paths in $\mathcal{R}_{n,n}$ with i diagonal steps, $0 \leq i \leq n$.)

Now, let $r_\ell(n,m)$, $0 \leq m \leq n$, be the number of restricted lattice paths in which diagonal steps have multiplicity ℓ, so $r_0(n,m) = c(n,m)$ and $r_1(n,m) = r(n,m)$. More generally, we have the 'additive' property (compare [17; §2]):

$$r_\ell(n,m) = r_\ell(n-1,m) + \ell r_\ell(n-1,m-1) + r_\ell(n,m-1) \ , \ 1 \leq m \leq n \ ,$$

subject to

$$r_\ell(n,0) = 1 \ ; \ r_\ell(n-1,n) = 0 \quad , \ n \geq 1 \ .$$

Then, returning to the use of generating functions, the first and last passage decomposition methods of [17; §2], give the 'multiplicative' results: -

$$R_\ell(x) = \sum_{n\geq 0} r_\ell(n,n)x^n = 1 + xR_\ell(x)(\ell+R_\ell(x)) \tag{11}$$

and

$$\sum_{n\geq 0} r_\ell(n+m-1,n)x^n = (R_\ell(x))^m \quad , \ m \geq 1 \ , \tag{12}$$

from which (8) and (10) may also be deduced (see the Appendix). It also follows that

$$r_\ell(n) = t_{\ell+1}(n) \quad , \ n \geq 0 \ ; \tag{13}$$

$$d(n,m) = 2r(n-2,n-m-1) \ , \ 2 \leq m < n \ . \tag{14}$$

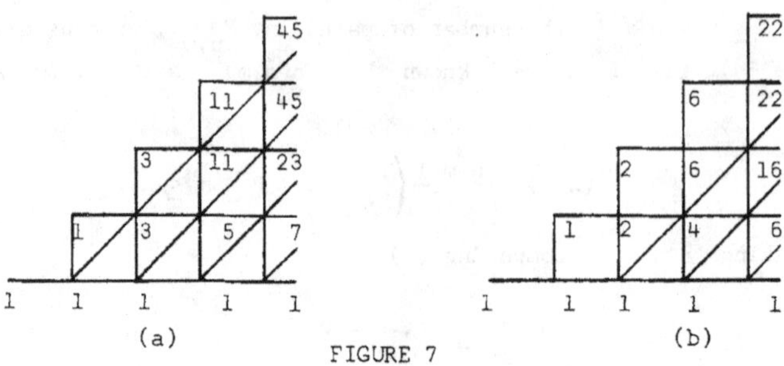

FIGURE 7

Further examples of lattice path enumerations involving the Schröder numbers are shown in Figures 7a, 7b and 8a and it is an amusing exercise to interpret these occurrences in terms of deque permutations and marked rooted planar trees by means of the correspondences established here. (It seems that the patterns in these figures do not continue; see, for example, Figure 8b. However, one curiosity of Figure 8b is that the diagonal entries are the column sums in Figure 7a).

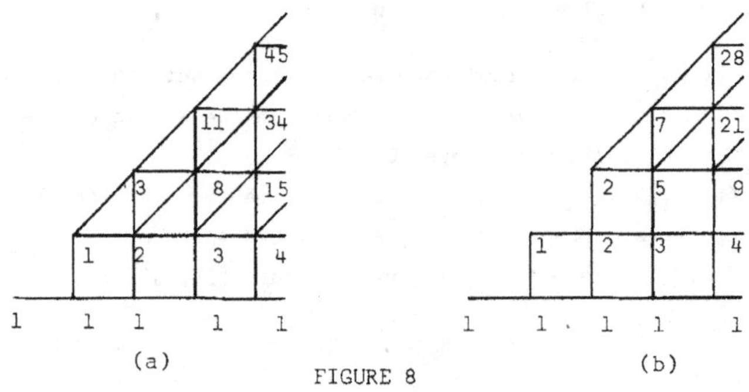

FIGURE 8

Lattice path enumeration problems have a large literature and the book [9] is devoted to this topic. Of special interest here are [1; pp. 80-81], [4], [10], [17; §2] and [24].

APPENDIX: LAGRANGE'S INVERSION THEOREM.

Suppose that the generating function $A(x)$ satisfies the functional equation

$$A(x) = \sum_{n\geq 0} a_n x^n = 1 + xH(A(x)) , \tag{15}$$

where $H(t)$ is a polynomial in t. Further, if $G(t)$ is a polynomial in t, write

$$G(A(x)) = \sum_{n\geq 0} g_n x^n .$$

Then, as a special case of *Lagrange's inversion theorem* (see [1; pp. 148-153] and [27; pp. 132-133]) we have

$$g_n = \text{coefficient of } t^{n-1} \text{ in } \frac{1}{n} G'(1+t)(H(1+t))^n , \quad n \geq 1 . \tag{16}$$

As a part of this theorem, we have the existence and uniqueness of solutions of the functional equation (15). (For further information, see [4; §3] and [23; pp. 126-127.)

Example 1 : $A(x) = T_1(x)$; $H(t) = t^2$.

In this case, $a_n = C_n, n \geq 0$, and, for $m \geq 1$, if $G(t) = t^m, g_m = c(n+m-1,n)$, $n \geq 0$ (compare (14)). The expressions (3) and (8) then follow from (16).

Example 2: $A(x) = T_k(x)$; $H(t) = t(k+t-1)$.

In this case, $a_n = t_k(n)$, $n \geq 0$. So, from (16), with $G(t) = t$,

$$\begin{aligned} t_k(n) &= \text{coefficient of } t^{n-1} \text{ in } \frac{1}{n}(1+t)^n(k+t)^n , \quad n \geq 1 , \\ &= \frac{1}{n}\sum_{i=0}^{n-1} \binom{n}{i}\binom{n}{i+1} k^{i+1} , \quad n \geq 1 , \end{aligned} \tag{17}$$

and then (5) follows on rearranging.

Example 3: $A(x) = T_{\ell+1}(n) = R_\ell(x)$; $H(t) = t(\ell+t)$

In this case, $a_n = t_{\ell+1}(n) = r_\ell(n,n)$, $n \geq 0$, and, for $m \geq 1$, if $G(t) = t^m$, $g_n = r_\ell(n+m-1,n)$, $n \geq 0$ (compare (14)).

Now $H(1+t) = \ell(1+t) + (1+t)^2$, so

$$(H(1+t))^n = \sum_{i=0}^{n} \binom{n}{i} \ell^i (1+t)^i (1+t)^{2(n-i)} = \sum_{i=0}^{n} \binom{n}{i} \ell^i (1+t)^{2n-i} , \quad n \geq 0 .$$

Hence, from (16),

$$\begin{aligned} r_\ell(n+m-1,n) &= \text{coefficient of } t^{n-1} \text{ in } \frac{m}{n} \sum_{i=0}^{m} \binom{n}{i} \ell^i (1+t)^{2n+m-1-i} , \quad n,m \geq 1 , \\ &= \frac{m}{n} \sum_{i=0}^{n} \binom{n}{i}\binom{2n+m-1-i}{n-1} \ell^i \end{aligned} \tag{18}$$

which, on rearrangement, noting (8), gives

$$r_\ell(n+m-1,n) = \sum_{i=0}^{n} \binom{2n+m-1-i}{i} c(n+m-1-i,n-i)\ell^i , \quad n \geq 0 , \quad m \geq 1 . \tag{19}$$

Taking $\ell = 0$ in (18), we recover (8), (compare Example 1) while taking $\ell = 1$ in (19) gives (10) of which (1) is a special case. (Note also, from (14), we may find $d(n,m)$ explicitly.)

Example 4: Davenport-Schinzel (D.-S.) sequences

In [12], it is shown, by a correspondence, that the number of (n,3)-D.-S. sequences of greatest length is the same as the number $t_2(n-1)$ of rooted planar trees with n-1 edges, namely $C_{n-1}, n \geq 1$. It is also shown there that the normal (n,3)-D.-S. sequences of greatest length are in correspondence with the face maps with n vertices and that if f_n is the number of these then (compare Figures 3 and 4):

$$F(x) = \sum_{n\geq 1} f_n x^n = x + \sum_{m\geq 2} (F(x))^m . \tag{20}$$

The remark is then made that, on applying Lagrange's theorem to (20), "the resulting expression is a finite summation with alternating signs which is undesirable for calculation" (see [12; p. 171]). This is, however, not the case if Lagrange's theorem is used in the form (16) above and indeed using this gives an explicit expression (21) for f_{n+1} which identifies it as the Schröder number $s_n, n \geq 0$. (This identification also follows on comparing the functional equations (20) and (4) for $k = 2$ (or (11) for $\ell = 1$) bearing in mind (1) and (2) and may further be established through combinatorial correspondences.)

For, writing $F(x) = xA(x)$, it follows from (20) that $A(x)$ satisfies (15) with $H(t) = t(2t-1)$. Hence, from (16) (compare (17) for $k=2$): -

$$\begin{aligned} f_{n+1} &= \text{coefficient of } t^{n-1} \text{ in } \frac{1}{n}(1+t)^n(1+2t)^n &&, n \geq 1 , \\ &= \sum_{i=0}^{n-1} \frac{1}{n}\binom{n}{i}\binom{n}{i+1} 2^i &&, n \geq 1 , \qquad (21) \\ &= \frac{1}{2}t_2(n) = \frac{1}{2}r_n = s_n &&, n \geq 1 . \end{aligned}$$

REFERENCES.

References having occurrences of the Schroder numbers are marked with an asterisk.

[1]*. L. Comtet, *Advanced Combinatorics,* (D. Reidel, Dordrecht, 1974), esp. pp. 56-57, 80-81.

[2]. R. Donaghey and L.W. Shapiro, "Motzkin Numbers", *J. Combinatorial Theory, Ser. A,* 23(1977), 291-301.

[3]*. A. Erdélyi and I.M.H. Etherington, "Some problems of non-associative combinations, I, II", *Edinburgh Math. Notes,* 32(1940),1-12, esp. p.6.

[4]*. I.M. Gessel, "A factorization for formal Laurent series and lattice path enumeration", *J. Combinatorial Theory, Ser. A,* 28(1980), 321-337, esp. p. 329.

[5]. H.W. Gould, *Research Bibliography of Two Special Number Sequences,* rev. ed., (Combinatorial Research Institute, Morgantown, W. Va., 1977).

[6]*. D.E. Knuth, *The Art of Computer Programming, Vol. 1; Fundamental Algorithms,* 2nd ed., (Addison Wesley, Reading, Ma., 1973), esp. pp. 235-239, 398, 533-534, 589.

[7]*. G. Kreweras, "Sur les partitions non croisées d'un cycle", *Discrete Math.,* 1(1972), 333-350, esp. p. 345.

[8]*. J.S. Lew, "Polynomial enumeration of multidimensional lattices", *Math. Systems Theory,* 12(1979), 253-270.

[9]. S.G. Mohanty, *Lattice Path Counting and Applications,* (Academic Press, New York, 1979).

[10]. L. Moser and W. Zayachkowski, "Lattice paths with diagonal steps", *Scripta Math.,* 26(1963), 223-229, esp. pp. 227, 228.

[11]*. Th. Motzkin, "Relations between hypersurface crossratios and a combinatorial formula for partitions of a polygon, for permanent preponderance and for non-associative products", *Bull. Amer. Math. Soc.,* 54(1948), 352-360, esp. p.359.

[12]*. R.C. Mullin and R.G. Stanton, "A map theoretic approach to Davenport-Schinzel sequences", *Pacific J. Math.,* 40(1972), 167-172.

[13]*. J. Riordan, *Combinatorial Identities,* (Wiley, New York, 1968), esp. pp. 149, 151, 168.

[14]. J. Riordan, "Enumeration of plane trees by branches and endpoints", *J. Combinatorial Theory, Ser. A,* 19(1975), 214-222.

[15]. D.G. Rogers, "The enumeration of a family of ladder graphs I : Connective relations", *Quart. J. Math. (Oxford)* (2), 28(1977), 421-431.

[16]*. D.G. Rogers, "The enumeration of a family of ladder graphs II : Schröder and Super connective relations", *Quart. J. Math. (Oxford)* (2), 31(1980),

[17]*. D.G. Rogers, "A Schröder triangle: three combinatorial problems", in *Combinatorial Mathematics V: Proceedings of the Fifth Australian Conference. Lecture Notes in Mathematics, Vol. 622,* (Springer-Verlag, Berlin, 1977), pp. 175-196.

[18]*. D.G. Rogers, "Pascal's triangle, Catalan numbers and renewal arrays",*Discrete Math.,* 22(1978), 301-311.

[19]*. D.G. Rogers, "Eplett's identity for renewal arrays", *Discrete Math.,* to appear.

[20]*. D.G. Rogers and L.W. Shapiro, "Some correspondences involving the Schröder number", in *Combinatorial Mathematics: Proceedings of International Conference, Canberra, 1977, Lecture Notes in Mathematics, Vol. 686*, (Springer-Verlag, Berlin, 1978), pp. 267-276.

[21]. E. Schröder, "Vier Kombinatorische Probleme", *Z. Math. Phys.,*15(1870), 361-376.

[22]*. N.J.A. Sloane, *A Handbook of Integer Sequences,* (Academic Press, New York, 1973).

[23]*. R.P. Stanley, "Generating functions" in *MAA Studies in Combinatorics* (Math. Assoc. Amer., Washington, D.C., 1978), pp. 100-141, esp. pp. 126, 129.

[24]. R.G. Stanton and D.D. Cowan, "Note on a "square" functional equation", *SIAM Review,* 12(1970), 277-279.

[25]*. H.N.V. Temperley and D.G. Rogers, "A note on Baxter's generalization of the Temperley-Lieb operators", in *Combinatorial Mathematics: Proceedings of International Conference, Canberra, 1977. Lecture Notes in Mathematics, Vol. 686,* (Springer-Verlag, Berlin, 1978), pp. 240-247.

[26]. J.H.van Lint, *Combinatorial Theory Seminar, Eindhoven University of Technology. Lecture Notes in Mathematics, Vol. 382,* (Springer-Verlag, Berlin, 1974), esp. pp. 21-27.

[27]. E.T. Whittaker and G.N. Watson, *A Course of Modern Analysis,* 4th ed., (Cambridge University Press, 1946).

68, Liverpool Road,
Watford, Herts., WD18DN,
ENGLAND .

Mathematics Department, Howard University,
Washington, D.C. 20059
U.S.A.

05 B 05

(05 B 10)

(05 B 30)

(62 K 10)

GRAECO-LATIN AND NESTED ROW AND COLUMN DESIGNS

DEBORAH J. STREET

In this paper certain balanced incomplete block designs (BIBD) and partially balanced incomplete block designs (PBIBD), constructed with the help of the theory of cyclotomy, are used to give some Graeco-Latin designs and some nested row and column designs.

The first section consists of a summary of notation and definitions useful in the remainder of the paper, section 2 contains the construction of the Graeco-Latin designs and section 3 those of the balanced and partially balanced nested row and column designs.

1. NOTATION AND A PRELIMINARY RESULT

The notation described below is that of Preece [3].

Consider a (v,b,r,k,λ) BIBD. It has two *constraints*, namely the blocks and the treatments, which occur at *levels* b and v respectively. A design may, of course, have several constraints. For instance, two mutually orthogonal Latin squares (Graeco-Latin squares) have four constraints: the rows, the columns and the two sets of treatments.

After ordering a design's constraints, we can define an incidence matrix of the ath constraint with respect to the bth by

$$N_{ab} = (n_{ij}),$$

where n_{ij} is the number of times the ith level of the ath constraint occurs with the jth level of the bth constraint. N_{ab} is $k_a \times k_b$, where there are k_i levels of the ith constraint, and $N_{ba} = N_{ab}^T$.

For example, if we regard the blocks as the first constraint and the treatments as the second in a (v,b,r,k,λ) BIBD then N_{21} is the usual incidence matrix and

$$N_{21}N_{21}^T = (r-\lambda)I + \lambda J,$$

where I is the identity matrix and J is the all ones matrix.

We include a family of supplementary difference sets (sds) given in [7] as they will be used several times in the remainder of the paper.

Lemma 1. *Let* $p^n = 2mf+1$ *be a prime power. Let* f *be odd and let* x *be a primitive root of* $GF(p^n)$. *Denote the cyclotomic classes with* $e = 2m$ *by*

$$C_i = \{x^{es+i} \mid s = 0,1,\ldots,f-1\}, \qquad i = 0,1,\ldots,e-1.$$

Let $i_0 = 0, i_1, \ldots, i_{m-1}$ *be a complete set of residues mod* m *such that* $0 \le i_j \le e-1$ for *every* j *and let* A *be a subset of* $\{0,1,\ldots,m-1\}$. *Then the* m *sets*

$$T_h = \bigcup_{j \in A} C_{i_j - i_h}, \qquad h = 0,1,\ldots,m-1$$

are $m - \{2mf+1\,;\, tf\,;\, t(tf-1)/2\}$ *sds, where* $t = |A|$.

2. GRAECO-LATIN DESIGNS

The first designs of this type we consider are reviewed in Preece [3].

Let the first constraint be blocks and the second and third constraints be sets of treatments. Then the designs of interest satisfy the following:

(i) $N_{i1}N_{i1}^T = (r_i - \lambda_i)I + \lambda_i J, \quad i = 2,3;$

(ii) $N_{23} = J$; and

(iii) $N_{21}N_{31}^T = kJ.$

Thus each set of treatments is arranged as a $(v_i, b, r_i, k, \lambda_i)$ BIBD $(i = 2,3)$ where $v_2 = r_3$ and $v_3 = r_2$.

Seberry [4] gave a family of these designs, constructed using cyclotomy, with $v_2 = r_3 = p+1$, $r_2 = v_3 = p$, $b = 2p$, $k = (p+1)/2$, for p a prime power. Below we give two more families with these parameters which are not isomorphic to each other nor to those of Seberry [4].

Theorem 2. (a) *If* $p \equiv 3 (4)$ *is a prime power then there exists a Graeco-Latin design satisfying* (i),(ii) *and* (iii) *above with* $v_2 = r_3 = p+1$, $r_2 = v_3 = p$, $b = 2p$, $k=(p+1)/2$ *and for which each of the underlying BIBDs is non-resolvable.*
(b) *If* $p = 6f+1 = 4x^2+27$ (f odd) *is a prime power then there exists a Graeco-Latin design satisfying conditions* (i),(ii) *and* (iii) *above with* $v_2 = r_3 = p+1$, $r_2 = v_3 = p$, $b = 2p$, $k = (p+1)/2$ *and for which one of the underlying BIBDs is resolvable and the other is a 2-multiple of a* $(p,p,(p+1)/2,(p+1)/2,(p+1)/4)$ *BIBD.*

Proof. (a) Consider a (v_2,b,r_2,k,λ_2) BIBD generated from the initial blocks $\{\infty\} \cup C_0$, $\{0\} \cup C_0$ (where C_0 is the set of quadratic residues in GF(p); that is, the factorization with $e = 2$ is used) and a (v_3,b,r_3,k,λ_3) BIBD with initial blocks $\{0\} \cup C_1$, $\{0\} \cup C_0$. Then,

$$N_{21} = \left[\begin{array}{c|c} A & A+I \\ 11\cdots 1 & 00\cdots 0 \end{array}\right], \qquad N_{31} = \left[\begin{array}{c|c} B+I & A+I \end{array}\right]$$

where A is the incidence matrix of C_0, B of C_1, so $A+B+I = J$ and $A^T = B$.

$$N_{21}N_{31}^T = \begin{bmatrix} A(J - A^T) + (A+I)(A^T+I) \\ kj \end{bmatrix}$$

where j is a row vector of ones. Now,

$$A(J - A^T) + (A+I)(A^T+I) = AJ + A + A^T + I = kJ$$

so $N_{21}N_{31}^T$ is of the required form.

To obtain $N_{23} = J$ we arrange the blocks as $(\infty 0, 1x, x^2x^3, \ldots, x^{p-3}x^{p-2})$ and $(00, 1x^{p-3}, x^2 1, \ldots, x^{p-3}x^{p-5})$ or $(\infty 0, 1x, x^2x^3, \ldots, x^{p-3}x^{p-2})$ and $(00, 1x^2, x^2x^4, \ldots, x^{p-3}1)$. In the first case, by considering the differences of corresponding positions, we obtain $(x-1)C_0$ and $(x^2-1)C_1$ and $\{-\infty, 0\}$ which gives the result if $(x+1) \in C_0$. The second pair of blocks give the result if $(x+1) \in C_1$.

(b) These designs are constructed in a similar way; however we use the Hall difference sets (see Storer [6]) for the resolvable design.

Consider a $(v_2, b, r_2, k, \lambda_2)$ BIBD generated from the initial blocks $\{0\} \cup C_0 \cup C_1 \cup C_3$ and $\{\infty\} \cup C_2 \cup C_4 \cup C_5$ ($e = 6$ here) and a $(v_3, b, r_3, k, \lambda_3)$ BIBD with initial block $\{0\} \cup C_0 \cup C_2 \cup C_4$ twice. Then,

$$N_{21} = \left[\begin{array}{c|c} A+I & B \\ 00\cdots 0 & j \end{array}\right], \qquad N_{31} = \left[\begin{array}{c|c} X+I & X+I \end{array}\right]$$

where A is the incidence matrix of $C_0 \cup C_1 \cup C_3$, $A + B + I = J$ and X is the incidence matrix of $C_0 \cup C_2 \cup C_4$. It is obvious that $N_{21}N_{31}^T$ is of the required form.

To obtain $N_{23} = J$, we write the blocks as either
$(00,\ 1x^{p-7},\ x^6 1, \ldots,\ x^{p-7}x^{p-13},\ xx^2,\ x^7x^8, \ldots,\ x^{p-6}x^{p-5},\ x^3x^4, \ldots,\ x^{p-4}x^{p-3})$ and
$(\infty 0,\ x^2x^{p-5},\ x^8x^2, \ldots,\ x^{p-5}x^{p-11},\ x^4x^{p-3},\ x^{10}x^4, \ldots,\ x^{p-3}x^{p-9},\ x^{p-2}1, \ldots,\ x^{p-8}x^{p-7})$
or $(00,\ 1x^6,\ x^6x^{12}, \ldots,\ x^{p-7}1,\ xx^2,\ x^7x^8, \ldots,\ x^{p-6}x^{p-5},\ x^3x^4, \ldots,\ x^{p-4}x^{p-3})$ and
$(\infty 0,\ x^2x^8,\ x^8x^{14}, \ldots,\ x^{p-5}x^2,\ x^4x^{10},\ x^{10}x^{16}, \ldots,\ x^{p-3}x^4,\ x^{p-2}1, \ldots,\ x^{p-8}x^{p-7})$,
depending on whether $(x^6-1)/(x-1)$ is a square or a non-square. The verification is the same as for part (a).

The second type of Graeco-Latin designs we consider are also discussed by Preece [2,3].

Again we let blocks be the first constraint and the two sets of treatments be the second and third constraints. Then the designs of interest satisfy the following:

(i) $N_{i1}N_{i1}^T = (r-\lambda)I + \lambda J, \quad i = 2,3;$

(ii) $N_{23}N_{23}^T = aI + bJ, \quad N_{32}N_{32}^T = cI + dJ$ for some a, b, c, d; and

(iii) $N_{21}N_{31}^T = N_{31}N_{21}^T = xI + yJ$ for some x and y.

Designs satisfying (i),(ii) and (iii) have also been called BIBDs for two sets of treatments.

Preece [2] gave a list of 59 such designs with $v \le 20$, $b \le 80$ satisfying the additional restrictions $r = v \pm 1$ and $N_{23} = N_{32} = J + (r-v)I$. These designs were subdivided further into two types, with properties as follows:

Type a : $N_{21} = N_{31}$, so $x = r - \lambda$ and $y = \lambda$;

Type b : $N_{21} = [A_1 | A_2]$, $N_{31} = [A_2 | A_1]$ and $A_1 \ne A_2$. Thus $x = r - \lambda - v$ and $y = \lambda + 1$.

Preece [3] shows that a type b design is preferable when $r = v - 1$ and a type a design is when $r = v + 1$. Many of the designs Preece gave are examples of more general constructions and these appear below.

Theorem 3. *Let* $v = 2mf + 1$ *be a prime power with* f *odd. Let* t *be an even number such that* $2 \le t \le m$. *Then there exist BIBDs for two sets of treatments, of type* a, *with the following parameters:*

(a) (v, mv, mtf, tf, t(tf-1)/2); *and*
(b) (v, mv, m(tf+1), tf+1, t(tf+1)/2).

Proof. (a) We use the design of Lemma 1. The initial blocks are, with $A = \{0,1,2,\ldots,t-1\}$,

$$T_h = \bigcup_{j=0}^{t-1} C_{i_j - i_h}, \qquad h = 0,1,\ldots,m-1.$$

We will write $(C_i, C_j, \ldots, C_k)$ to represent the block

$(x^i, x^{i+2m}, \ldots, x^{i+2m(f-1)}, x^j, x^{j+2m}, \ldots, x^{j+2m(f-1)}, \ldots, x^k, x^{k+2m}, \ldots, x^{k+2m(f-1)})$.

The initial blocks for the design with two sets of treatments are

$$(\underline{C_{i_0-i_h}}\, C_{i_1-i_h},\ \underline{C_{i_1-i_h}}\, C_{i_0-i_h},\ \ldots,\ \underline{C_{i_{t-2}-i_h}}\, C_{i_{t-1}-i_h},\ \underline{C_{i_{t-1}-i_h}}\, C_{i_{t-2}-i_h})$$

where $h = 0,1,\ldots,m-1$ and the treatments of the first set are underlined.

By construction $N_{21} = N_{31}$ and $N_{21}N_{21}^T = (r-\lambda)I + \lambda J$. To show that (ii) is satisfied we show that $N_{23} = (t/2)(J-I)$: again consider the differences of corresponding positions. This gives (where & denotes strong union)

$$\mathop{\&}_{h=0}^{m-1}\ \mathop{\&}_{s=0}^{t/2-1} (x^{i_{2s+1}-i_{2s}}-1)(C_{i_{2s}-i_h} \cup C_{i_{2s}-i_h+m})$$

$$= \mathop{\&}_{s=0}^{t/2-1} (x^{i_{2s+1}-i_{2s}}-1) \mathop{\&}_{h=0}^{m-1} (C_{i_{2s}-i_h} \cup C_{i_{2s}-i_h+m}).$$

Now

$$\mathop{\&}_{h=0}^{m-1} (C_{i_{2s}-i_h} \cup C_{i_{2s}-i_h+m}) = GF(v)\backslash\{0\}$$

as $i_{2s}-i_h \equiv i_{2s}-i_k(2m)$ implies $i_h \equiv i_k(m)$, a contradiction for $k \neq h$ and $i_{2s}-i_h \equiv i_{2s}-i_k+m(2m)$ again implies $i_h \equiv i_k(m)$. This completes the proof of (a).

(b) The proof proceeds as in (a) using the initial blocks $S_h = \{0\} \cup T_h$.

Example. Let $f = m = 3$, $t = 2$. Then there is a (19, 57, 18, 6, 5) BIBD for two sets of treatments, of type a, with the three initial blocks ($x=2, e=6, i_j=j, 0\le j\le 2$)

$$(\underline{1}\ 2,\ \underline{7}\ 14,\ \underline{11}\ 3,\ \underline{2}\ 1,\ \underline{14}\ 7,\ \underline{3}\ 11),$$

$$(\underline{10}\ 1,\ \underline{13}\ 7,\ \underline{15}\ 11,\ \underline{1}\ 10,\ \underline{7}\ 13,\ \underline{11}\ 15)$$

$$(\underline{10}\ 5,\ \underline{13}\ 16,\ \underline{15}\ 17,\ \underline{5}\ 10,\ \underline{16}\ 13,\ \underline{17}\ 15).$$

No design with these parameters is given by Preece [2].

Theorem 4. *Let* $v = 2mf + 1$, f *odd, be a prime power. Let* t *be an integer such that* $1 \le t \le m$. *Then there exist BIBDs for two sets of treatments, of type* b, *with the following parameters:*

(a) $(v,\ 2mv,\ 2mtf,\ tf,\ t(tf-1))$;

(b) $(v,\ 2mv,\ 2m(tf+1),\ tf+1,\ t(tf+1))$; *and*

(c) $(v+1,\ 2mv,\ mv,\ mf+1,\ m^2f)$.

Proof. (a) Again we use the design of Lemma 1. Thus, putting $A = \{0,1,\ldots,t-1\}$, the initial blocks for the first treatment set are

$$T_h = \bigcup_{j=0}^{t-1} C_{i_j-i_h}, \qquad h = 0,1,\ldots,m-1.$$

Define

$$S_h = \bigcup_{j=0}^{t-1} C_{i_j-i_h+m}, \qquad h = 0,1,\ldots,m-1.$$

These also give $m - \{2mf+1;\ tf;\ t(tf-1)/2\}$ sds and are the initial blocks for the second treatment set. Using the notation of Theorem 3, the initial blocks for the design are

$(\underline{C_{i_0-i_h}\ C_{i_0-i_h+m}},\ \underline{C_{i_1-i_h}\ C_{i_1-i_h+m}},\ldots,\ \underline{C_{i_{t-1}-i_h}\ C_{i_{t-1}-i_h+m}})$ and

$(\underline{C_{i_0-i_h+m}\ C_{i_0-i_h}},\ \underline{C_{i_1-i_h+m}\ C_{i_1-i_h}},\ldots,\ \underline{C_{i_{t-1}-i_h+m}\ C_{i_{t-1}-i_h}})$, $h=0,1,\ldots,m-1$.

Let the incidence matrix of S_i be A_i, that of T_i be B_i. Then $A_i+B_i+D_i = J-I$ and the D_i are incidence matrices of $m-\{2mf+1;\ 2(m-t)f;\ (m-t)(2(m-t)f-1)\}$ sds. Then,

$$N_{21} = (B_0B_1\ldots B_{m-1}\mid A_0A_1\ldots A_{m-1}) \text{ and } N_{31} = (A_0A_1\ldots A_{m-1}\mid B_0B_1\ldots B_{m-1}),$$

and

$$\begin{aligned} N_{21}N_{31}^T = N_{31}N_{21}^T &= \sum_{i=0}^{m-1}(B_iA_i^T+A_iB_i^T) \\ &= -\sum_{i=0}^{m-1}(B_iB_i^T+A_iA_i^T) + \sum_{i=0}^{m-1}(J-I-D_i)(J-I-D_i^T) \\ &= t^2f(J-I), \text{ establishing (iii).} \end{aligned}$$

To verify (ii) we consider differences of corresponding positions. This gives

$$\mathop{\&}_{h=0}^{m-1}(x^m-1)T_h \ \&\ \mathop{\&}_{h=0}^{m-1}(1-x^m)T_h = \mathop{\&}_{h=0}^{m-1}(x^m-1)\left[\bigcup_{j=0}^{t-1}\{C_{i_j-i_h}\cup C_{i_j-i_h+m}\}\right].$$

Now C_a occurs when $i_j-i_h \equiv a(2m)$ or $i_j-i_h+m \equiv a(2m)$. As $i_0,i_1,\ldots,i_{m-1}$ represent a complete set of residues modulo m, $i_0+a,\ i_1+a,\ldots,\ i_{m-1}+a,\ i_0+a+m,\ldots,\ i_{m-1}+a+m$ represent a complete set of residues modulo 2m. Thus there are t values of j which satisfy the congruence relations above and so $N_{23} = t(J-I)$.

(b) The proof is similar to that for (a) using the two sets of initial blocks $\{0\}\cup T_h$, $\{0\}\cup S_h$, $h=0,1,\ldots,m-1$.

(c) The proof is similar to that for (a) using the two sets of initial blocks $\{\infty\}\cup T_h$, $\{0\}\cup S_h$, $h=0,1,\ldots,m-1$.

Theorem 5. *Let* p *be a prime or a prime power and write* $p = ef+1$. *Then there exist BIBDs for two sets of treatments, of type* a *(and of type* b *if* e *is even), with the following parameters:*

(a) (p, ep, ef, f, f-1);

(b) (p, ep, e(f+1), f+1, f+1); *and, if* $e=2$ *and* $p\equiv 3(4)$

(c) (p+1, 2p, p, f+1, f).

Proof. (a) Let $C_i = \{x^{es+i}\mid s=0,1,\ldots,f-1\}$, $i=0,1,\ldots,e-1$, where x is a generator of $GF(p)^*$. In the notation of Theorem 3, the initial blocks for the type a design are $(\underline{C_0}\ C_0),(\underline{C_1}\ C_1),\ldots,(\underline{C_{e-1}}\ C_{e-1})$ and for the type b design (e = 2m, say) are $(\underline{C_0}\ C_m),(\underline{C_1}\ C_{m+1}),\ldots,(\underline{C_{m-1}}\ C_{2m-1}),(\underline{C_m}\ C_0),\ldots,(\underline{C_{2m-1}}\ C_{m-1})$. The proof is similar to those of Theorems 3 and 4.

(b) As in (a), using the initial blocks $\{0\} \cup C_i$, $i = 0,1,\ldots,e-1$ for both of the underlying BIBDs.

(c) As in (a), using the initial blocks $(\underline{\infty}\ 0,\ \underline{C_0}\ C_1)$, and $(0\ \underline{\infty},\ \underline{C_1}\ C_0)$

3. NESTED ROW AND COLUMN DESIGNS

Singh and Dey [5] introduced balanced incomplete block designs with nested rows and columns (BIBRC). If we let the first constraint be blocks, the second rows, the third columns and the fourth treatments then these designs satisfy the following:

(i) each treatment occurs at most once in a block;

(ii) each block is arranged as an array with p rows and q columns; and

(iii) $pN_{42}N_{42}^T + qN_{43}N_{43}^T - N_{41}N_{41}^T = \lambda(J - I) + (p + q - 1)rI$, where r is the replication number and $\lambda(v - 1) = r(p - 1)(q - 1)$.

We denote such a design by BIBRC (v,b,r,p,q,λ).

Singh and Dey [5] considered the construction of such designs using the method of differences; below we give some families of BIBRC designs constructed by this method in both the balanced and partially balanced cases.

Theorem 6. *Let* $v = 2mf + 1$, f *odd, be a prime power. Let* t *be an integer such that* $1 \le t \le m$. *Then there exists a BIBRC* (v, mv, mtf, t, f, t(t-1)(f-1)/2).

Proof. We use the design of Lemma 1. Let $A = \{0,1,\ldots,t-1\}$ and write the hth initial block, $h = 0,1,\ldots,m-1$, as

$$\begin{array}{cccc} x^{i_0-i_h} & x^{i_0-i_h+2m} & \cdots & x^{i_0-i_h+2m(f-1)} \\ x^{i_1-i_h} & x^{i_1-i_h+2m} & \cdots & x^{i_1-i_h+2m(f-1)} \\ \vdots & \vdots & & \vdots \\ x^{i_{t-1}-i_h} & x^{i_{t-1}-i_h+2m} & \cdots & x^{i_{t-1}-i_h+2m(f-1)} \end{array}.$$

Clearly each block has t rows and f columns.

To verify (iii) we define λ_R, λ_C and λ_E to be the number of times that two treatments appear in the same row, the same column and the same block but not in the same row or column respectively. Now the set of ith rows from the m initial blocks form an $m-\{2mf+1;\ f;\ (f-1)/2\}$ sds so $\lambda_R = t(f-1)/2$. Also,

$$\bigcup_{j=0}^{m-1} \{C_{i_0-i_j} \cup C_{i_0-i_j+m}\} = GF(2mf+1)\setminus\{0\}$$

so by considering the differences of corresponding positions we see that $\lambda_C = t(t-1)/2$. Using the fact that we started with the initial blocks of an $m-\{2mf+1;\ tf;\ t(tf-1)/2\}$ sds gives $\lambda_E = t(tf-1)/2 - \lambda_R - \lambda_C = t(t-1)(f-1)/2$. The result follows as $(t-1)\lambda_R + (f-1)\lambda_C - \lambda_E = \lambda = (t-1)(f-1)/2$.

Example. Let $f = m = t = 3$. Then there is a BIBRC (19, 57, 27, 3, 3, 6) with the three initial blocks as follows:

$$\begin{matrix} 1 & 7 & 11 \\ 10 & 13 & 15 \\ 5 & 16 & 17 \end{matrix}\ ,\quad \begin{matrix} 2 & 14 & 3 \\ 1 & 7 & 11 \\ 10 & 13 & 15 \end{matrix}\quad \text{and}\quad \begin{matrix} 4 & 9 & 6 \\ 2 & 14 & 3 \\ 1 & 7 & 11 \end{matrix}\ .$$

In an analogous manner the partially balanced incomplete block designs of Morgan, Street and Wallis [1] can be used to give nested row and column designs. Before giving these, however, we introduce some notation. For more details see [1].

As in Storer [6], the cyclotomic number (i,j) is the number of ordered pairs s,t such that

$$x^{es+i} + 1 = x^{et+j}, \qquad (0 \le s,\ t \le f-1)$$

where x is a primitive root of $GF(p^n)$ and $p^n = ef+1$. If there is doubt as to which factorization of $p^n - 1$ is being used, it is specified by writing $(i,j)_e$.

As in [1], let $p^n = \alpha\beta\gamma + 1$ be an odd prime power where n, α, β, γ are positive integers with $\alpha,\beta,\gamma \ge 2$. Let x be a primitive root of $GF(p^n)$ and define

$$C_i = \{x^{s\alpha+i} \mid s = 0,1,\ldots,\beta\gamma-1\},\quad i = 0,1,\ldots,\alpha-1,$$

$$D_i = \{x^{s\alpha\beta+i} \mid s = 0,1,\ldots,\gamma-1\},\quad i = 0,1,\ldots,\alpha\beta-1.$$

Then $C_i = \bigcup_{j=0}^{\beta-1} D_{i+j\alpha}$. Let $B_j = \bigcup_{h=0}^{\alpha-1} D_{h+j\alpha}$, $j = 0,1,\ldots,\alpha\beta-1$ and let $E_j = \bigcup_{h=1}^{t} D_{a_h+j}$, $j = 0,1,\ldots,\alpha\beta-1$, where $0 \le a_1 < a_2 < \ldots < a_t \le \alpha\beta-1$.

Designs which can be constructed using these sets are given below. The values of λ_i are not explicitly given as these depend on which cyclotomic class numbers such as $x^{a_j-a_i}-1$ are in; it is easy to show, however, that these numbers are independent of the particular pair of ith associates chosen.

Theorem 7. *The β sets* B_j, $j = 0,1,\ldots,\beta-1$ *or the β sets* $E_{j\alpha}$, $j = 0,1,\ldots,\beta-1$ *may be used as the blocks of a PBIBRC (m), where* $v = p^n$, $b = \beta p^n$, $m \le \alpha$ *and each associate class consists of a cyclotomic class* C_i *or a union of such classes. The parameters of the designs are given below when* $m = \alpha$ *so that each cyclotomic class is an associate class.*

	Initial Blocks	r	*Number of rows*	*Number of cols*	*Parity Conditions*
(a)	$E_{j\alpha}$	$t\beta\gamma$	t	γ	$\beta\gamma$ *even*
(b)	B_j	$\alpha\beta\gamma$	α	γ	β *even*, γ *odd*
(c)	B_j	$\alpha\beta\gamma$	α	γ	γ *even*

Proof. (a) The blocks are laid out in the same manner as in the proof of Theorem 6. We define λ_{iR}, λ_{iC} and λ_{iE} to be the number of times that two treatments, which are ith associates, occur in the same row, the same column, or in the same block but not in the same row or column respectively.

The set of jth rows from the β initial blocks form a PBIBD (α) (Theorem 2, [1]) so $\lambda_{iR} = \sum_{d=1}^{t} \sum_{j=0}^{\beta-1} (0, j\alpha+i-a_d)_{\alpha\beta}$ (where each cyclotomic class C_i is an associate class).

To establish that λ_{iC} is a constant, independent of the particular pair of ith associates chosen, consider rows f and g in each of the initial blocks. The differences of corresponding components of these rows give $(x^{a_f-a_g}-1)C_{a_f}$ showing that λ_{iC} is constant. The result now follows.

The proofs for parts (b) and (c) are similar.

We now consider balanced incomplete block designs with nested rows and columns for two sets of treatments. Thus we introduce a fifth constraint, a second treatment set, to the designs above. Then the designs of interest have each set of treatments arranged as a BIBRC (v,b,r,p,q,λ) and have $N_{45}N_{45}^T = aI + bJ$, $N_{54}N_{54}^T = cI + dJ$ for some a, b, c, d.

Theorem 8. *Let* $v = 2mf+1$, f *odd, be a prime power. Let* t *be an even integer such that* $2 \le t \le m$. *Then there exists a BIBRC* $(v, mv, mtf, t, f, t(t-1)(f-1)/2)$ *for two sets of treatments.*

Proof. Use the design of Theorem 6 and arrange the hth initial block, $h = 0,1,\ldots,m-1$, as (where the treatments of the first set are underlined)

$$\begin{array}{llll}
\underline{x^{i_0-i_h}} \quad x^{i_1-i_h}, & \ldots, & \underline{x^{i_0-i_h+2m(f-1)}} & x^{i_1-i_h+2m(f-1)} \\
\underline{x^{i_1-i_h}} \quad x^{i_0-i_h}, & \ldots, & \underline{x^{i_1-i_h+2m(f-1)}} & x^{i_0-i_h+2m(f-1)} \\
\vdots \qquad \vdots & & \vdots & \vdots \\
\underline{x^{i_{t-2}-i_h}} \quad x^{i_{t-1}-i_h}, & \ldots, & \underline{x^{i_{t-2}-i_h+2m(f-1)}} & x^{i_{t-1}-i_h+2m(f-1)} \\
\underline{x^{i_{t-1}-i_h}} \quad x^{i_{t-2}-i_h}, & \ldots, & \underline{x^{i_{t-1}-i_h+2m(f-1)}} & x^{i_{t-2}-i_h+2m(f-1)}
\end{array}$$

As in the proof of Theorem 3, $N_{45} = t/2(J - I)$.

REFERENCES

[1] Elizabeth J. Morgan, Anne Penfold Street and Jennifer Seberry Wallis, Designs from cyclotomy, *Combinatorial Math. IV*, Proc. Fourth Australian Conf., Lecture Notes in Math. 560, 158-176 (Springer-Verlag, Berlin, Heidelberg, New York, 1976).

[2] D.A. Preece, Some balanced incomplete block designs for two sets of treatments, *Biometrika* 53 (1966), 497-506.

[3] D.A. Preece, Non-orthogonal Graeco-Latin designs, *Combinatorial Math. IV*, Proc. Fourth Australian Conf., Lecture Notes in Math. 560, 7-26 (Springer-Verlag, Berlin, Heidelberg, New York, 1976).

[4] Jennifer Seberry, A note on orthogonal Graeco-Latin designs, *Ars Comb.* 8 (1979), 85-94.

[5] M. Singh and A. Dey, Block designs with nested rows and columns, *Biometrika* 66 (1979), 321-326.

[6] Thomas Storer, *Cyclotomy and Difference Sets.* (Markham Publishing Co., Chicago, 1967).

[7] Deborah J. Street, Bhaskar-Rao designs from cyclotomy, *J. Austral. Math. Soc.* 29 (1980), 425-430.

Department of Mathematical Statistics,
University of Sydney,
N.S.W.

NOTE ADDED IN PROOF :
I thank the referee for strengthening Theorem 5(c).

05 C 99

05 C 05

05 C 40

CONSTRAINED SWITCHINGS IN GRAPHS

R. Taylor

This paper investigates realizations of a given degree sequence, and the way in which they are related by switchings. The results are given in the context of simple graphs, multigraphs and pseudographs. We show that we can transform any connected graph to any other connected graph of the same degree sequence, by switchings which are constrained to connected graphs. This is done for certain labelled graphs, the result for unlabelled graphs following as a corollary. The results are then extended to infinite degree sequences.

1. INTRODUCTION

The *degree sequence* of a finite graph, denoted $\underset{\sim}{d} = (d_1, \ldots, d_n)$, is the list of the degrees of all the vertices of the graph, conventionally beginning with the maximum degree and arranged non-increasingly.

A graph G is a *realization* of a degree sequence $\underset{\sim}{d}$ if the collection of the degrees of the vertices of G is the same as the collection of terms in $\underset{\sim}{d}$. A *labelled* realization of a degree sequence $\underset{\sim}{d} = (d_1, \ldots, d_n)$, $d_1 \geq d_2 \geq \ldots \geq d_n$ is a graph whose vertices are labelled $v_1, \ldots, v_n$ with the restriction that the degree of each vertex v_i is $d(v_i) = d_i$. Unless otherwise stated any labelled realization of a degree sequence is to be considered as having this restriction.

We shall use the classification of graph types introducted by Eggleton and Holton in [5], and shall confine our attention to graphs of type $\tau = (0,0,1)$, $(0,0,\infty)$, $(0,\infty,\infty)$, that is, simple graphs, multigraphs and pseudographs respectively.

A *switching* is a transformation of a graph that eliminates two edges and introduces two new ones. For simple graphs and multigraphs a switching involves two independent edges, say (u,v) and (x,y), and transforms the graph by eliminating these edges and introducing new edges (u,x) and (v,y). This is illustrated in Figure 1.1.

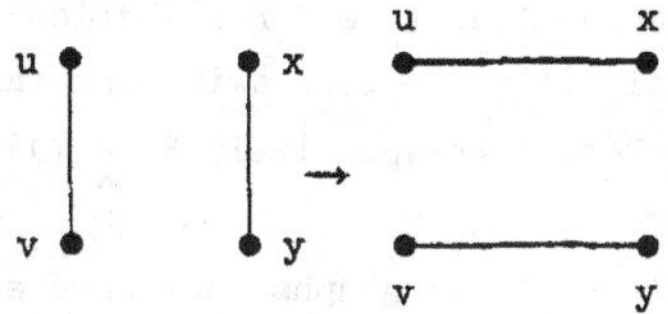

Figure 1.1

Algebraically we may represent this operation as $[(u,v), (x,y)] \rightarrow [(u,x),(v,y)]$. In the case of simple graphs, the switching is only permitted if neither of the edges (u,x) and (v,y) is already present. No such restriction applies to multigraphs. For pseudographs the loops allow more kinds of switchings; these are shown below in Figure 1.2.

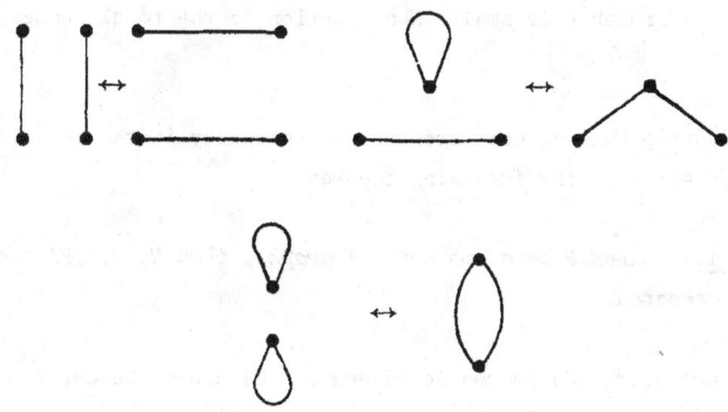

Figure 1.2

To formalize the way in which the various realizations of a degree sequence are related by switchings, the following graph has been introduced by Eggleton and Holton [4].

<u>Definition</u>. The *graph of type* τ *realizations* (respectively the graph of labelled type τ realizations) of a degree sequence $\underset{\sim}{d} = (d_1, \ldots, d_n)$, $d_1 \geq d_2 \geq \ldots \geq d_n$ is a graph whose vertices are identified with the type τ

realizations (respectively, the labelled type τ realizations) of $\underset{\sim}{d}$, and where two vertices are adjacent if and only if there is a switching that sends one graph into the other. This is denoted by $R(\underset{\sim}{d},\tau)$ (respectively $R_\ell(\underset{\sim}{d},\tau)$).

Often we are interested only in graphs that have a certain property and so motivated by this, we make the following definition.

Definition. Let P be a property which a graph may possess, then $R(\underset{\sim}{d},\tau,P)$ (respectively $R_\ell(\underset{\sim}{d},\tau,P)$) is the subgraph of $R(\underset{\sim}{d},\tau)$ (respectively, $R_\ell(\underset{\sim}{d},\tau)$) induced by those vertices that correspond to graphs with property P.

An interesting question to ask about the graph of realizations is for which P, if any, is it connected. Properties for which $R(\underset{\sim}{d},\tau,P)$ is connected are said to be *complete* (see [2]). If a property is complete then we may find all the graphs of a given degree sequence with that property by switching only through graphs with the property. This is more desirable than testing all the realizations of a degree sequence to see which have the property, particularly if the number of realizations with the property is small in proportion to the total number of realizations.

A relationship between completeness of a property in the labelled and unlabelled cases is given in the following theorem.

Theorem 1.1. *Let P be a property of graphs, then $R_\ell(\underset{\sim}{d},\tau,P)$ connected implies $R(\underset{\sim}{d},\tau,P)$ connected.*

Proof. Let $R_\ell(\underset{\sim}{d},\tau,P)$ be connected and let G_1 and G_2 be any two unlabelled realizations of $\underset{\sim}{d}$ with property P. Label them with vertices v_i in any way subject to the restriction that $d(v_i) = d_i$. Denote the labelled forms by $G_1^{\ell_1}$ and $G_2^{\ell_2}$. Now since $R_\ell(\underset{\sim}{d},\tau,P)$ is connected, there must exist a sequence of switchings σ, that transforms $G_1^{\ell_1}$ to $G_2^{\ell_2}$ and goes through only graphs with property P. If we now drop the labels on all graphs involved in the switchings we see that σ transforms G_1 to G_2 and only through graphs with property P. Note that many of the graphs in this transformation sequence may be isomorphic. Since G_1 and G_2 were arbitrary realizations of $\underset{\sim}{d}$ with property P we may conclude that $R(\underset{\sim}{d},\tau,P)$ is connected.

In general, however, note that $R(\underset{\sim}{d},\tau,P)$ connected does not imply $R_\ell(\underset{\sim}{d},\tau,P)$ connected, as is shown by Example 3.1 at the end of Section 3.

2. KNOWN RESULTS

In this section we shall review some of the known results about switchings. The authors state their results about unlabelled graphs, however the proofs in fact give the corresponding stronger result on labelled graphs.

Theorem 2.1. *Let* $\underset{\sim}{d} = (d_1, \ldots, d_n)$, $d_1 \geq \ldots \geq d_n$ *be a degree sequence, then* $R_\ell(\underset{\sim}{d}, \tau)$ *is connected for* $\tau = (0,0,\infty), (0,\infty,\infty), (0,0,1)$.

Proof. When the graphs are of types $(0,\infty,\infty)$ and $(0,0,\infty)$ see Eggleton and Holton [6] and Hakimi [7], when the graphs are of type $(0,0,1)$ the result is stated without proof by Eggleton in [3]. We present a proof here and also refer the reader to the paper [8] by Eggleton and Holton, in this volume, for an independent demonstration of the result. Our proof is by induction on the number of terms in $\underset{\sim}{d} = (d_1, \ldots, d_n)$.

$n = 3$. In this case there are only four graphic degree sequences and each has a unique labelled realization. These are shown in Figure 2.1.

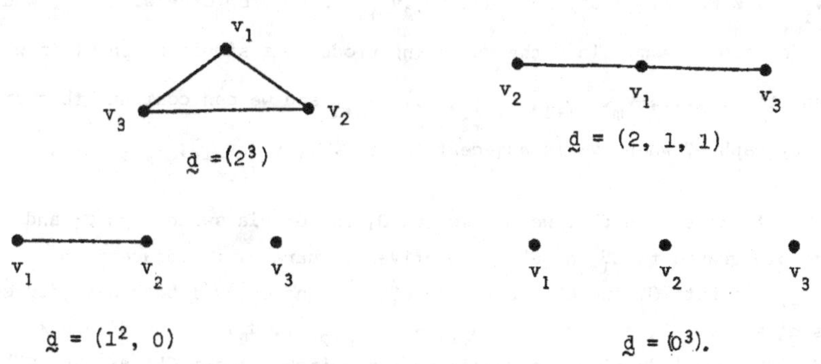

Figure 2.1

So let us assume the result holds whenever $n \leq k - 1$.

$n = k$. Let $\underset{\sim}{d} = (d_1, \ldots, d_k)$, $d_1 \geq d_2 \geq \ldots \geq d_k$ be a degree sequence and let G_1, G_2 be any two labelled simple realizations of $\underset{\sim}{d}$, where G_1, G_2 have vertices $v_1, \ldots, v_k$ with $d(v_i) = d_i$. We show how to switch on G_1 and G_2 to obtain graphs in which v_1 is adjacent to $v_2, \ldots, v_{s+1}$ where $s_1 = d(v_1) = d_1$. Take G_1 for example and assume that v_1 is adjacent to $v_2, \ldots, v_m, v_{i_{m+1}}, \ldots, v_{i_{s+1}}$ in G_1 with $v_{m+1} \neq v_{i_j}$ $\forall$ $j > m$. Now every vertex adjacent to v_{m+1} cannot also be adjacent to $v_{i_{m+1}}$ for this would yield the contradiction $d(v_{m+1}) < d(v_{i_{m+1}})$, since v_1 is adjacent to $v_{i_{m+1}}$ but not v_{m+1}. Thus there exists some vertex v_ℓ with $(v_{m+1}, v_\ell) \in EG_1$ and $(v_{i_{m+1}}, v_\ell) \notin EG_1$. Then we make the switching $[(v_1, v_{i_{m+1}}), (v_\ell, v_{m+1})] \to [(v_1, v_{m+1}), (v_\ell, v_{i_{m+1}})]$ as indicated in Figure 2.2.

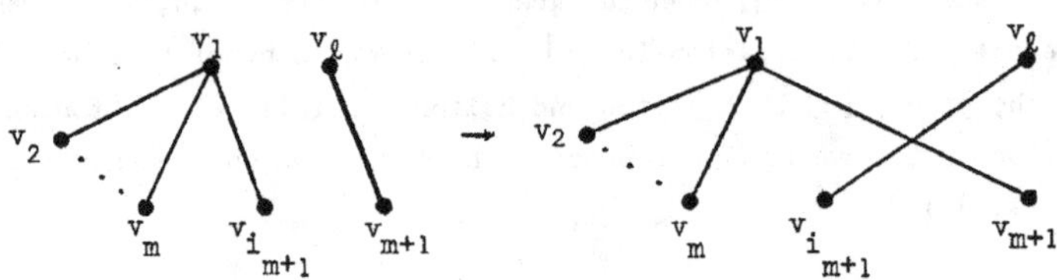

Figure 2.2

Now $(v_\ell, v_{i_{m+1}}) \notin EG_1$ by assumption, and $(v_1, v_{m+1}) \notin EG_1$ for otherwise v_{m+1} would equal v_{i_j} for some $j > m$. Thus the switching produces a simple graph G_1' in which v_1 is adjacent to $v_2, \ldots, v_m, v_{m+1}, v_{i_{m+2}}, \ldots, v_{s+1}$ and we can continue this process to obtain a graph G_1'' where v_1 is adjacent to $v_2, \ldots, v_m, v_{m+1}, v_{m+2}, \ldots, v_{s+1}$.

It follows then that we may switch G_1 and G_2 via switchings σ_1 and through simple graphs to G_1'' and G_2'', respectively, where v_1 is adjacent to $v_2, \ldots, v_{s+1}$ in both G_1'' and G_2''. But then $G_1''\backslash\{v_1\}$ and $G_2''\backslash\{v_1\}$ both have degree sequences $\underset{\sim}{d}' = (d_2 - 1, d_3 - 1, \ldots, d_{s+1} - 1, d_{s+2}, ., d_k)$. Now $\underset{\sim}{d}'$ has $k - 1$ terms and so by our induction hypothesis we can switch between $G_1''\backslash\{v_1\}$ and $G_2''\backslash\{v_1\}$ via a sequence of switchings θ through simple graphs. It follows that this same sequence of switchings will transform G_1'' into G_2'' through simple graphs.

$$G_1 \xrightarrow{\sigma_1} G_1'' \xrightarrow{\theta} G_2'' \xleftarrow{\sigma_2} G_2$$

Figure 2.3

Thus the sequence of switchings defined by $\sigma_1 \theta \sigma_2^{-1}$ (acting from the left) transforms G_1 into G_2 via simple graphs (See Figure 2.3). Then since G_1 and G_2 were arbitrary labelled realizations of $\underset{\sim}{d}$ we can conclude that $R_\ell(\underset{\sim}{d},(0,0,1))$ is connected whenever $\underset{\sim}{d}$ has k terms. Hence by induction we have the result for all $\underset{\sim}{d}$.

Corollary 2.3. *$R(d,\tau)$ is connected for $\tau = (0,0,\infty),(0,\infty,\infty),(0,0,1)$.*

More recently Colborn [2] and Syslo [11] have independently shown the following result.

Theorem 2.4. *$R_\ell(\underset{\sim}{d},(0,0,1),P)$ is connected where $P \equiv$ "Tree".*

The natural generalization of this last result is to replace the property of being a tree with that of being connected. We explore this matter in the next section.

3. SWITCHINGS THROUGH CONNECTED GRAPHS

In this section we show that we can switch between two connected realizations of a degree sequence, through connected graphs. This is done for multigraphs, pseudographs and simple graphs.

Theorem 3.1. *$R_\ell(\underset{\sim}{d},(0,0,\infty),P)$ is connected where $P \equiv$ "Connected".*

Proof. Let $\underset{\sim}{d} = (d_1, \ldots, d_n)$, $d_1 \geq \ldots \geq d_n$. We shall assume $n \geq 4$ since if $n \leq 3$ then $\underset{\sim}{d}$ has at most one realization (see Hakimi [9]) and the theorem is satisfied trivially. Also we assume that $\sum_{i=1}^{n} d_i \geq 2(n-1)$ since if $\underset{\sim}{d}$ has a connected realization G, then G must have at least n - 1 edges. So let $\sum_{i=1}^{n} d_i = 2(n-1+m)$ where m is a non-negative integer. We shall prove the theorem by induction on m.

m = 0. Here $\sum_{i=1}^{n} d_i = 2(n-1)$, so that every connected realization of $\underset{\sim}{d}$ is a simple tree. Thus $R_\ell(\underset{\sim}{d},(0,0,\infty),P) \cong R_\ell(\underset{\sim}{d},(0,0,1),P')$ where $P' \equiv$ "Tree", and the result then follows by Theorem 2.4. Let us assume then that we have the result for $m = k \geq 0$. We shall now show that the result holds for $m = k + 1$.

m = k + 1. Let G_1 and G_2 be any two labelled connected realizations of $\underset{\sim}{d}$ where $\sum_{i=1}^{n} d_i = 2(n-1+k+1) = 2(n+k)$. Take G_1 for example and assume that $(v_1,v_2) \notin EG_1$. We show that we may switch G_1 through connected graphs so as to make (v_1,v_2) an edge in the transformed graph. Since G_1 is connected there must be some path of minimum length between v_1 and v_2. Let $(v_1, a_1, \ldots, a_r)$, with $a_r = v_2$ and $r \geq 2$, be such a path. (Note that many of the edges may be multiple, but for the purposes of the proof we need consider only one edge between any two vertices). Now there is at least one edge between a_1 and a_2, and since $d(v_1) \geq d(a_1)$ not all edges from v_1 can be incident with a_1. So let $(v_1, p) \in EG_1$ where $p \neq a_1$. We then note that $p \neq a_2, \ldots, a_r$, or else this would contradict the fact that the path $(v_1, a_1, \ldots, a_r)$ was a path of minimum length between v_1 and v_2. Then we can switch $[(p,v_1),(a_{r-1},v_2)] \rightarrow [(p,a_{r-1}),(v_1,v_2)]$ and so transform G_1 to a graph in which v_1 is adjacent to v_2. Thus we can switch G_1 and G_2 by switchings σ_1 and σ_2 through connected graphs to G_1' and G_2' respectively, with $(v_1,v_2) \in EG_1' \cap EG_2'$.

We now wish to switch G_i', $i = 1,2$ to obtain graphs in which (v_1,v_2) is not a cut edge (that is a bridge). This will allow us to apply our induction hypothesis to the subgraphs formed by removing the edge (v_1,v_2). So assume that (v_1,v_2) is a cut edge of G_1' and let $G_1' \backslash (v_1,v_2)$ have components A and B. Now since $\sum_{i=1}^{n} d_i \geq 2n$, then either A or B has a multiple edge or an edge on a cycle. So let (p,q) be a single edge where either (p,q) lies on a cycle or $G_1' \backslash (v_1,v_2)$ has other edges between p and q. We shall assume without loss of generality that $(p,q) \in EA$ and also that $v_2 \in VA$ (the case $v_1 \in VA$ can be treated similarly). Since $d_1 \geq d_2 \geq 2$ and we have $n \geq 4$ vertices, there must be a vertex $a \neq v_2$ which is adjacent to v_1. This is illustrated in Figure 3.1.

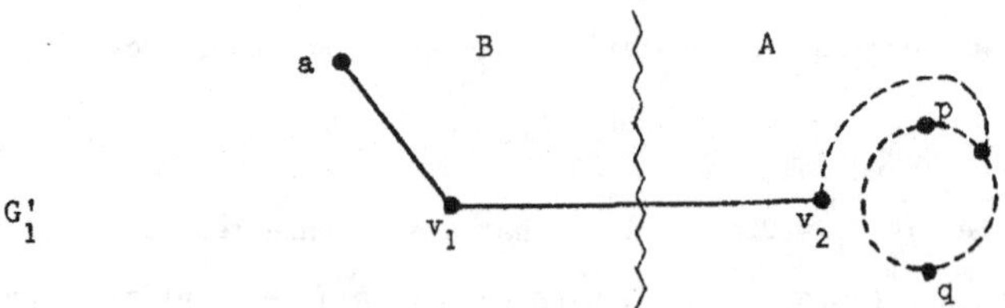

Figure 3.1

Here the dotted line incident with v_2 is any path that leads to the cycle in A, which must exist since A is connected. The dotted lines from p to q are either paths or single edges. We now switch $[(a,v_1),(p,q)] \to [(a,q),(v_1,p)]$, and observe that the resultant graph is still connected and (v_1,v_2) is not a cut edge. It follows therefore that G_1' and G_2' can be transformed via switchings θ_1 and θ_2, respectively, to G_1'' and G_2'' in which (v_1,v_2) is not a cute edge of either graph.

Let $G_1''' = G_1''\backslash(v_1,v_2)$ and $G_2''' = G_2''\backslash(v_1,v_2)$, then G_1''' and G_2''' are both connected realizations of the degree sequence $\underset{\sim}{d}' = (d_1 - 1, d_2 - 1, d_3, \ldots, d_n) = (d_1', \ldots, d_n')$. Thus $\sum_{i=1}^{n} d_i' = (\sum_{i=1}^{n} d_i) - 2 = 2(n + k - 1)$.

We may apply the induction hypothesis to transform G_1''' into G_2''' by a sequence of switchings γ, through connected graphs. It then follows that these same switchings when applied to the subgraph of G_1'' will transform it via connected graphs to G_2''. In Figure 3.2 we illustrate the various graphs and their relationships by switchings.

$$G_1 \xrightarrow{\sigma_1} G_1' \xrightarrow{\theta_1} G_1'' \xrightarrow{\gamma} G_2'' \xleftarrow{\theta_2} G_2' \xleftarrow{\sigma_2} G_2$$

Figure 3.2

Thus the switching sequence $\sigma_1\theta_1\gamma\theta_2^{-1}\sigma_2^{-1}$ transforms G_1 to G_2 via connected graphs. Since G_1 and G_2 were arbitrary connected realizations of $\underset{\sim}{d}$ we may conclude that $R_\ell(\underset{\sim}{d};(0,0,\infty),P)$ is connected whenever $\sum_{i=1}^{n} d_i = 2(n - 1 + k + 1)$. We therefore have the result for m = k + 1, and by induction for all m, hence for all $\underset{\sim}{d}$.

<u>Corollary 3.1</u>. *$R(d,(0,0,\infty),P)$ is connected where $P \equiv$ "Connected".*

<u>Theorem 3.2</u>. *$R_\ell(d,(0,\infty,\infty),P)$ is connected where $P \equiv$ "Connected".*

<u>Proof</u>. The proof operates by taking two connected realizations of $\underset{\sim}{d}$ and transforming them to eliminate as many loops as possible. Then we apply Theorem 3.1. So let G_1 and G_2 be any two labelled connected pseudographic realizations of $\underset{\sim}{d}$. Take G_1 for example and for every pair of independent loops that occur perform a switching of a type indicated in Figure 3.3.

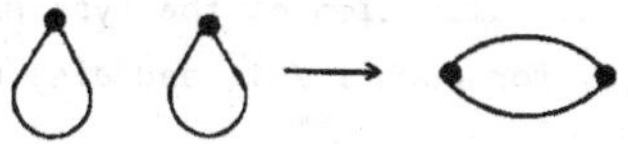

Figure 3.3

This clearly leaves G_1 connected. If we perform all the possible switchings of this type any loops in the resultant graph must all be incident with a single vertex. Now if the result has any edges independent of the loops we may switch as indicated in Figure 3.4

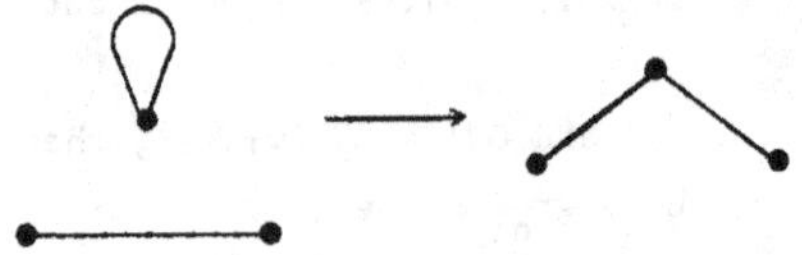

Figure 3.4

It is obvious that this switching preserves connectivity. If we exhaust all possible switchings of this type we must finally produce a graph G_1' where either (a) G_1' is a multigraph (no loops) or (b) G_1' has say $m \geq 1$ loops on one vertex and no edges independent of the loops.

Now in a muligraph every edge incident with v_1 is also incident with some other vertex and we may conclude that case (a) occurs if an only if $d_1 \leq \sum_{i=1}^{n-1} d_i$. In case (b), every edge is incident with v_1, and further we know that v_1 carries $m \geq 1$ loops so case (b) occurs if and only if $d_1 > \sum_{i=2}^{n-1} d_i$.

So if $d_1 \leq \sum_{i=2}^{n-1} d_i$ it follows that we may switch G_1 and G_2 via switchings σ_1 and σ_2 through connected graphs to multigraphs G_1' and G_2' . We may then apply Theorem 3.1 and transform G_1' into G_2' through connected multigraphs with say the switching sequence θ_1. Thus the sequence of switchings defined by $\sigma_1\theta_1\sigma_2^{-1}$ transforms G_1 to G_2 via connected pseudographs.

If on the other hand $d_1 > \sum_{i=1}^{n-1} d_i$, then by switching to eliminate loops we may transform G_1 and G_2 to graphs G_1' and G_2' of the form indicated in Figure 3.5. Note that $m = \frac{1}{2}(d_1 - \sum_{i=1}^{n-1} d_i)$ and is therefore dependent purely on the degree sequence.

But there is a unique labelled realization of the type shown in Figure 3.5 It has d_i edges between v_i and v_1 for each $i \neq 1$, and also m loops on v_1.

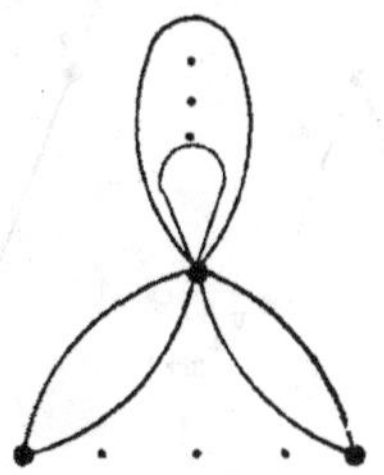

Figure 3.5

Hence in either case (a) or (b) we see that G_1 can be transformed into G_2 by a sequence of switchings through connected graphs. The theorem follows since G_1 and G_2 were any two labelled connected pseudographic realizations of $\underset{\sim}{d}$.

Corollary 3.2. *$R(d,(0,\infty,\infty),P)$ is connected, where $P \equiv$ "Connected".*

Theorem 3.3. *$R_\ell(d,(0,0,1),P)$ is connected where $P \equiv$ "Connected".*

Proof. Let $\underset{\sim}{d} = (d_1, d_2, \ldots, d_n)$, $d_1 \geq d_2 \geq \ldots \geq d_n$. The proof is by induction on n. We take any two realizations of d and transform them so that the neighbourhood of the vertex v_n is the same in each. The induction hypothesis can then be applied to the induced subgraphs formed by the removal of v_n from each graph. Care must be taken to ensure that after each switching the graphs remain both simple and connected.

$n \leq 3$. In this case the result is easily shown by taking cases. So assume that the result holds for $n \leq k - 1$, for some $k \geq 4$.

$n = k$. Let G_1 and G_2 be any two labelled connected realizations of $\underset{\sim}{d}$. We will switch on these graphs to obtain new graphs in which v_n is adjacent to $v_1, \ldots, v_s$ where $s = d_n = d(v_n)$. Suppose that v_n is adjacent to $v_1, \ldots, v_m$, $v_{i_{m+1}}, \ldots, v_{i_s}$ in G_1 where $v_{i_j} \neq v_{m+1}$ for all $j \geq m + 1$. Now every edge adjacent to v_{m+1} cannot also be adjacent to $v_{i_{m+1}}$ since this would imply $d(v_{m+1}) < d(v_{i_{m+1}})$ (note that $i_{m+1} \geq m + 1$). We can therefore choose a vertex v_ℓ which is adjacent to v_{m+1} but not $v_{i_{m+1}}$.

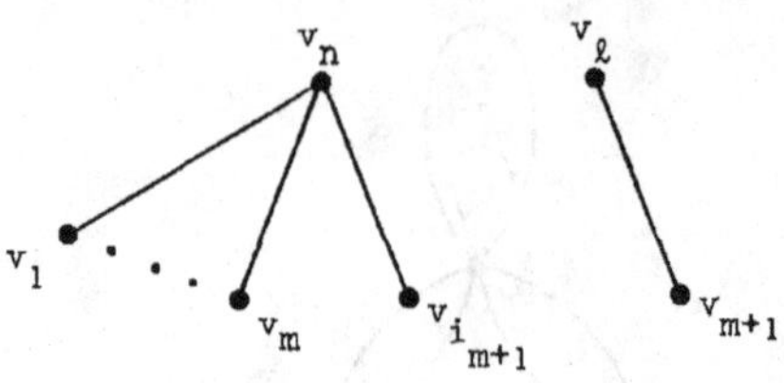

Figure 3.6

G_1 is connected, so must contain at least one of the following kinds of path.

1. Between v_n and v_ℓ, independent of v_{m+1} and $v_{i_{m+1}}$.

2. " v_n and v_{m+1}, independent of v_ℓ and $v_{i_{m+1}}$.

3. " $v_{i_{m+1}}$ and v_ℓ , independent of v_n and v_{m+1}.

4. " $v_{i_{m+1}}$ and v_{m+1}, independent of v_n and v_ℓ.

Case 1. (See Figure 3.7)

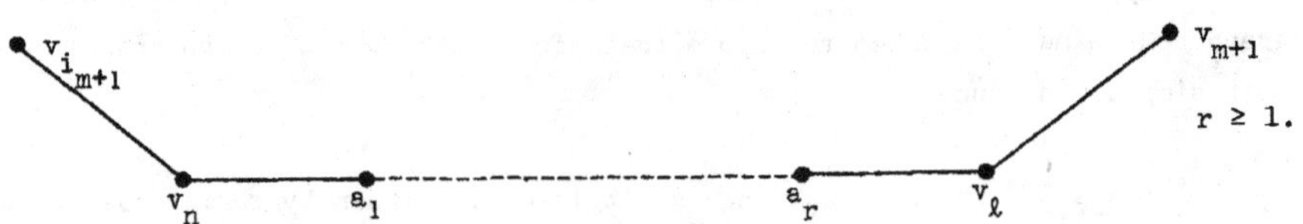

Figure 3.7

We perform the following switching $[(v_{i_{m+1}},v_n),(v_\ell,v_{m+1})] \to [(v_{i_{m+1}},v_\ell),(v_n,v_{m+1})]$. Observe that G_1 remains connected and simple, since $(v_\ell,v_{i_{m+1}}) \notin EG_1$ and $(v_n,v_{m+1}) \notin EG_1$.

Case 2. (See Figure 3.8)

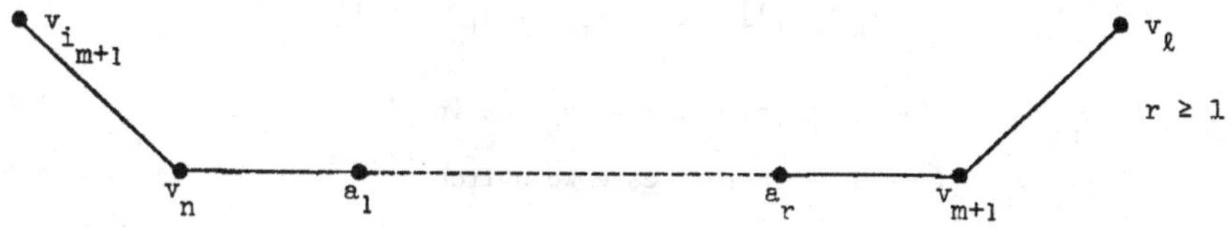

Figure 3.8

We distinguish two subcases,

(a) If $(v_{i_{m+1}}, a_r) \in EG_1$ we switch
$[(v_{i_{m+1}}, v_n), (v_{m+1}, v_\ell)] \to [(v_{i_{m+1}}, v_\ell), (v_n, v_{m+1})]$, and observe that G_1 remains simple and connected.

(b) If $(v_{i_{m+1}}, a_r) \notin EG_1$ we switch
$[(v_{i_{m+1}}, v_n), (a_r, v_{m+1})] \to [(v_{i_{m+1}}, a_r), (v_n, v_{m+1})]$ and again the resultant graph is simple and connected.

Case 3. (See Figure 3.9)

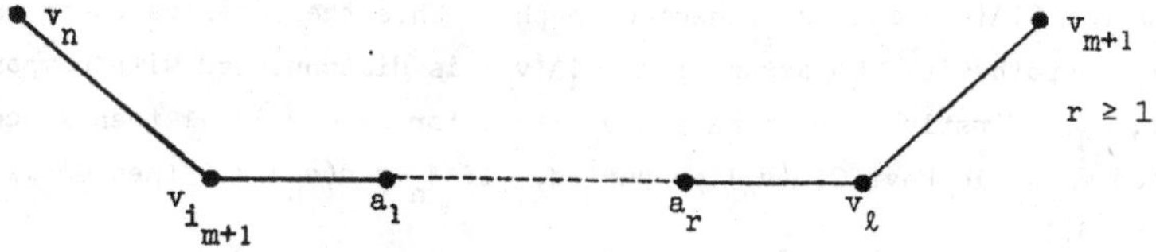

Figure 3.9

Since $d(v_{m+1}) \geq d(v_{i_{m+1}})$ either

(a) v_{m+1} is adjacent to one of $v_{i_{m+1}}, a_1, \ldots, a_r$, in which case we switch $[(v_n,v_{i_{m+1}}),(v_\ell,v_{m+1})] \to [(v_n,v_{m+1}),(v_{i_{m+1}},v_\ell)]$.

(b) v_{m+1} is adjacent to some vertex v_p not in Figure 3.9 and $(v_p,v_{i_{m+1}}) \notin EG_1$. In this case we switch $[(v_n,v_{i_{m+1}}),(v_{m+1},v_p)] \to [(v_n,v_{m+1}),(v_{i_{m+1}},v_p)]$.

Case 4. (See Figure 3.10)

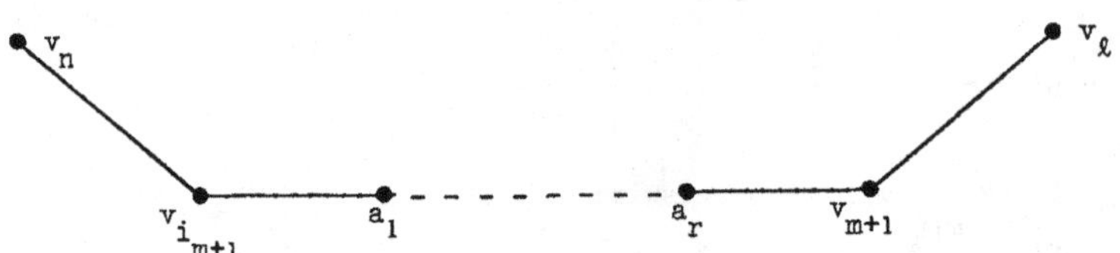

Figure 3.10

Here we switch $[(v_n,v_{i_{m+1}}),(v_\ell,v_{m+1})] \to [(v_n,v_{m+1}),(v_{i_{m+1}},v_\ell)]$.

The overall conclusion is that we can switch G_1 to another simple connected graph in which v_n is adjacent to $v_1, v_2, \ldots, v_m, v_{m+1}, v_{i_{m+2}}, \ldots, v_{i_s}$. Continuing this process we obtain a graph in which v_n is adjacent to $v_1, v_2, \ldots, v_s$. It follows therefore that there exists two sequences of switchings σ_1 and σ_2, which transform G_1 and G_2 to connected graphs G_1' and G_2' in which v_n is adjacent to $v_1, \ldots, v_s$. Further, the transformed graphs between G_i and G_i' in the sequences are all connected and simple. We now switch on G_1' to transform $G_1' \backslash \{v_n\}$ into a connected graph. This then enables us to use the induction hypothesis. So assume that $G_1' \backslash \{v_n\}$ is disconnected with components $H_1, \ldots, H_d$. Firstly we must have $d(v_n) \geq 2$, for if $d(v_n) = 1$ then since G_1' is connected we would have $G_1' \backslash \{v_n\}$ connected. If $d_n = d(v_n) = 2$ then G_1' is illustrated in Figure 3.11.

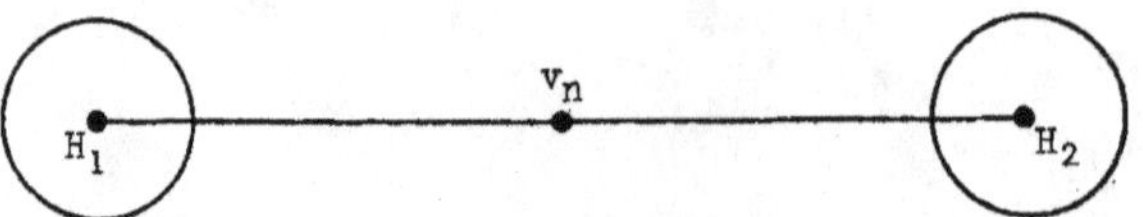

Figure 3.11

H_1 is connected and can have at most one vertex of degree 1 since $d_i \geq 2$ for $i = 1, \ldots, n$. Thus H_1 cannot be a tree and must contain a cycle, similarly for H_2. We can then select edges $(a,b) \in EH_1$ and $(c,d) \in EH_2$ with (a,b) and (c,d) each on a cycle. The switching $[(a,b),(c,d)] \rightarrow [(a,c),(b,d)]$ will give a simple connected graph in which v_n is not a cut vertex. Observe also that the neighbourhood of v_n is unaffected by this switching.

The case $d_n = d(v_n) \geq 3$ is illustrated in Figure 3.12 where H_i and H_j are arbitrary components of $G_1'\backslash\{v_n\}$. Since every vertex in H_i has degree at least two, H_i cannot be a tree. Consequently we may select edges $(a,b) \in EH_i$ and $(c,d) \in EH_j$ with (a,b) and (c,d) each on cycles. Switching as before transforms G_1' so that $G'\backslash\{v_n\}$ is switched into a graph with one fewer component. Observe again that the neighbourhood of v_n is unchanged. We may continue this process to obtain a graph G_1'' in which $G_1''\backslash\{v_n\}$ is connected.

We can therefore switch G_1' and G_2' through simple connected graphs by switchings θ_1 and θ_2, say, to graphs G_1'' and G_1'', where v_n is adjacent to $v_1, \ldots, v_s$, and v_n is not a cut vertex of G_1'' or G_2''. But $G_1''\backslash\{v_n\}$ and $G_2''\backslash\{v_n\}$ are connected labelled graphs with common degree sequence $\underset{\sim}{d}' = (d_1 - 1, d_2 - 1, \ldots, d_s - 1, d_{s+1}, \ldots, d_{n-1})$ which has $n - 1 = k - 1$ terms. Thus by the induction hypothesis there must exist a sequence of switchings γ through simple connected graphs, which transforms $G_1''\backslash\{v_n\}$ into $G_2''\backslash\{v_n\}$. It follows then that this same sequence of switchings transforms G_1'' to G_2'' via simple connected graphs.

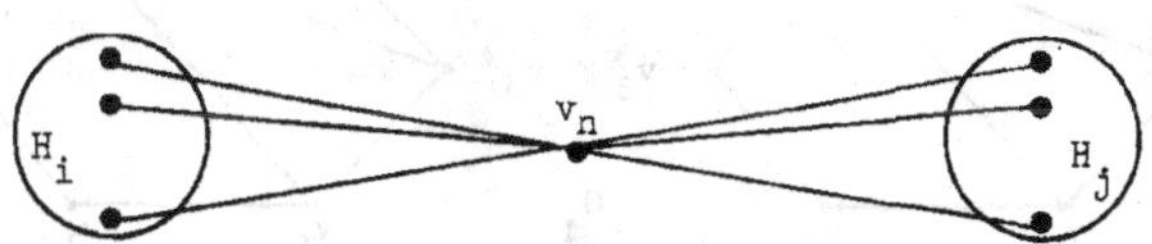

Figure 3.12

Thus the sequence of switchings $\sigma_1\theta_1\gamma\theta_2^{-1}\sigma_2^{-1}$ (see Figure 3.13) transforms G_1 to G_2 in the required manner. Since G_1 and G_2 were arbitrary labelled connected simple realizations of $\underset{\sim}{d} = (d_1, \ldots, d_k)$ we have the result for $n = k$.

$$G_1 \xrightarrow{\sigma_1} G_1' \xrightarrow{\theta_1} G_1'' \xrightarrow{\gamma} G_2'' \xleftarrow{\theta_2} G_2' \xleftarrow{\sigma_2} G_2$$

Figure 3.13

<u>Corollary 3.3.</u> *$R(\underset{\sim}{d},(0,0,1),P)$ is connected, where P ≡ "Connected".*

We are now in a position to give the example referred to in Section 1.

<u>Example 3.1.</u> Let $\underset{\sim}{d} = (4^2,3^4,2^6)$, and P be the property "Two components, each with degree sequence $(4,3^2,2^3)$", then $R(\underset{\sim}{d},(0,0,1),P)$ is connected but $R_\ell(d,(0,0,1),P)$ is not.

To show that $R(\underset{\sim}{d},(0,0,1),P)$ is connected, let G_1 and G_2 be any two unlabelled realizations of $\underset{\sim}{d}$ with property P. Let G_1 and G_2 have components A_1, A_2 and B_1, B_2 respectively. By Corollary 3.3 we can switch between A_1 and B_1 through connected graphs by some sequence of switchings σ say, and similarly between A_2 and B_2 by the sequence θ say. Thus the sequence of switchings $\sigma\theta$ transforms G_1 to G_2 via graphs with property P. Since G_1 and G_2 were arbitrary unlabelled realizations of $\underset{\sim}{d}$ with property P, we can conclude that $R(\underset{\sim}{d},(0,0,1),P)$ is connected.

To show that $R_\ell(\underset{\sim}{d},(0,0,1),P)$ is not connected consider the two labelled realization of $\underset{\sim}{d}$ shown in Figure 3.14.

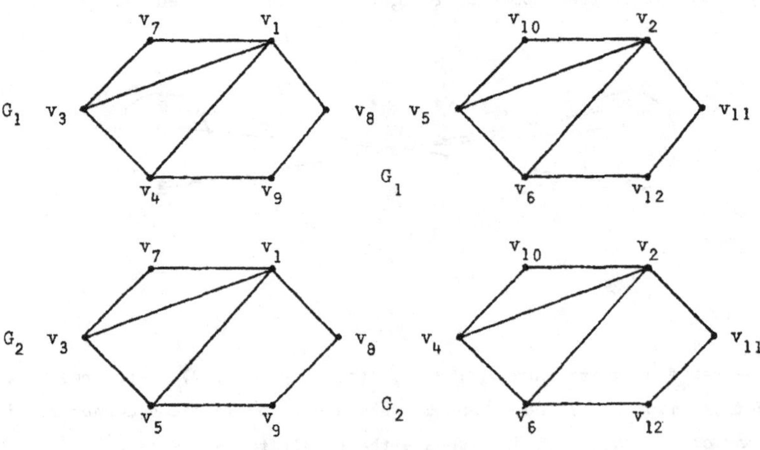

Figure 3.14

Note that in G_2, v_3 is in the same component as v_5 whilst in G_1, v_3 is in a different component. Consequently any sequence of switchings which transforms G_1 into G_2 must at some stage switch between the two components. It is easily shown however that every connected realization of $(4,3^2,2^3)$ has each edge on a cycle. Thus any switching between components forms one large component, violating P.

4. CONNECTIVITY k.

In this section we explore switchings constrained to graphs of connectivity k. In [2] Colbourn shows that connectivity k is an incomplete property for k = 0, 1 but adds that the general case has not been settled. Here we show that connectivity k is incomplete for all $k \geq 1$.

Let $k \geq 1$ be given. We form the two graphs G_1 and G_2 (see Figure 4.1).

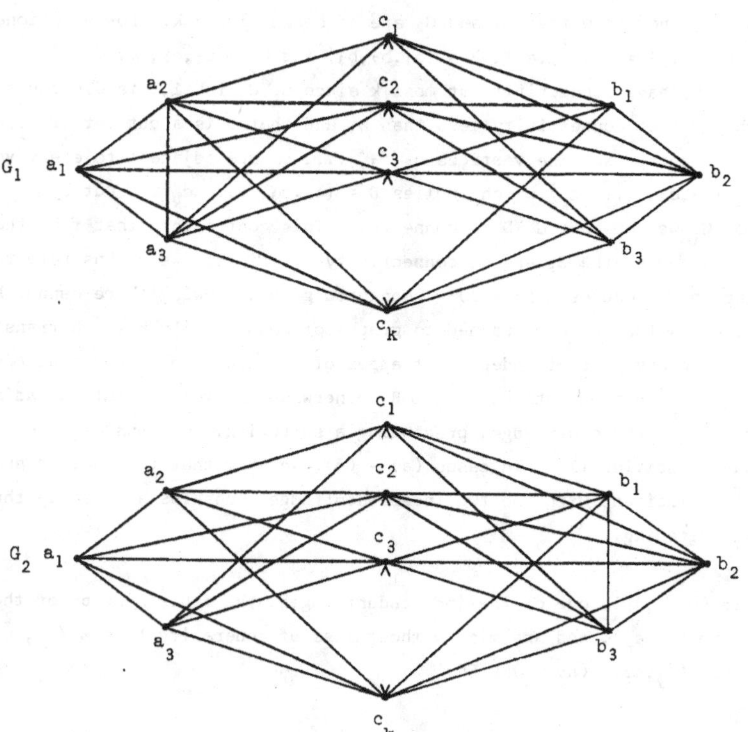

Figure 4.1

Note that these are not labelled realizations with the restriction mentioned in Section 1. The vertex set for each is $\{a_1,a_2,a_3\} \cup \{b_1,b_2,b_3\} \cup \{c_1, \ldots, c_k\}$. The edge sets are defined as follows, where the subscripts range over all values for which they are defined.

$$
\begin{array}{lll}
(c_i,c_j) \in EG_s & & \\
(c_i,a_j) \in EG_s & \text{unless} & (c_i,a_j) = (c_1,a_1) \\
(c_i,b_j) \in EG_s & & \\
(a_i,a_j) \in EG_1 & & \\
(b_i,b_j) \in EG_2 & & \\
(b_i,b_j) \in EG_1 & \text{unless} & (b_i,b_j) = (b_1,b_2) \\
(a_i,a_j) \in EG_2 & \text{unless} & (a_i,a_j) = (a_2,a_3).
\end{array}
$$

Observe that G_1 and G_2 share the degree sequence $\underset{\sim}{d} = ((k+5)^{k-2}, k+4, (k+2)^3, (k+1)^3)$. Also note that $G_1 \not\cong G_2$, since in G_2 we have a vertex of degree $k+1$ adjacent to two others of degree $k+1$. This is not so in G_1. We now show that G_1 and G_2 are of connectivity k. For notational convenience we set $A = \{a_1,a_2,a_3\}$, $B = \{b_1,b_2,b_3\}$ and $C = \{c_1,c_2, \ldots, c_k\}$. Clearly G_1 and G_2 have connectivity at most k since $G_i \backslash C$, $i = 1,2$ is disconnected. So assume that G_i has connectivity less than k, and that D is a cut set of G_i with less than k elements. Now the vertices $c_2,c_3, \ldots, c_k$ are adjacent to every vertex in G_i so $D \supseteq \{c_2,c_3, \ldots, c_k\}$ which implies $D = \{c_2,c_3, \ldots, c_k\}$. But by inspection of G_i we see that $G \backslash D$ is connected. This contradicts that fact that D is a cut set, and so the G_i are of connectivity k. Finally we demonstrate that any switching on G produces a $(k+1)$ - connected graph, showing there cannot be any sequence of switchings constrained to graphs of connectivity k which transforms G_1 into G_2. Take any pair of independent edges of G_1. The two independent edges cannot have more than two vertices in $C \cup B$, otherwise one vertex would be adjacent to both vertices on the other edge, precluding a switching. We consider the remaining cases individually. In cases (a) - (d), we show that the induced subgraph on the vertices incident with the two independent edges contains a triangle thus preventing a switching.

(a) If (c_i,c_j) is one of the independent edges, the other must be of the form (a_ℓ,a_m), and assuming without loss of generality that $\ell \neq 1$, note that (a_ℓ,c_i), $(a_\ell,c_j) \in EG_1$.

(b) If (a_i,c_j) and (a_ℓ,c_m) are the independent edges, then we may suppose $\ell \neq 1$ without loss of generality, and note that $(a_i,a_\ell),(c_j,a_\ell) \in EG_1$.

(c) If (a_i,c_j) and (a_ℓ,a_m) are the independent edges, then (a_i,a_ℓ), $(a_i,a_m) \in EG_1$.

(d) If (b_i,c_j) is one of the independent edges, the other must be of the form (a_ℓ,a_m). Then (c_j,a_ℓ), $(c_j,a_m) \in EG_1$ if $j \neq 1$ or if $\ell, m \neq 1$.

(e) If (b_i,b_j) is one of the independent edges, the other must be of the form (a_ℓ,a_m), in which case there are no further edges between the pairs of vertices concerned.

(f) If (b_i,c_1) and (a_1,a_m) are the independent edges, then (c_1,a_m) is the only further edge between the pairs of vertices concerned.

The only cases where we may perform a switching on the given edges are (e) and (f). However any switching in these cases introduces an edge between a vertex in A and a vertex in B, which has the effect of increasing the connectivity of G_1. Thus we have shown that for any k, connectivity k is not complete.

As far as we know, the corresponding question for multigraphs and pseudographs is open.

5. INFINITE DEGREE SEQUENCES.

In this section we extent the results on finite degree sequences to infinite degree sequences. We define an infinite degree sequence $\underset{\sim}{d}$ by $\underset{\sim}{d} = (d_1,d_2, \ldots, d_n, \ldots)$ where the terms d_i are non-negative integers. Labelled realizations of $\underset{\sim}{d}$ are defined analagously to the finite case, i.e. with the restriction that v_i has degree $d(v_i) = d_i$ for all i. We now introduce some definitions that are useful in relating results about finite degree sequences to infinite degree sequences.

Definition. Two labelled graphs of the same degree sequence G_1 and G_2 are said to have *finite difference* if $S = S(G_1,G_2) = \{(v_i,v_j); (v_i,v_j) \in EG_1 \Delta EG_2\}$ is a finite set, where Δ indicates symmetric difference.

Definition. We say that two unlabelled graphs G_1 and G_2 have *finite difference* if there exist labellings of G_1 and G_2, so that their labelled forms have finite difference.

Theorem 5.1. *Any two labelled realizations of* $\tilde{d}$ *of type* $\tau = (0,0,1), (0,0,\infty), (0,\infty,\infty)$ *are joined in* $R_\ell(\tilde{d},\tau)$*, the graph of realizations of* $\tilde{d}$*, by a finite sequence of switchings if and only if they have finite difference.*

Proof. Firstly we prove sufficiency. Let G_1 and G_2 be two labelled realizations of $\tilde{d}$ with finite difference. Thus S is finite, and also S_v is finite, where $S_v(G_1,G_2) = \bigcup_{(v_i,v_j)\in S} \{v_i \cup v_j\}$. Observe that

$G_1\backslash S_v \cong G_2\backslash S_v$, by the definition of S. Furthermore, we claim that $\langle S_v\rangle_{G_1}$ and $\langle S_v\rangle_{G_2}$ have the same degree sequence, where $\langle S_v\rangle_{G_i}$ is the induced subgraph of G_i on the vertices S_v. To show this it is sufficient to show that any vertex $a \in S_v$ has the same degree in $\langle S_v\rangle_{G_1}$ as it does in $\langle S_v\rangle_{G_2}$. So assume without loss of generality that a has larger degree in $\langle S_v\rangle_{G_1}$ than in $\langle S_v\rangle_{G_2}$. Then there must be some $b \notin S_v$ with $(a,b) \in EG_1$ and $(a,b) \notin EG_2$. Thus $(a,b) \in EG_1\backslash EG_2$ and so $(a,b) \in S$ which implies $b \in S_v$, a contradiction. By Theorem 2.1 it now follows that there exists a sequence of switchings σ which transforms $\langle S_v\rangle_{G_1}$ into $\langle S_v\rangle_{G_2}$. Let G_1' be the result of the application of σ to the whole graph G_1. Since σ switches edges both of whose endpoints are in S_v, then $G_1'\backslash S_v \cong G_1\backslash S_v \cong G_2\backslash S_v$, while, by definition of σ, $\langle S_v\rangle_{G_1'} \cong \langle S_v\rangle_{G_2}$. Let $(a,b) \in EG_1'$, and assume firstly that a and b are in S_v. Then $(a,b) \in EG_2$ since $\langle S_v\rangle_{G_1'} \cong \langle S_v\rangle_{G_2}$. Assume on the other hand that one of a or b is not in S_v. Then $(a,b) \notin S$ by definition of S_v. Also σ does not involve the edge (a,b) since one of the endvertices is not in S_v, consequently $(a,b) \in EG_1$. We may conclude therefore, that $(a,b) \in EG_2$. Thus $EG_2 \supseteq EG_1'$ and by a similar argument we may show $EG_1' \supseteq EG_2$, whence $G_1' \cong G_2$. This completes the demonstration of sufficiency.

We now prove necessity. Let G_1 and G_2 be any two labelled realizations of $\tilde{d}$ that are joined in the graph of realizations by a finite sequence of switchings, σ. Let U be the collection of edges involved in any switching in σ. Then since σ is finite, U is finite. Take any edge $(a,b) \in S(G_1,G_2)$, and assume without loss of generality that $(a,b) \in EG_1\backslash EG_2$. If (a,b) were not involved in any switching in σ, then it would remain throughout the switching sequence σ and this would imply $(a,b) \in EG_2$, a contradiction. Thus (a,b) must be involved in a switching and so $(a,b) \in U$. Therefore we may conclude that $U \supseteq S(G_1,G_2)$ and so cardinality S $\leq$ cardinality U $< \infty$. So G_1 and G_2 have finite difference.

Corollary 5.1. *Any two realizations of* $\tilde{d}$ *of type* $\tau = (0,0,1), (0,0,\infty), (0,\infty,\infty)$ *are joined by a finite sequence of switchings if and only if they have finite difference.*

We now show that it is by no means obvious to discern whether or not there exists a sequence of switchings between two realizations of a given infinite degree sequence. This is done by giving two realizations of the same degree sequence which appear to be almost identical, but which have no finite sequence of switchings between them. The following example is a slight modification of one given by Billington.

Example 5.1. (D. Billington [1]) Consider the two realizations G_1 and G_2 of $d = (1^2, 2^\infty)$ shown in Figure 5.1.

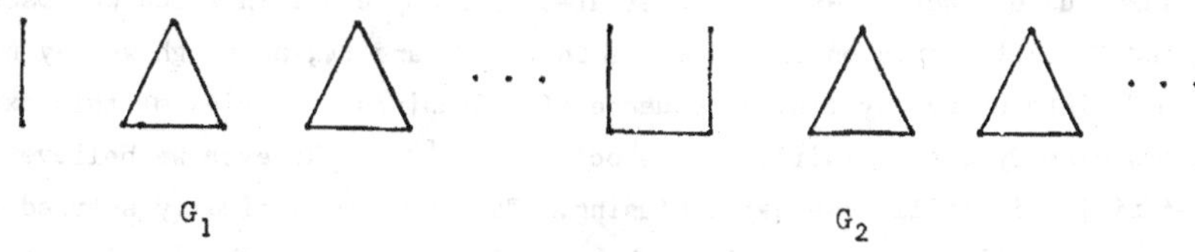

Figure 5.1

Assume there is a finite sequence of switchings σ, which transforms G_1 into G_2. Thus G_1 and G_2 have finite difference, by Corollary 5.1, and so there exist labellings ℓ_1 and ℓ_2 so that $G_1^{\ell_1}$ and $G_2^{\ell_2}$ have finite difference. Note therefore that $S_v(G_1^{\ell_1}, G_2^{\ell_2})$ is finite. We let S_v' be the set of vertices defined by $a \in S_v' \Leftrightarrow a$ is joined by some path in $G_1^{\ell_1}$ or $G_2^{\ell_2}$ to some vertex in S_v.

For any $x \notin S_v$, we have $(x,y) \in EG_1^{\ell_1}$ if and only if $(x,y) \in EG_2^{\ell_2}$.

Thus $a \in S_v' \Leftrightarrow a$ is joined by some path in $G_1^{\ell_1}$ and $G_2^{\ell_2}$ to some vertex in S_v.

Now any vertex in S_v can be joined by a path to at most two other vertices in $G_1^{\ell_1}$ and so $|S_v'| \le 3|S_v| < \infty$. Consider the induced subgraphs $\langle S_v' \rangle_{G_1} \ell_1$ and $\langle S_v' \rangle_{G_2} \ell_2$. Both contain all the vertices in S_v and so must have common degree sequence $(1^p, 2^q)$, for some $p \in \{0,1,2\}$ and q a non-negative integer.

From our discussion of S_v' it follows that $\langle S_v' \rangle_{G_1} \ell_1$ and $\langle S_v' \rangle_{G_2} \ell_2$ contain a collection of connected components of the graphs $G_1^{\ell_1}$ and $G_2^{\ell_2}$. Also the component P_4 in G_2 must occur in $\langle S_v' \rangle_{G_2} \ell_2$, since P_4 contains an edge between vertices of degrees one and two, whilst no such edge exists in G_1. Similarly we can show that the component P_2 must occur in $\langle S_v' \rangle_{G_1} \ell_1$. Thus $\langle S_v' \rangle_{G_1} \ell_1 \cong P_2 \cup rC_3$ and

$\langle S_v{}' \rangle_{G_2} \ell_2 \cong P_4 \cup sC_3$ for some r, $s \in \mathbb{N}$. We can conclude therefore that $q \equiv 0 (\mathrm{mod}\ 3)$ and also that $q \equiv 2 (\mathrm{mod}\ 3)$, a contradiction. Consequently there can be no finite sequence of switchings which transforms G_1 into G_2.

Conditions under which two unlabelled graphs are sufficiently similar so that one may be transformed into the other by a finite sequence of switchings are discussed by Eggleton and Holton in [6], [7] and [8]. The authors define the term associates and claim that a graph may be transformed into another graph precisely when the two graphs are associates. In [6] two graphs are defined to be associates if they differ at only a finite number of vertices. This definition appears inadequate however, as is demonstrated by Example 5.1 in which we observe that G_1 and G_2 differ only on the vertices in the P_2 and P_4, although we may not transform G_1 into G_2 by any finite sequence of switchings. In view of this example the authors clarify the definition of associates in [7] . However we believe their definition is still somewhat confusing. This matter is finally settled in [8], where two realizations are said to be associates whenever there exists a degree preserving bijection between their vertices which identifies all but finitely many of their edges. Note that this definition is equivalent to the definition of finite difference we use here.

We now extend the results of Section 3 to infinite degree sequences.

Theorem 5.2. *Any two connected labelled realizations of an infinite degree sequence* $\underset{\sim}{d}$ *of type* $\tau = (0,0,1), (0,0,\infty), (0,\infty,\infty)$ *are joined in the graph of realizations by a finite sequence of switchings through connected graphs if and only if they have finite difference.*

Proof. We prove the sufficiency first. Let G_1 and G_2 be any two labelled connected realizations of $\underset{\sim}{d}$ with finite difference. Then $S_v(G_1,G_2)$ is finite. Consider the induced subgraphs $\langle S_v \rangle_{G_1}$ and $\langle S_v \rangle_{G_2}$. We now enlarge the set S_v so that the induced subgraphs on this larger collection of vertices are connected. So let $\langle S_v \rangle_{G_1}$ and $\langle S_v \rangle_{G_2}$ have components $A_1, \ldots, A_n$ and $B_1, \ldots, B_m$, respectively. Since G_1 is connected there must exist a collection of paths P_i, $i = 1, \ldots, k$ with the following properties.

(i) For each i the endvertices of P_i are in S_v but all other vertices on P_i are not.

(ii) The collection of paths connects the graph $\langle S_v \rangle_{G_1}$.

We then let $S_v' = S_v \cup \{v: v \in P_i \text{ for some } i\}$. Thus $\langle S_v' \rangle_{G_1}$ is connected. Further note that the number of components of $\langle S_v' \rangle_{G_2}$ is at most equal to the number of components of $\langle S_v \rangle_{G_2}$. This follows from the fact that if $a \notin S_v$ the $(u,a) \in EG_1$ implies $(u,a) \in EG_2$ for all a. Now consider $\langle S_v' \rangle_{G_2}$. As before we may enlarge S_v' to a set S_v'', with $\langle S_v'' \rangle_{G_2}$ connected and where the number of components of $\langle S_v'' \rangle_{G_1}$ is at most equal to the number of components in $\langle S_v' \rangle_{G_1}$, and so is connected.

Now since $S_v'' \supseteq S_v$, by the same reasoning as used in Theorem 5.1, we know that $\langle S_v'' \rangle_{G_1}$ and $\langle S_v'' \rangle_{G_2}$ have the same degree sequence and also that $G_1 \backslash S_v'' \cong G_2 \backslash S_v''$. Since S_v'' is finite we can conclude from Theorems 3.1, 3.2 and 3.3 that there exists a sequence of switchings σ which transforms $\langle S_v'' \rangle_{G_1}$ into $\langle S_v'' \rangle_{G_2}$ through connected graphs. Now if we switch an induced subgraph of a connected graph through connected graphs, the whole graph must remain connected throughout the sequence of switchings. So then σ transforms G_1 to G_2, as in the proof of Theorem 5.1. This completes the demonstration of sufficiency. The necessity follows as a corollary to Theorem 5.1.

Corollary 5.2. *Any two connected realizations of an infinite degree sequence* $\underset{\sim}{d}$*, of type* $\tau = (0,0,1),(0,0,\infty),(0,\infty,\infty)$ *are joined in the graph of realizations by a finite sequence of switchings through connected graphs if and only if they have finite difference.*

REFERENCES

[1] D. Billington, Private Communication.

[2] C.J. Colbourn, Graph enumeration, Dept. of Computer Science, University of Waterloo, *Research Report* CS-77-37 (1977).

[3] R.B. Eggleton, Graphic sequences and graphic polynomials: a report, in *Infinite and Finite Sets,* Vol. 1, ed. A. Hajnal *et al*, Colloq. Math. Soc. J. Bolyai 10, (North Holland, Amsterdam, 1975) 385-392.

[4] R.B. Eggleton and D.A. Holton, Path realizations of multigraphs, I. The disconnected case, Dept. of Mathematics, University of Melbourne, *Research Report* 33 (1978).

[5] R.B. Eggleton and D.A. Holton, Graphic sequences *Comb. Maths. VI,* Proc. Sixth Aust. Conf., Lecture Notes in Maths. 748 (Springer-Verlag, 1979) 1-10.

[6] R.B. Eggleton and D.A. Holton, The graph of type $(0,\infty,\infty)$ realizations of a graphic sequence, *Comb. Math. VI,* Proc. Sixth Aust. Conf., Lecture Notes in Maths. 748 (Springer-Verlag, 1979) 40-54.

[7] R.B. Eggleton and D.A. Holton, Pseudographic realizations of an infinitary degree sequence, *Comb. Math. VII,* Proc. Seventh Aust. Conf., Lecture Notes in Maths. 829 (Springer-Verlag, 1980) 94-109.

[8] R.B. Eggleton and D.A. Holton, Simple and multigraphic realizations of degree sequences, *this volume.*

[9] S.L. Hakimi, On realizability of a set of integers as degrees of the vertices of a linear graph I, *J. Soc. Indust. Appl. Math.* 10 (1962) 492-506.

[10] S.L. Hakimi, On realizability of a set of integers as degrees of the vertices of a linear graph II, uniqueness, *J. Soc. Indust. Appl. Math.* 11 (1963) 135-147.

[11] M.M. Syslo, Private Communication.

Department of Mathematics
University of Melbourne
Parkville, Victoria 3052

ONE-FACTORISATIONS OF WREATH PRODUCTS

W.D. WALLIS

Some sufficient conditions are proven for the existence of one-factorisations of wreath products of graphs. That these conditions are not necessary is then established by examples.

1. INTRODUCTION

We assume familiarity with the basic ideas of graph theory. A *factorisation* of a graph G is a decomposition of G into edge-disjoint spanning subgraphs or *factors*. A *one-factor* is a spanning subgraph in which every vertex has valency 1, and a *one-factorisation* is a factorisation in which every factor is a one-factor; similarly a *two-factor* is a spanning union of vertex-disjoint cycles, and a *two-factorisation* is a factorisation into two-factors.

The best-known results on one-factorisations are proofs that a few families of graphs are always one-factorable, namely the complete graph K_n when n is even, the complete regular m-partite graph $K_{n,n,\ldots,n}$ when mn is even, and the graph $C_m[\overline{K}_n]$ (as defined below) when mn is even - see [1,5,7]. Of more interest is the proof in [3] that if G is one-factorable and H is regular then the cartesian product $G \times H$ is one-factorable. Subsequently Kotzig [4] has generalised this result; and more recent discussions of 1-factorisation of cartesian products appear in [6] and [8].

In this paper we consider the *wreath product* or *composition* of two graphs: if G is a graph with vertices $p_1, p_2, \ldots, p_v$, and H is any graph, then the wreath product G[H] consists of the disjoint union of v copies $H_1, H_2, \ldots, H_v$ of H, to which are added all the edges joining vertices in H_i to vertices in H_j if and only if p_i is adjacent to p_j in G.

Various properties of the wreath product are known. For example, if $G = G_1 \cup G_2$ is a factorisation, and $\overline{K}$ is the empty graph on the vertices of H, then

$$G[H] = G_1[H] \cup G_2[\overline{K}], \tag{1}$$

and the two graphs on the right are edge-disjoint. Similarly, if G_1 and G_2 are vertex-disjoint, then $G_1[H]$ and $G_2[H]$ are vertex-disjoint. Another important property is

$$(G[H])[J] = G[H[J]]. \tag{2}$$

The proofs of the "G[H]" case in Theorem 1, and of Theorem 2, were separately discovered by the author and by P.E. Himelwright and J.E. Williamson, in unpublished papers; and it is hoped that a joint version will appear [2]. Also the proof of Theorem 1 could be shortened by reference to [7]. In both cases I thought that a complete, self-contained paper would be preferable.

2. SOME FACTS ABOUT FACTORISATIONS

Lemma 1. *Suppose the graph* Y *is a union of edge-disjoint spanning subgraphs:*

$$Y = Y_1 \cup Y_2 \cup \ldots \cup Y_n.$$

If each Y_i *has a one-factorisation then so does* Y.

Proof. One takes all the one-factors in all the Y_i as one-factors in Y.

Lemma 2. *Suppose* X *is a regular graph which is the union of disjoint components:*

$$X = X_1 \cup X_2 \cup \ldots \cup X_n.$$

If every X_i *has a one-factorisation then so does* X.

Proof. Let k be the valency of X. Then each X_i is also regular of valency k. Suppose X_i has one-factorisation

$$X_i = X_{i1} \cup X_{i2} \cup \ldots \cup X_{ik}.$$

Then X has a one-factorisation in which factor j is

$$X_{1j} \cup X_{2j} \cup \ldots \cup X_{nj}.$$

3. A SUFFICIENT CONDITION FOR FACTORISATION

We commence with a Lemma concerning a different type of product. The cartesian product $G \times H$ of two graphs may be defined as follows: it consists of a copy of G in which each vertex is replaced by a copy of H, and each edge is replaced by a set of edges, one for each vertex of H, joining the corresponding vertices in the copies of H. Thus $G \times H$ is like G[H], but with specific one-factors rather than complete bipartite graphs replacing the edges of G.

Lemma 3. *The cartesian product* $K_2 \times C_k$ *of an edge with a cycle has a one-factorisation.*

Proof. Take the graph to have vertices $1,2,\dots,k,\hat{1},\hat{2},\dots,\hat{k}$ and edges $12, 23,\dots, k1, \hat{1}\hat{2}, \hat{2}\hat{3},\dots, \hat{k}\hat{1}, 1\hat{1}, 2\hat{2},\dots, k\hat{k}$. If k is even there is no problem; if k is odd then suitable factors are

$$1\hat{1},\ 23,\ 45,\dots,\ (k-1)k,\ \hat{2}\hat{3},\ \hat{4}\hat{5},\dots,\ (\widehat{k-1})\hat{k}$$
$$2\hat{2},\ 34,\ 56,\dots,\ k1,\ \hat{3}\hat{4},\ \hat{5}\hat{6},\dots,\ \hat{k}\hat{1}$$
$$12,\ \hat{1}\hat{2},\ 3\hat{3},\ 4\hat{4},\dots,\ k\hat{k}.$$

Theorem 1. *If* H *is a non-empty graph with a one-factorisation and* G *is a graph which can be factored into one-factors and two-factors, then* H[G] *and* G[H] *are one-factorable.*

Proof. Say G and H have g and 2q vertices respectively; denote the vertices of G by $1,2,\dots,g$. Suppose the factorisations are

$$G = G_1 \cup G_2 \cup \dots \cup G_m,$$

$$H = H_1 \cup H_2 \cup \dots \cup H_n,$$

where $G_1, G_2,\dots,G_\ell$ are edge-disjoint two-factors and $G_{\ell+1}, G_{\ell+2},\dots,G_m$ are edge-disjoint one-factors, as are the H_i. Write $G^2 = G_1 \cup G_2 \cup \dots \cup G_\ell$, $G^1 = G_{\ell+1} \cup G_{\ell+2} \cup \dots \cup G_m$, and denote by $\overline{K}$ the empty graph with the same vertices as G. Then

$$H[G] = H_1[G] \cup H_2[\overline{K}] \cup \dots \cup H_n[\overline{K}],$$

an edge-disjoint union of spanning subgraphs, and moreover,

$$H_1[G] = H_1[G^2] \cup G^*$$

where G^* is the disjoint union of $2q$ copies of G^1. Clearly each $H_i[\overline{K}]$ has a one-factorisation, as does G^*. So we consider $H_1[G^2]$. Since $H_1[G^2]$ is a disjoint union of copies of $K_2[G^2]$, it is sufficient to show that the latter graph has a one-factorisation.

In each of $G_1, G_2,\dots,G_\ell$, select a direction for every constituent cycle. Then define a_{ij} to be the vertex following j in the cycle containing it in G_i. The a_{ij} clearly form a latin rectangle of size $\ell \times g$, which can, of course, be extended to a latin square A of side g. Define F_i to be the graph (one-factor) on vertices $1,2,\dots,g,\hat{1},\hat{2},\dots,\hat{g}$, with j adjacent to $\hat{a}_{ij}$. If $K_2[G^2]$ consists of two copies of G^2, one on $1,2,\dots$ and the other on $\hat{1},\hat{2},\dots$, together with $K_{g,g}$, and if $\hat{G}_i$ denotes G_i with each vertex j replaced by $\hat{j}$, then

$$K_2[G^2] = (F_1 \cup G_1 \cup \hat{G}_1) \cup (F_2 \cup G_2 \cup \hat{G}_2) \cup \dots \cup (F_\ell \cup G_\ell \cup \hat{G}_\ell)$$
$$\cup F_{\ell+1} \cup F_{\ell+2} \cup \dots \cup F_m,$$

an edge-disjoint union. Now each $F_i \cup G_i \cup \hat{G}_i$ has a one-factorisation by Lemma 3, so $K_2[G^2]$ has a one-factorisation.

To factorise G[H], we first observe that

$$G[H] = G[\overline{K}_{2q}] \cup \overline{K}[H].$$

$\overline{K}[H]$ consists of g disjoint copies of H; if H_i^* is the union of g copies of H_i, one in each copy of H, then the H_i^* are one-factors whose union is $\overline{K}[H]$. Now by (1), $G[\overline{K}_{2q}]$ is the disjoint union of the $G_i[\overline{K}_{2q}]$. If G_i is a one-factor then obviously $G_i[\overline{K}_{2q}]$ is factorable, and if it is a two-factor then $G_i[\overline{K}_{2q}]$ is a disjoint union of graphs $C_k[\overline{K}_{2q}]$ for various k. So it is sufficient to factor $C_k[\overline{K}_{2q}]$. In order to do this we examine $C_k[\overline{K}_q]$.

We show that $C_k[\overline{K}_q]$ can be factored into two-factors. We denote the vertices of $C_k[\overline{K}_q]$ as $1_1, 1_2, \ldots, 1_q, 2_1, \ldots, k_q$, and say i_x is adjacent to j_y if and only if $j \equiv i \pm 1 \pmod k$. We write P_{ab} for the path (assuming k is odd)

$$1_a 2_{a+b} 3_a 4_{a+b} \ldots k_a$$

(where subscripts are treated as integers modulo q). Let $C_{ab,cd}$ be the union $P_{ab} \cup P_{cd}$ together with the edges $(k_a, 1_c)$ and $(k_c, 1_a)$; $C_{ab,ab}$ is a cycle of length k, while $C_{ab,cb}$ is a cycle of length 2k when $a \neq c$. If q is odd, say $q = 2t+1$, the bth factor in the required factorisation $(1 \leq b \leq q)$ is

$$C_{1b,(b-1)b} \cup C_{2b,(b-2)b} \cup \ldots \cup C_{tb,(b-t)b} \cup C_{(t+1)b,(b-t-1)b}.$$

If q is even, say $q = 2t$, then one factor is the union of the q cycles $C_{a0,a0}$ for $1 \leq a \leq q$ and the others have the form

$$C_{1b,(b-1)b} \cup C_{2b,(b-2)b} \cup \ldots \cup C_{tb,(b-t)b} ,$$

for $1 \leq b \leq q-1$. We obtain a similar factorisation of $C_k[\overline{K}_q]$ if k is even.

Now $C_k[\overline{K}_{2q}] = C_k[\overline{K}_q[\overline{K}_2]] = (C_k[\overline{K}_q])[\overline{K}_2]$ by (2), so $C_k[\overline{K}_{2q}]$ is a union of factors $J[\overline{K}_2]$ where J is a 2-factor, a union of disjoint cycles. When we show that $J[\overline{K}_2]$ can be one-factored we are finished; and by Lemma 2 it is enough to prove that $C_n[\overline{K}_2]$ has a one-factorisation. If n is even, C_n splits into four one-factors, and the result follows easily.

Suppose n is odd. We denote the vertices of $C_n[\overline{K}_2]$ as $1, \hat{1}, 2, \hat{2}, \ldots, n, \hat{n}$, where the vertices of copy i of $\overline{K}_2$ are $\{i, \hat{i}\}$. Then the four factors

$$F_1:\ 12,\hat{2}\hat{3},34,\hat{4}\hat{5},\ldots,(n-2)(n-1),\widehat{(n-1)}\hat{n},n\hat{1}$$

$$F_2:\ 1\hat{2},2\hat{3},3\hat{4},45,\hat{5}\hat{6},67,\ldots,(n-1)n,\hat{n}\hat{1}$$

$$F_3:\ \hat{1}2,\hat{2}3,\hat{3}4,\hat{4}5,\ldots,\widehat{(n-1)}n,\hat{n}1$$

$$F_4:\ \hat{1}\hat{2},23,\hat{3}\hat{4},4\hat{5},5\hat{6},6\hat{7},\ldots,(n-1)\hat{n},n1$$

are a one-factorisation of $C_n[\overline{K}_2]$. (This is illustrated in Figure 1 in the case of $C_7[\overline{K}_2]$.)

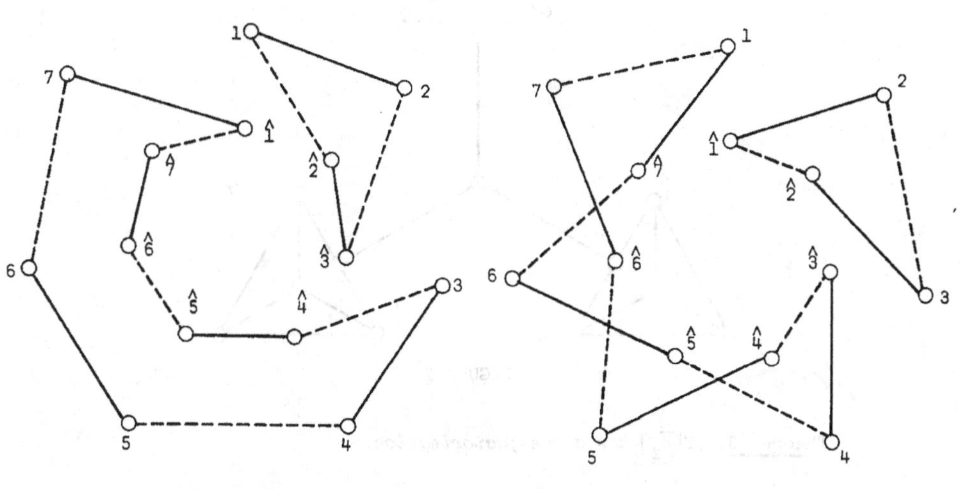

FIGURE 1. One-factorisation of $C_7[\overline{K}_2]$.

From the proof of the preceding Theorem, it is easy to see:

Theorem 2. *If* G *can be factored into one-factors and two-factors, then* $G[\overline{K}_{2q}]$ *has a one-factorisation.*

4. SOME EXAMPLES

We now give some examples to prove that the sufficient conditions of the preceding section are not necessary. They involve the particularly ugly graph U on 16 vertices which is connected and has valency 3, but which contains no one-factor (see Figure 2). We prove that both $U[\overline{K}_2]$ and $U[\overline{K}_3]$ have one-factorisations.

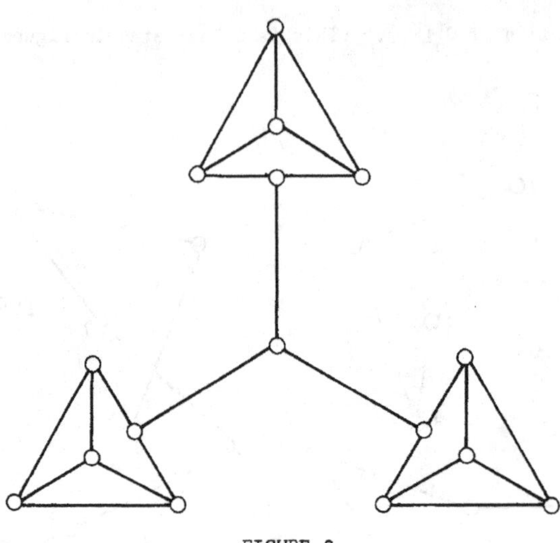

FIGURE 2

Theorem 3. $U[\overline{K}_2]$ *has a one-factorisation.*

Proof. We can consider U as the union of three graphs, U_1, U_2 and U_3, where U_1 is shown in Figure 3, U_2 is formed by rotating U_1 through 120° about the central point F, and U_3 is formed by rotating a further 120°. Then we decompose U_i into two subgraphs U_{i1} and U_{i2}; U_{11} and U_{12} are shown in Figure 3, and the others are decomposed similarly.

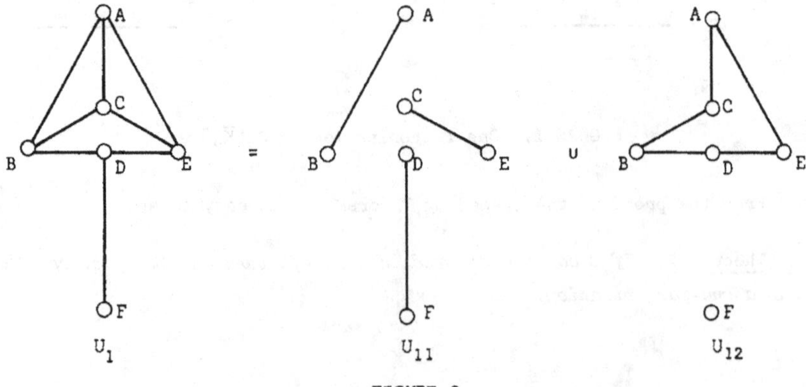

FIGURE 3

Since U_{i1} is a one-factor, we can write $U_{i1}[\overline{K}_2] = F_{i1} \cup F_{i2}$, a union of two one-factors. Since U_{i2} is a cycle (if F is deleted), we can write $U_{i2}[\overline{K}_2]$ as a union of four one-factors (except that they do not contain the two points corresponding to F), $F_{i3} \cup F_{i4} \cup F_{i5} \cup F_{i6}$. Then the six graphs

$$\begin{array}{l} F_{11} \cup F_{23} \cup F_{33} \\ F_{12} \cup F_{24} \cup F_{34} \\ F_{13} \cup F_{21} \cup F_{35} \\ F_{14} \cup F_{22} \cup F_{36} \\ F_{15} \cup F_{25} \cup F_{31} \\ F_{16} \cup F_{26} \cup F_{32} \end{array}$$

are a one-factorisation of $U[\overline{K}_2]$.

Corollary. *$U[K_{2q}]$ has a one-factorisation.*

Proof. Since $U[\overline{K}_2]$ has a one-factorisation, then clearly $U[K_2]$ has one also. But $U[K_{2q}] = (U[K_2])[K_q]$ by (2); it is well-known that K_q has a one-factorisation when q is even and a two-factorisation when q is odd, so the result follows from Theorem 1.

Theorem 4. *$U[\overline{K}_3]$ has a one-factorisation.*

Proof. We use the same notation U_1, U_2, U_3, and the labelling of U_1 as shown in Figure 3. In converting U to $U[\overline{K}_3]$, we assume that vertex X becomes {X1,X2,X3}. Clearly

$$U[\overline{K}_3] = U_1[\overline{K}_3] \cup U_2[\overline{K}_3] \cup U_3[\overline{K}_3].$$

Let $U_i(j)$ denote $U_i[\overline{K}_3]$ with all of F_1, F_2, F_3 deleted except for F_j. We shall exhibit an edge-disjoint decomposition

$$U_i[\overline{K}_3] = U_{i1} \cup U_{i2} \cup U_{i3}$$

where U_{ij} is the union of three disjoint one-factors of $U_i(j)$. Then it is clear that each of the following is a set of three one-factors of $U[\overline{K}_3]$:

$$\begin{array}{l} U_{11} \cup U_{22} \cup U_{33} \\ U_{12} \cup U_{23} \cup U_{31} \\ U_{13} \cup U_{21} \cup U_{32} \end{array}$$

Together they form a one-factorisation of $U[\overline{K}_3]$.

Here are the components of $U_1[\overline{K}_3]$. In each case, each row is a one-factor.

U_{11}	A1B2	A2D2	A3B1	B3C3	C1E3	C2E1	D1E2	D3F1
	A1B3	A2B1	A3D1	B2C2	C1E2	C3E1	D2F1	D3E3
	A1D3	A2B2	A3B3	B1C1	C2E2	C3E3	D1F1	D2E1
U_{12}	A1C2	A2C3	A3D2	B1E3	B2C1	B3E2	D1E1	D3F2
	A1D1	A2C1	A3C2	B1C3	B2E3	B3E1	D2F2	D3E2
	A1C1	A2D3	A3C3	B1E1	B2E2	B3C2	D1F2	D2E3
U_{13}	A1D2	A2B3	A3C1	B1C2	B2E1	C3E2	D1E3	D3F3
	A1C3	A2D1	A3B2	B1E2	B3C1	C2E3	D2F3	D3E1
	A1B1	A2C2	A3D3	B2C3	B3E3	C1E1	D1F3	D2E2

The components of $U_2[\overline{K}_3]$ and $U_3[\overline{K}_3]$ are formed by rotation.

It might be as well to conclude with an example of a wreath product which does not have a one-factorisation.

Theorem 5. $U[K_3]$ *has no one-factorisation.*

Proof. $U[K_3]$ is a regular graph of valency 11, so any one-factorisation contains 11 factors. If we denote the vertices as we did in $U[\overline{K}_3]$, then $S = \{A1,A2,A3,B2,\ldots,E3\}$ contains 15 vertices. Since 15 is odd, each one-factor of $U[K_3]$ contains at least one edge with one endpoint in S and one outside. This edge must be of the form D_iF_j. Only nine such edges exist, so eleven factors are impossible.

This argument is based on the fact that $U[K_n]$ has valency $4n-1$, and that the number of edges D_iF_j is n^2; so, for n odd, a factorisation is impossible when $n^2 < 4n-1$. But the only case ruled out is $n = 3$. It is quite possible that $U[K_5]$ has a one-factorisation, and in fact I conjecture that it has one.

REFERENCES

[1] M. Behzad, G. Chartrand and J.K. Cooper Jr., The colour numbers of complete graphs, *J. London Math. Soc.* 42 (1967), 226-228.

[2] P.E. Himelwright, W.D. Wallis and J.E. Williamson, On one-factorisations of compositions of graphs (to appear).

[3] P.E. Himelwright and J.E. Williamson, On 1-factorability and edge-colorability of cartesian products of graphs, *Elem. Math.* 29 (1974), 66-68.

[4] A. Kotzig, Problems and recent results on 1-factorizations of cartesian products of graphs, *Proc. Ninth South-eastern Conference on Combinatorics, Graph Theory and Computing.* (Utilitas Math. Publ., Winnipeg, 1978), 457-460.

[5] R. Laskar and W. Hare, Chromatic numbers for certain graphs, *J. London Math. Soc.* (2) 4 (1972), 489-492.

[6] E. Mahmoodian, On edge-colorability of cartesian products of graphs, *Canad. Math. Bull.* (to appear).

[7] E.T. Parker, Edge-coloring numbers of some regular graphs, *Proc. Amer. Math. Soc.* 37 (1973), 423-424.

[8] W.D. Wallis, A one-factorisation of a cartesian product, *Utilitas Math.* (to appear).

Department of Mathematics
University of Newcastle
New South Wales 2308

DIVISIBLE SEMISYMMETRIC DESIGNS

Peter Wild

1. Symmetric 2-Designs and Semisymmetric Designs

A $2\text{-}(v,k,\lambda)$ design is an incidence structure of v points and b blocks such that

(i) every block is incident with k points $(0 < k < v)$

(ii) any two points are incident with λ common blocks

(iii) no two blocks are incident with the same set of points.

From Fisher's Inequality (Fisher [1]) we have that $b \geq v$ and further $b = v$ if and only if any two blocks are incident with λ common points.

A 2-design with $b = v$ is called a symmetric 2-design. Although infinitely many symmetric 2-designs with $\lambda = 1$ (i.e. projective planes) are known, for each $\lambda \geq 2$ only finitely many symmetric 2-designs are known.

Associated with the Buekenhout diagram $\circ\overset{C}{-}\circ\overset{C^*}{-}\circ$ is a class of connected incidence structures satisfying the following two properties

(i) any two points are incident with 0 or 2 common blocks

(ii) any two blocks are incident with 0 or 2 common points.

These structures are called Semibiplanes and clearly include all symmetric 2-designs with $\lambda = 2$ (i.e. biplanes). Whereas only finitely many biplanes are known, infinite families of semibiplanes have been found (see Hughes [2]).

In general we may define a Semisymmetric Design as a connected incidence structure satisfying for some $\lambda \geq 2$ the following two properties.

(i) any two points are incident with 0 or λ common blocks

(ii) any two blocks are incident with 0 or λ common points.

<u>Result 1</u>. Let S be a semisymmetric design. Then

(i) there is an integer k such that every block of S is incident with k points, and every point of S is incident with k blocks

(ii) the number of points, v, of S is equal to the number of blocks

(iii) $v \geq \frac{k(k-1)}{\lambda} + 1$; if equality holds and $k > \lambda$ then S is a symmetric 2-design.

<u>Proof</u>: (i) Let (p,x) be an incident point-block pair of S. Any other block incident with p meets x in $\lambda-1$ other points. Also any other point incident with x is joined to p by $\lambda-1$ other blocks. Hence the number of blocks incident with p equals the number of points incident with x. Now connectivity implies the result.

(ii) this follows by counting the number of incident point-block pairs in two ways.

(iii) each point p is incident with k blocks, and each of these blocks contains $k-1$ other points. Any point on one common block with p is on λ common blocks. Hence there are $\frac{k(k-1)}{\lambda}$ points on a common block with p. Thus $v \geq \frac{k(k-1)}{\lambda} + 1$. Also, if $v = \frac{k(k-1)}{\lambda} + 1$ then any two points are incident with λ common blocks.

Thus a semisymmetric design (or SSD) has three parameters (v,k,λ).

2. Divisible Semisymmetric Designs

A special class of semisymmetric designs consists of those which are divisible, i.e. there is a partition of the points into classes such that two points from the same class are on 0 blocks, and two points from different classes are on λ blocks.

Result 2. Let S be a divisible $SSD(v,k,\lambda)$. Then

(i) there are $m = \frac{v}{n}$ classes each of size $n = v - \frac{k(k-1)}{\lambda}$

(ii) $n \leq \frac{k}{\lambda}$ and $v \leq \frac{k^2}{\lambda}$.

Proof. (i) there are $\frac{k(k-1)}{\lambda}$ points lying on λ common blocks with any point p. p and the remaining $v - \frac{k(k-1)}{\lambda} - 1$ points make up the class containing p. Thus $n = v - \frac{k(k-1)}{\lambda}$ and there are $\frac{v}{n}$ classes.

(ii) a point p is on λ blocks with each of the n points of a class not containing p. Hence there are at least $n\lambda$ blocks incident with p (since no block contains two points of a class). Thus $k \geq n\lambda$, and $v = \frac{k(k-1)}{\lambda} + n \leq \frac{k(k-1)}{\lambda} + \frac{k}{\lambda} = \frac{k^2}{\lambda}$.

Example: Consider a projective space P of dimension $r \geq 2$, with $q+1$ points on a line. Let p be a point and x a hyperplane of P. Define the following structure S. The points of S are the points of P besides p and the points on x. The blocks of S are the hyperplanes of P besides x and the hyperplanes through p.

Incidence in S is the natural incidence of P.

Each point ℓ of S is incident with q^{r-1} blocks of S (the q^{r-1} hyperplanes of P through ℓ but not through p). Two points of S are on 0 blocks of S if they lie on a line through p, and are on q^{r-2} blocks otherwise (the q^{r-2} hyperplanes of P through the two points but not through p).

Similarly any two blocks of S contain 0 or q^{r-2} points of S. Thus S is a divisible semisymmetric design, a class of the division consisting of the points of S lying on a line through p.

If $p \in x$ S has parameters (q^r, q^{r-1}, q^{r-2}) and $n=q$.

If $p \notin x$ S has parameters $(q^r-1, q^{r-1}, q^{r-2})$ and $n=q-1$.

3. Dual Properties

In this section we see that divisible SSD's have a very nice structure, in that their duals are also divisible and these divisions determine a tactical decomposition of the structure. Also we see that divisible SSD's are closely related to symmetric 2-designs.

<u>Lemma</u>: Let S be a divisible SSD. Then the dual of S is also divisible, i.e. there is a partition of the blocks of S into classes such that two blocks from the same class meet in 0 points, and two blocks from different classes meet in λ points.

<u>Proof</u>: Let x be a block of S. Let $p_1,\dots,p_k$ be the points on X. Let $C_1,\dots,C_k$ be the point classes containing $p_1,\dots,p_k$ respectively. Let $q \in C_1\setminus\{p_1\}$. q lies on λ blocks with each of $p_2,\dots,p_k$ and each block through q and meeting x, meets x in λ of $p_1,\dots,p_k$. Thus there are $k-1$ blocks through q and meeting x, and so exactly one block through q and not meeting x.

Hence the $k(n-1)$ points of $C_1\setminus\{p_1\},\dots,C_k\setminus\{p_k\}$ each lie on exactly one of the $n-1$ blocks not meeting x. It follows that these $n-1$ blocks are disjoint and together with x form a class of a division.

If a point of a class C is incident with a block of a class D, then each point of C is incident with exactly one block of D. Thus the point and block classes form a tactical decomposition of S with $(C,D) = (D,C) = 0$ or $(C,D) = (D,C) = 1$, where (C,D) is the number of points of class C on each block of D, and (D,C) is defined dually.

Hence if S is a divisible $SSD(v,k,\lambda)$, we may define a new structure $C(S)$ whose points are the point classes of S and whose blocks are the block classes of S, with point class C incident with block class D if and only if $(C,D) = 1$.

<u>Theorem</u>: Let S be a divisible $SSD(v,k,\lambda)$ with m classes of n points each. Then

(i) if $k = \lambda n$, $C(S)$ consists of k points and k blocks and every point is incident with every block.

(ii) if $k > \lambda n$ $C(S)$ is a symmetric 2-$(m,k,\lambda n)$ design.

<u>Proof</u>: Clearly each block of $C(S)$ is incident with k points of $C(S)$ (the k classes containing the k points on a block of the block class). If $k = \lambda n$, then $v = \frac{k^2}{\lambda}$ and $m = k$, and $C(S)$ is as described. If $k > \lambda n$, consider two classes C_1 and C_2. Let $p \in C_1$. p lies on λ blocks with each of the n points of C_2. These λn blocks are distinct (since no block contains two points of a class) and represent λn block classes incident with C_1 and C_2.

These are the only block classes incident with C_1 and C_2, since any such block class contains a block incident with p. Thus $C(S)$ is a symmetric $2\text{-}(m,k,\lambda n)$ design.

4. Examples from Hadamard Matrices and Balanced Weighing Matrices

An $SSD(v,k,\lambda)$ with $v = \frac{k(k-1)}{\lambda} + 2$ is necessarily divisible with $n = 2$ (to each point corresponds a unique point not on any block with it). Consider an incidence matrix associated with the division for such an SSD, i.e. such that the two points of a class correspond to consecutive rows, and the two blocks of a class correspond to consecutive columns. This matrix is partitioned into 2×2 blocks. Each block is one of the following three $\begin{bmatrix}0&0\\0&0\end{bmatrix}$, $\begin{bmatrix}1&0\\0&1\end{bmatrix}$, and $\begin{bmatrix}0&1\\1&0\end{bmatrix}$.

If we replace each block $\begin{bmatrix}0&0\\0&0\end{bmatrix}$ by 0,

each block $\begin{bmatrix}1&0\\0&1\end{bmatrix}$ by 1,

and each block $\begin{bmatrix}0&1\\1&0\end{bmatrix}$ by -1,

we get a $(0,1,-1)$ matrix M such that $MM^t = M^tM = kI$. If $k = 2\lambda$, M is a Hadamard matrix, and if $k > 2\lambda$, M is a balanced weighing matrix. Clearly this process may be reversed.

5. A Construction for Divisible SSD's Using Quadrics.

Let $\mathcal{A}$ and $\mathcal{B}$ be sets of symmetric $r\times r$ matrices over $GF(q)$, q odd, r odd, such that $|\mathcal{A}| = |\mathcal{B}| = t$, and

$A_1 - A_2$ is non-singular for all $A_1 \neq A_2 \in \mathcal{A}$,
and $B_1 - B_2$ is non-singular for all $B_1 \neq B_2 \in \mathcal{B}$.
For $A \in \mathcal{A}$ and $B \in \mathcal{B}$ put $m_{AB} = \det(I-4AB)$.
For $x \in GF(q)$ let

$$f(x) = \begin{cases} 0 & \text{if } x = 0 \\ 1 & \text{if } x \text{ is a non-zero square} \\ -1 & \text{if } x \text{ is a non-square.} \end{cases}$$

Put $h_{AB} = f(m_{AB})$ and suppose $H = (h_{AB})$ is a Hadamard matrix. Put $Z_{AB} = (I-4AB)^{-1}$ for $A \in \mathcal{A}$ and $B \in \mathcal{B}$. Define a structure S in the following way.

The points of S are all triples (a_1,a,A) where $a_1 \in GF(q)$, a is an r-tuple of elements of $GF(q)$ and $A \in \mathcal{A}$. The blocks of S are all triples (b_1,b,B) where $b_1 \in GF(q)$, f is an r-tuple of elements of $GF(q)$ and $B \in \mathcal{B}$. $(a_1,a\ A)$ is incident with (b_1,b,B) if and only if

$$a_1 + b_1 + aZ^t_{AB}b^t + bAZ^t_{AB}b^t + aBZ_{AB}a^t = 0\ .$$

Then S is a divisible $SSD(tq^{r+1},tq^r,tq^{r-1})$.

<u>Proof</u>: Two points (a_1,a,A) and (x_1,a,A) with $a_1 \neq x_1$ are on no common blocks.

Two points (a_1,a,A) and (x_1,x,X) with $A \neq X$ are on a common block whenever there is a solution for b,B to the equation

$$a_1-x_1 + (aZ_{AB}^t-xZ_{XB}^t)b^t + b(AZ_{AB}^t-XZ_{XB}^t)b^t$$

$$+ aBZ_{AB}a^t - xBZ_{XB}x^t = 0.$$

Now $AZ_{AB}^t-XZ_{XB}^t = Z_{XB}(A-X)Z_{AB}^t = Z_{AB}(A-X)Z_{XB}^t$ is non-singular.

Put $y = b + \frac{1}{2}(aZ_{AB}^t-xZ_{AB}^t)(Z_{XB}(A-X)Z_{AB}^t)^{-1}$.

Substituting for b in the above equation, and simplifying we have

$$a_1-x_1 - \frac{1}{4}(a-x)(A-X)^{-1}(a-x)^t + y(Z_{XB}(A-X)Z_{AB}^t)y^t = 0.$$

There is a one-to-one correspondence between the solutions for b,B to the first equation and the solutions for y,B to this equation.

For each B, this equation corresponds to a quadric Q in r-dimensional projective space over $GF(q)$, and solutions for b correspond to points of Q outside the hyperplane meeting Q in the quadric corresponding to the equation $y(Z_{XB}(A-X)Z_{AB}^t)y^t = 0$.

If $a_1-x_1 - \frac{1}{4}(a-x)(A-X)^{-1}(a-x)^t = 0$, then Q is a cone for all B, and for each B there are q^{r-1} solutions for b.

If $a_1-x_1 - \frac{1}{4}(a-x)(A-X)^{-1}(a-x)^t \neq 0$, then Q is an elliptic quadric (so that there are $q^{r-1} - q^{\frac{r-1}{2}}$ solutions for b) or an hyperbolic quadric (so that there are $q^{r-1} + q^{\frac{r-1}{2}}$ solutions for b) depending on whether $\det(Z_{XB}(A-X)Z_{AB}^t)$ is a square or non-square. (See Primrose [3] for properties of quadrics). Since the matrix H defined above is a Hadamard matrix an elliptic quadric occurs equally many times as a hyperbolic quadric, as B varies. Thus in any case the points (a_1,a,A) and (x_1,x,X) are on tq^{r-1} common blocks.

Thus two points are on 0 or tq^{r-1} common blocks. Dual arguments show that two blocks meet in 0 or tq^{r-1} common points, and S is a semisymmetric design. Also S is divisible, two points (a_1,a,A) and (x_1,x,X) being in the same class if and only if $a=x$ and $A=X$.

REFERENCES

[1] R.A. Fisher, An examination of the different possible solutions of a problem in incomplete blocks, Ann. Eugenics 10 (1940) 52-75.

[2] D.R. Hughes, Biplanes and Semibiplanes, Lecture Notes in Mathematics No. 686, Springer-Verlag, Berlin Heidelberg New York 1978, 55-58.

[3] E.J.F. Primrose, Quadrics in finite geometries, Proc. Cambridge Phil. Soc. 47 (1951) 299-304.

Department of Pure Mathematics, The University of Adelaide, Box 498, G.P.O. Adelaide, South Australia, 5001.

05C99

08B99

05C15

GRAPHS AND UNIVERSAL ALGEBRAS

SHEILA OATES-WILLIAMS

A method of associating graphs with universal algebras is exhibited and the possibility of using this relationship to investigate the properties of conjunctions of graphs is considered.

1. SHALLON ALGEBRAS

In her Ph.D. thesis [6], Caroline Shallon introduced the following type of universal algebra associated with a graph.

Definition 1. *Let* Γ *be a directed graph with vertices* $V(\Gamma) = \{a_1,\ldots,a_n\}$ *and edge set* $E(\Gamma) \subseteq V(\Gamma) \times V(\Gamma)$. *Then the Shallon algebra,* $A(\Gamma)$, *has as elements* $\{0,a_1,\ldots,a_n\}$ *with* 0 *as a 0-ary operation and a binary operation defined by* $a_ia_j = a_i$ *if* $(a_i,a_j) \in E(\Gamma)$ *and all other products zero. If* Γ *is not directed then* $A(\Gamma)$ *can still be defined if we regard an undirected edge as two directed edges in opposite directions.*

Clearly any universal algebra with a zero and a binary operation such that any product is either zero or equal to the first factor is an $A(\Gamma)$ for some Γ.

It is not difficult to see that these algebras are not in general semi-groups; in fact this will occur only when the connected components of Γ are either complete graphs with loops at each vertex or single points.

Recall that a variety of universal algebras is the class of all universal algebras of the same (finitary) type satisfying a given set of laws, or, equivalently, (see [1]) a class of universal algebras closed under taking subalgebras, homomorphic images and cartesian products. Let us consider these operations as applied to Shallon algebras:

Any subset of a Shallon algebra which contains zero is clearly a subalgebra, and if non-trivial is a Shallon algebra corresponding to the spanning subgraph on the vertices in the subset.

Any non-trivial homomorphic image of a Shallon algebra is also a Shallon algebra. However most Shallon algebras arising from connected graphs have only trivial homomorphic images, one of the few exceptions being $A(C_4)$ which has $A(P_3)$ as homomorphic image, (see [6], Theorem 7.11).

At first sight the situation with cartesian products looks even worse as the direct product of two Shallon algebras is not necessarily a Shallon algebra. For example, if we take the direct product $A(P_2) \times A(P_2)$ where the paths have vertices a_1, a_2 ; b_1, b_2, then $(a_1,a_2)(b_1,0) = (a_1,0)$ which is neither zero nor the first factor. However we do have the following result which is proved in [5].

Lemma. *Let* $A_1,\dots,A_n$ *be Shallon algebras obtained from the graphs* $\Gamma_1,\dots,\Gamma_n$, *then the relation* ρ *on* $A_1 \times \dots \times A_n$ *defined by* $(a_1,\dots,a_n)\ \rho\ (b_1,\dots,b_n)$ *if either there exists* i,j *such that* $a_i = b_j = 0$ *or* $a_i = b_i$ $(i=1,\dots,n)$ *is a congruence relation and* $(A_1 \times \dots \times A_n)/\rho$ *is a Shallon algebra corresponding to the graph* $\Gamma_1 \wedge \dots \wedge \Gamma_n$. (Here $\Gamma_1 \wedge \Gamma_2$ denotes the graph whose vertex set is $V(\Gamma_1) \times V(\Gamma_2)$ and such that $((x_1,x_2),(y_1,y_2)) \in E(\Gamma_1 \wedge \Gamma_2)$ if and only if $(x_1,y_1) \in E(\Gamma_1)$ and $(x_2,y_2) \in E(\Gamma_2)$. This particular operation has a multitude of names in the literature. In this paper, following Harary and Wilcox, [2], it will be called the *conjunction*.)

2. APPLICATIONS OF VARIETY THEORY

First note that laws of a variety of Shallon algebras are of two types, $w(x_1,\dots,x_n) = 0$ or $w_1(x_1,\dots,x_n) = w_2(x_1,\dots,x_n)$. In each case the words will consist of a product of the x_i with appropriate bracketting. The following result is easily verified:

Lemma. *With the notation of the previous lemma,* $(A_1 \times \dots \times A_n)/\rho$ *satisfies a law of the type* $w = 0$ *if and only if one of the* A_i *satisfies this law.*

From now on all graphs considered will be undirected and without loops .

Theorem 1 *Let* $c(\Gamma)$ *denote the size of the maximum clique in a finite graph* Γ. *Then*

$$c(\Gamma_1 \wedge \dots \wedge \Gamma_r) = \min\{c(\Gamma_i)\}.$$

Proof. Consider the law $w_n = 0$ where w_n is defined recursively as follows:

$$w_2(x_1,x_2) = x_2x_1$$

$$w_n(x_1,\dots,x_n) = x_n(x_{n-1}(x_n(x_{n-2}(\dots(x_n(x_1(w_{n-1})\dots).$$

w_n has the property that each (unordered) pair of variables occurs at least once in adjacent positions. Now consider what happens when elements of a Shallon algebra $A(\Gamma)$ are substituted for the variables in w_n. At each stage in the evaluation, because of the nature of the binary operation in Shallon algebras, we have that a term within a given set of brackets is either zero or equal to whatever was substituted for the first variable within the brackets. It follows that w_n will reduce to zero if

(i) any variable is replaced by zero

or (ii) two distinct variables are replaced by the same element (since then we have a term ..g(g...).. and $g^2 = 0$)

or (iii) two distinct variables are replaced by elements corresponding to non-adjacent vertices in Γ.

Also if $\{a_1,\ldots,a_n\}$ are the vertices of K_n then $w(a_1,\ldots,a_n) = a_n \neq 0$ so $w = 0$ is not a law in $A(K_n)$.

Putting these comments together, we see that $w_n = 0$ is a law in $A(\Gamma)$ if and only if Γ does not contain a subgraph isomorphic to K_n, that is, if $c(\Gamma) < n$. The theorem now follows immediately from the lemma.

Akin to the above result is the conjecture of Hedetniemi [3] that $\chi(\Gamma_1 \wedge \ldots \wedge \Gamma_r) = \min(\chi(\Gamma))$. (Here, as usual, $\chi(\Gamma)$ is the chromatic number of Γ). It is readily verified that $\chi(\Gamma_1 \wedge \ldots \wedge \Gamma_r) \leq \min(\chi(\Gamma))$. Again, we can use varietal techniques to establish the truth of the conjecture in a simple case.

Theorem 2. *Let $\Gamma_1,\ldots,\Gamma_r$ be non-bipartite graphs, then $\Gamma_1 \wedge \ldots \wedge \Gamma_r$ is not bipartite.*

Proof. This time we consider the law $v_n = 0$, $n \geq 1$, where $v_n(x_1,\ldots,x_n) = x_1(x_2(x_3(\ldots x_{2n}(x_{2n+1}\, x_1)\ldots)$. It is easily verified that $v_n = 0$ is a law in $A(C_{2m+1})$ if and only if $m > n$, so that a graph is bipartite if and only if it satisfies all the laws $v_n = 0$.

Since each of the Γ_i above is non-bipartite it contains C_{2r_i+1} for some r_i. But then $A(\Gamma_i)$ fails to satisfy $v_n = 0$ for $n \geq r_i$. If $m = \max_i(r_i)$ then $A(\Gamma_1 \wedge \ldots \wedge \Gamma_r)$ fails to satisfy $v_m = 0$ and so is not bipartite.

Corollary. *If $\min \chi(\Gamma_i) = 3$ then $\chi(\Gamma_1 \times \ldots \times \Gamma_r) = 3$.*

The above laws can also be used to establish a result of Miller's [4] which shows that Hedetniemi's conjecture is false in the infinite case.

Theorem 3. *The graph Γ which is the conjunction of the infinite set of graphs $\{C_{2r+1} : r \in N\}$ is bipartite.*

Proof. We use the laws $v_n = 0$ of theorem 2. Since $A(C_{2r+1})$ satisfies $v_n = 0$ for $r > n$, Γ satisfies $v_n = 0$ for all n. Hence Γ is bipartite.

3. REMARKS

1. Except perhaps for theorem 3, the above method of proof does not yield a shorter method of proof of the results used as illustrations than would a direct argument. However, we feel it is worthwhile drawing attention to the existence of this technique.

2. The strong direct product $\Gamma_1 \overline{\times} \Gamma_2$ of (undirected) graphs is defined by

$$V(\Gamma_1 \overline{\times} \Gamma_2) = V(\Gamma_1) \times V(\Gamma_2)$$

$E(\Gamma_1 \overline{\times} \Gamma_2) = \{\{(u_1,u_2),(v_1,v_2)\} | (u_1 = v_1 \wedge \{u_1,u_2\} \in E(\Gamma_2)) \vee (\{u_1,v_1\} \in E(\Gamma_1) \wedge \wedge \{u_2,v_2\} \in E(\Gamma_2)) \vee (\{u_1,v_1\} \in E(\Gamma_1) \wedge u_2 = v_2)\}$. This product can also be related to a quotient of a directed product of algebras associated with the original graphs provided we change the definition of our algebras so that $a_i^2 = a_i$ always.

REFERENCES

[1] G. Birkhoff, On the structure of abstract algebras, *Proc. Cambridge Phil. Soc.* 31(1935), 433-454.

[2] Frank Harary and Gordon W. Wilcox, Boolean operations on graphs, *Math. Scand.* 20(1967), 41-51.

[3] Stephen T. Hedetniemi, Homomorphisms of graphs and automata, *University of Michigan Technical Report,* Project 03105-44-T, 1966.

[4] Donald J. Miller, The categorical product of graphs, *Can. J. Math.* 20(1968), 1511-1521.

[5] Sheila Oates-Williams, Murskii's algebra does not satisfy MIN, *Bull. Austral. Math. Soc.* 22(1980), 199-203.

[6] Caroline Ruth Shallon, Non-finitely based binary algebras derived from lattices (Ph.D. thesis, University of California, Los Angeles, 1979).

Department of Mathematics

University of Queensland

St. Lucia

Queensland 4067

05B30

05B45

UNIVERSAL FABRICS

SHEILA OATES-WILLIAMS AND ANNE PENFOLD STREET

A fabric is said to be k-universal if it exhibits every possible $k \times k$ square coloured black and white. Here we investigate the possibility of using pseudo-random sequences and arrays to construct such fabrics.

1. INTRODUCTION

A fabric consisting of two sets of strands, the warp and the weft, may be represented by a pattern of black and white squares; a white square indicating where the weft passes over the warp, and a black square indicating the converse situation. In their paper [3] (to which the reader is referred for more detailed definitions) Grünbaum and Shephard introduced the concepts of isonemal and mononemal for fabrics, and also, what we are considering here, k-universal fabrics.

Definition *A fabric is strongly* k*-universal if it contains every possible* $k \times k$ *block coloured black and white and* k*-universal if it contains a representative of each orbit of the symmetry group on such blocks (where the allowable symmetries are rotations, reflections, colour interchange, and combinations of these).*

Grünbaum and Shephard also require their universal fabrics to be isonemal, but we will not necessarily impose that restriction.

There are really two stages to the problem, the first being to find a rectangle of black and white squares which, when used to tile the plane, yield all possible $k \times k$ squares, and the second being to ensure this gives a fabric, (here the criterion given by Clapham [2] is very useful).

For the remainder of the paper we will work with arrays of 0s and 1s rather than arrays of black and white squares.

2. PSEUDO-RANDOM ARRAYS

Definition *A pseudo-random sequence of length* $2^n - 1$ *is a sequence of* 0s *and* 1s *in which every possible sequence of* 0s *and* 1s *of length* n *occurs as a subsequence except the all zero one.*

Example 1. 000100110101111 is a pseudo-random sequence of length 15.

Such sequences can be generated from an initial sequence of $n - 1$ 0s and a 1 by a recurrence relation derived from a polynomial of degree n irreducible over GF[2]. The polynomial used for the above example is $x^4 + x + 1$ which corresponds to the recurrence relation

$$a_{i+4} = a_{i+1} + a_i$$

Details of this may be found in MacWilliams and Sloane [4]. From there also come the following results on pseudo-random arrays. We do not need the technical properties of pseudo-random arrays here, only the fact that they have the window property, that is, for an appropriate k and ℓ every possible $k \times \ell$ block of 0s and 1s, except the all-zero block, occurs when these arrays are used to tile the plane. The precise statement of the result is as follows.

Theorem [4]. *Let* $m = k\ell$ *be such that* $x = 2^k - 1$ *and* $y = (2^m - 1)/(2^k - 1)$ *are relatively prime. Then the* $x \times y$ *array obtained by writing a pseudo-random sequence of length* $2^m - 1$, *derived from an irreducible polynomial, along the diagonals has the* $k \times \ell$ *window property.*

Example 2. The sequence in Example 1 yields the array

01111
00110
01001

(where the entries are inserted in the order

$b_{11}, b_{22}, b_{33}, b_{14}, b_{25}, b_{31}, b_{12}, b_{23}, b_{34}, b_{15}, b_{21}, b_{32}, b_{13}, b_{24}, b_{35}$).

This has the 2×2 window property.

It is also shown in [4] that the entries in the columns in such an array satisfy a recurrence relation; for instance, in the above example we have

$$b_{i+2,j} = b_{i,j} + b_{i+1,j} .$$

As a result such an array always contains an $x \times (\ell-1)$ block of 0s, arising from any $k \times \ell$ block whose first $\ell - 1$ columns are all zero. Thus these arrays have two failings as far as our requirements are concerned, they have no all-zero blocks, and they are not suitable for weaving (since the strands corresponding to the all-zero columns would just drop off). The first problem is easily rectified by inserting an extra column of zeros to give a $x \times \ell$ block of zeros. We have two ways of tackling the second, which we illustrate by reference to the array of Example 2. This has now become

001111
000110
001001.

If we add two rows to this as follows -

001111
000110
001001
001111
110000

we obtain a block that tiles the plane giving all possible 2×2 blocks and also gives a fabric that hangs together (as is seen by applying Clapham's criterion, [2]).

This, of course, is far from being isonemal. To obtain an isonemal fabric we can use a result of Grünbaum and Shephard [3], which shows that any non-monochrome $p \times q$ block, with p and q relatively prime, can be embedded in an isonemal fabric of period $2pq$. In this case we would have to use the block

1001111
0000110
1001001
1001111

to ensure having all possible 2×2 blocks, thus obtaining a fabric of period 56, not nearly as good as the one of period 10 given by Grünbaum and Shephard [3].

However, of the two methods the latter is the easier to generalise, the difficulty with the former lying in the correct choice of the row that disposes of the columns of zeros, as checking Clapham's criterion in the general case would seem a formidable task.

So, for a k-universal fabric we first of all require that $x = 2^k - 1$ and $y = (2^{k^2} - 1)/(2^k - 1)$ be relatively prime, so that MacWilliam and Sloane's method applies. Since y leaves remainder k on division by x, this is true if and only if x and k are relatively prime. This is certainly true if k is prime or a power of 2, but, unfortunately, is not true in general, for instance if $k = 6$, $x = 63$. We then add an extra k columns (including a column of 0s) and an extra $k - 1$ rows so that the block as written has all $k \times k$ subsquares. If $x + k - 1$ and $y + k$ are relatively prime we get a fabric of period $2(x + k)(y + k) \leq 4.2^{k^2}$. If not, a few extra rows or columns may have to be added, but we will still obtain a fabric of period considerably less than the approximately $k^2 \times 2^{k^2}$ that is obtained by juxtaposing the 2^{k^2} different squares of size $k \times k$ and then applying Grünbaum and Shephard's method.

3. DE BRUIJN SEQUENCES

It might be hoped that De Bruijn sequences, of length 2^{k^2} which contain every possible sequence of length 2^{k^2} would be usable in a similar fashion to give an immediate construction of a block containing all $k \times k$ squares of 0s and 1s. Of course, since 2^{k^2} cannot be factored non-trivially into two relatively prime integers we have the immediate problem of how to specify the arrangement of the sequence along the diagonals. However, even if we could produce a rule for this, we still cannot obtain the only 4×4 square, namely

0100
0111
1110
0010

which tiles the plane so as to give all 2×2 squares, by any arrangement of a De Bruijn sequence of length 16 along its diagonals, since nowhere in it does the sequence 0100 occur, which must occur in any De Bruijn sequence. That this is indeed the only such square may be verified by examination of the list of all pantactic squares given in [1].

4. MISCELLANEOUS

One of the standard methods of constructing De Bruijn sequences, namely the Euler walk technique, may be applied in an analogous fashion to obtain an $k \times 2^{k^2}$ strip, which, when used to tile the plane, yields all possible $k \times k$ squares. The $2^{k(k-1)}$ $k \times (k-1)$ blocks of 0s and 1s are taken as the vertices of a graph and an edge is drawn from A to B if there is a $k \times k$ block whose first $k-1$ columns give A and whose last $k-1$ columns give B. Every vertex then has in-degree 2^k and out-degree 2^k so an Euler walk exists and this yields the required strip. We illustrate the method in the case $k = 2$.

<u>Example 3.</u> Here we have four 2×1 blocks

0 1 0 1
0 0 1 1

which we denote by A,B,C,D. All edges exist so the graph is the complete directed graph on four vertices. An example of an Euler walk is AABBCCBDDCDBACADA which yields the strip

0011001110110001
0000110111100101

This is obviously wasteful, it should be possible to produce a 3×8 strip with the same properties. Such a strip is

00110110
00110011
11000110.

However, it is not clear how this particular construction can be generalised to larger k.

We have made little progress with the k-universal case. The lack of symmetry seems to make a systematic construction harder. We do have the following example

Example 4. The 3×3 square

$$\begin{matrix} 1 & 1 & 0 \\ 1 & 1 & 0 \\ 1 & 0 & 1 \end{matrix}$$

is 2-universal since its four 2×2 subspaces

$$\begin{matrix} 11, & 10, & 11, & 10 \\ 11 & 10 & 10 & 01 \end{matrix}$$

belong to the four different orbits that the 16 2×2 squares fall into under the action of the symmetry group. However it does not weave, since it has a column of 1s. The 4×4 square

$$\begin{matrix} 1100 \\ 1100 \\ 1011 \\ 0011 \end{matrix}$$

obtained from the above square by adjoining an extra row and column does weave, but, of course is not even mononemal.

REFERENCES

[1] C.J. Bouwkamp, P.Jannsen and A. Koene, Note on pantactic squares, *Math. Gaz.* 54 (1970), 348-351.

[2] C.R.J. Clapham, When a fabric hangs together, *Bull. London Math. Soc.* 12(1980), 161-164.

[3] Branko Grünbaum and Geoffrey C. Shephard, Satins and twills: an introduction to the geometry of fabrics, *Math. Mag.* 53(1980), 139-161.

[4] F. Jessie MacWilliams and Neil J.A. Sloane, Pseudo-random sequences and arrays, *Proc. IEEE,* 64(1976), 1715-1729.

Department of Mathematics
University of Queensland
St. Lucia
Queensland 4067

seems to have a ... [illegible] ... order. We ... have the following example.

Example [illegible]

[illegible]

REFERENCES

[1] [illegible]

[2] [illegible] 141-164.

[3] [illegible]

[4] [illegible]

Department of Math...
[illegible]